ASTECHNOVA International Energy Conference 2017

E3S Web of Conferences Volume 43 (2018)

Yogyakarta, Indonesia
1 November 2017

Editors:

I. Sunarno
A. P. Sasmito
L. P. Hong

ISBN: 978-1-5108-6593-8

Printed from e-media with permission by:

Curran Associates, Inc.
57 Morehouse Lane
Red Hook, NY 12571

Some format issues inherent in the e-media version may also appear in this print version.

Printed by Curran Associates, Inc. (2018)

For additional information, please contact EDP Sciences – Web of Conferences
at the address below.

EDP Sciences – Web of Conferences
17, Avenue du Hoggar
Parc d'Activité de Courtabœuf
BP 112
F-91944 Les Ulis Cedex A
France

Phone: +33 (0) 1 69 18 75 75
Fax: +33 (0) 1 69 28 84 91

contact-edps@webofconferences.org

Additional copies of this publication are available from:

Curran Associates, Inc.
57 Morehouse Lane
Red Hook, NY 12571 USA
Phone: 845-758-0400
Fax: 845-758-2633
Email: curran@proceedings.com
Web: www.proceedings.com

TABLE OF CONTENTS

Dept. of Nuclear Engineering and Engineering Physics
Universitas Gadjah Mada

Preface

Energy-Food-Water Nexus is an essential factor of human life. This energy-food-water relationship is need to be taken into account in the researches and innovation for sustainable future. This is the theme of our 6th International Energy Conference: ASTECHNOVA 2017 held by Department of Nuclear Engineering and Engineering Physics (DNEEP), Universitas Gadjah Mada.

ASTECHNOVA 2017 received 42 extended abstracts and after peer review process from International Scientific and National Scientific reviewer, there are 41 extended abstract which continue to next step for submitting the full paper. In order to increase the quality of the paper, ASTECHNOVA 2017 committee conducted similarity crosscheck with software, and peer-review of full paper by scientific reviewers.

Finishing of all the review process, we have 33 encouraging papers, coming from 12 different institutions, 8 different countries, 83 participants, and 31 papers have been presented at the conference. Throughtout the keynote speech and presentation from participants, ideas sharing between researchers especially in the same field of research can be achieved. The discussions also provide the general participants to explore what are the recent issues and challenges related to Energy-Food-Water Nexus field.

Let me close this preface with the highest appreciation for all of your participation, fruitful discussion and let keep our network strong for contribution in energy field development.

Thank you very much.

Best regards,

Dr. Rachmawan Budiarto
ASTECHNOVA 2017 General Chair

Dept. of Nuclear Engineering and Engineering Physics
Universitas Gadjah Mada

ABOUT THE CONFERENCE

ASTECHNOVA 2017
INTERNATIONAL ENERGY CONFERENCE

"Innovative Sustainable System in Energy - Food - Water Nexus"

ABOUT THE CONFERENCE

ASTECHNOVA 2017

INTERNATIONAL ENERGY CONFERENCE

"Innovative Sustainable System in Energy - Food - Water Nexus"

Short Description

With huge population growth and unstable economic problem, nowadays, the challenges on securing the needs of energy, water, and food (E-W-F) become a big issue. The connection of the E-W-F aspects offer ascent to the requirement for handling the difficulties with a nexus approach. The discussion between the research progresses between E-W-F aspects is vital for better comprehension the complicity of E-W-F nexus and adopting coordinated strategies for the next progress.

The ASTECHNOVA 2017 is to host communication and dissemination of proposed new approaches in dealing with E- W-F nexus issues, as studying complexity and knowledge management. The papers on the issues relating to technology innovations and policy implementation of the E- W-F nexus s are welcomed. This event is the third international energy conference. Since 2014 till 2016 the conference was held successfully and attended by several participants and experts from Japan, Cuba, Cambodia, Vietnam, Philippines, Nepal, South Africa, China, Japan, Cuba and USA. Map of reviewer and presenter of Astechnova 2014 – 2016 is shown as follow:

Department of Nuclear Engineering and Engineering Physics, Universitas Gadjah Mada

Phone/Fax : +62 274 580882 - Email : astechnova@ugm.ac.id

www.tf.ugm.ac.id/astechnova

The ASTECHNOVA 2017 -International Energy Conference- will systematically examine the following topics

- Oil and Gas
- New and Renewable Energy
- Nuclear Engineering
- Green Technologies
- Energy Management and Services
- Environmental Compliance
- Plant Engineering and Instrumentation & Control
- Energy efficiency
- Nexus from the Energy perspective related to water and food
- Future clean energy technologies and systems under water and food constrains
- Land, climate and energy constraints on water and food sectors
- Future urban energy systems with considerations of water and food security
- Technology innovations for supporting global secure access to clean energy and water
- Other relevant topics on energy – water – food nexus.

Rationale

The 6th ASTECHNOVA 2017 is one of contribution of Department of Nuclear Engineering and Engineering Physics, Universitas Gadjah Mada (DNEEP) for energy-related researches and publication. In this context, by promoting research innovation in energy-related researches, DNEEP can reach the global network. This conference will be part of big steps in DNEEP to provide the ideas exchange, fruitful discussion, and to faciliate the researchers to present their work as part of energy-community information circle. DNEEP believes that by holding the ASTECHNOVA 2017 - is a good opportunity to open more networks for researchers and practitioners from academia and industry whose comments and suggestions will help DNEEP to excel as a centre of excellent in energy system in the near future.

In ASTECHNOVA 2017, not so different with previous ASTECHNOVA, there will be the scientific members who theirs expertise and networks will be essential points in energy innovation. The scientific members; Prof. Sunil Chirayath (Texas A&M University, USA), Dr. Liem Peng Hong (NAIS Co., Inc, Japan), Prof. Agus Pulung Sasmito (McGill University, Canada), Prof. Christophe Demaziere (Chalmers University of Technology, Sweden), Hiroshige Kikura (Tokyo Institute of

Technology), Prof. Sunarno, Prof. Phillip Beeley (Khalifa University, United Arab Emirates), Prof. Yonghee Kim (KAIST, South Korea), Dr. Jim Kuijper (NUCLIC, Netherlands) Dr.-Ing. Sihana (Universitas Gadjah Mada, Indonesia), Dr. Gea O.F. Parikesit (Universitas Gadjah Mada, Indonesia), Dr. Totok Soehartanto (Institut Teknologi Sepuluh Nopember, Indonesia) and Dr. Jarot Setyowiyoto (Universitas Gadjah Mada, Indonesia), Prof. Dr. Ing. Ir. Misri Gozan, M.Tech (Universitas Indonesia, Indonesia), Prof. John Z. Zhai (University of Colorado Boulder).

Aims

1. To facilitate researchers and experts both from industry and academia to give the state of the art of innovation in energy system,
2. To present recent achievements and results of advanced technology developments in energy system,
3. To discuss and exchange ideas and experiences in the field of technology developments in energy system.

Event

The ASTECHNOVA 2017 - International Energy Conference - is proposed to take two days in November 2017 with 100 participants world wide that will present their papers, i.e.

Date	: 1 November 2017
Time	: 07:30 – 16:30 with five parallel scientific sessions
Venue	: Eastparc Hotel Yogyakarta

Important dates for participants for this event

Abstract submission deadline	: 11.07.2017
Notification of acceptance	: 07.08.2017
Full paper submission	: 04.09.2017

The detail schedule of ASTECHNOVA -International Energy Conference-, as follows:

The Grand Ballroom	
Time	**Schedule**
07:30– 08:00	Registration
08:00 – 08:10	**Welcome Remark** by General Chair of ASTECHNOVA 2017
08:10 – 08:20	**Opening** by Vice Rector for Research and Community Service of Universitas Gadjah Mada
08:20 – 08:30	**Souvenirs** for Keynote Speaker, Main Speakers, Volunteer Reviewers and Moderators by Head of - Department of Nuclear Engineering and Engineering Physics UGM
08:30 – 08:45	**Keynote Speech** by **Ambassador/Senior Advisor for Political, Legal and Security Affairs, Director of Non-Aligned Movement Centre for South-South Technical Cooperation (NAM CSSTC) – Prianti Gagarin Djatmiko-Singgih**
08:45 – 09:00	**Coffee break**
09:00 – 10:50	**Plenary Session 1 "Developing Energy - Food - Water Nexus for Today's and Future Civilization"** **Main Speaker :** - Ambassador of Bolivarian Republic of Venezuela – **Gladys F. Urbaneja Duran** - Federal Democratic Republic of Nepal – **Sundar Bahadur Khadka** - Republic of Cuba – **Barbara Hernandez Martinez** **Chair :** Rachmawan Budiarto
11:00 – 12:00	**Plenary Session 2 "Energy for Human Life"** **Main Speaker :** - Tokyo Institute of Technology – **Hiroshige Kikura** "Status in Fukushima and Decommissioning Project" - Department of Nuclear Engineering and Engineering Physics UGM – **Gea Oswah Fatah Parikesit** "Energy, Culture, and Humanities: A Quantitative Study on the Indonesian Art of Wayang Kulit" **Moderator :** Haryono Budi Santosa
12:00 – 13:00	**Lunch** at Grand Ballroom

Department of Nuclear Engineering and Engineering Physics, Universitas Gadjah Mada
Phone/Fax : +62 274 580882 - Email : astechnova@ugm.ac.id
www.tf.ugm.ac.id/astechnova

Scientific Parallel Session

Time	Orchid Room	Tulip Room	Magnolia Room	Carnation Room	Dahlia Room
	New and Renewable Energy	**Oil and Gas, Plant Engineering and Instrumentation & Control**	**Green Technologies, Energy Efficiency and Other**	**Nuclear Engineering and Other**	**New and Renewable Energy**
	Chair : Sihana	**Chair : Awang Wardhana**	**Chair : Sentagi Sesotya Utami**	**Chair : Alexander Agung**	**Chair : Nazrul Effendy**
13:00-13:15	#6 Analysis Of Characteristics Of Pendulum Oscillation In PLTGL-SB (Juniarko Prananda)	#2 Sustainable Innovation Of Processing Bagasse To Sodium Ligno sulphonate Surfactant Utilized In Enhanced Oil Recovery (Rini Setiati)	#21 Life Cycle Assessment for Canning Crab Product (Portunus Pelagius) Product in Indonesia (Putri Amelia)	#14 Longitudinal and shear wave velocities measurement in compacted bentonite for water content (Shun Kimura, et.al.)	#35 Study of Supervisory Control Implementation In A Small Scale Variable Speed Wind Turbine (Katherin Indriawati)
13:15-13:30	#12 Response to Pontoon and Pendulum Motion at Wave Energy Converter Based on Pendulum System (Teuku Arvisa)	#7 Design of PLC Integrated Process Simulation Software for Steam Generator Using Matlab Simulink (Ikhtiander Ikhtiander)	#11 Demand Simulation for Water, Food Irrigation, and Energy from Micro Hydro Power Plant in Sungai Bayang, Bayang Utara, Pesisir Selatan, West Sumatera (Siti Nurlaila Indriani, et.al.)	#33 Analysis Of Maintenance Task Of Secondary Cooling Process System At G.A. Siwabessy Research Reactor Using Reliability Centered Maintenance Method (Winda Sarmita)	#9 Implementation of Maximum Power Point Tracking (MPPT) Technique on Solar Tracking System Based on Adaptive Neuro-Fuzzy Inference System (ANFIS) (Imam Abadi, et.al.)
13:30-13:45	#39 Design And Implementation Soft-switching MPPT SEPIC Converter Using P&O Algorithm (Ony Qudsi, et.al.)	#26 An Application of Statistical Modelling on Power Generating Asset Performance Analysis at the Asam-Asam Steam Power Plant of South Kalimantan (Aqli Mursadin)	#10 The Effect Of Inlet Curvature And Flange On Wind Turbine Diffuser Performance (Muhammad Nasir, et.al.)	#37 Arc Plasma Sintering; Novel Technology for Producing of Zirconia Enhanced ODS Alloy (Arbi Dimyati)	#4 Performance Test Of Solar-Powered Ice Maker: Case Study in South Lampung (Putri Wullandari, et.al.)

Time	Orchid Room	Tulip Room	Magnolia Room	Carnation Room	Dahlia Room
	New and Renewable Energy	Oil and Gas, Plant Engineering and Instrumentation & Control	Green Technologies, Energy Efficiency and Other	Nuclear Engineering and Other	New and Renewable Energy
	Chair : Sihana	Chair : Awang Wardhana	Chair : Sentagi Sesotya Utami	Chair : Alexander Agung	Chair : Nazrul Effendy
13:45-14:00	#1 Synthesis of Fatty Acid Methyl Ester from Soybean Oil Using Electrolysis with Kaolinite Acid as Catalyst (Rudy Syah Putra, et.al)	#34 Adsorption of CO_2 and Methane on Covalent Organic Polymer (Lee Siew-Pei, et.al.)	#15 Analysis of Palm-Oil-Based Bio-diesel in Indonesia as Technological Innovation System (Dintani Naimah)	#22 Determination LMWL Wonosobo Area By Using Nuclear Technology, Case Study: Hydrogeology Study for Aqua Danone CSR Program (Agus Wijatna, et.al)	#38 Design and Implementation Buck Converter for 540WP Solar Charger Using Fuzzy Logic Control (Sutedjo, et.al.)
14:00-14:15	#20 Al-Alginate as Acid Catalyst for FAME Synthesis Using Electrolysis Process (Rudy Syah Putra)	#13 Design of New Ultrasonic Transducer with Two Elements for Flow Rate Measurement Using Ultrasonic Doppler Method (Jevin Tanius Owen , Ari Hamdani)	#5 Analog tachometer as an indicator machine weathering of motorcycle machine based on moving coil (Dhanis Woro Fittri, et.al.)	#23 Prognostic Methods on Accelerator's Anode Voltage Regulator (Robby Christian, Balza Achmad)	#19 Analysis of Back Surface Field (BSF) Performance in P-Type And N-Type Monocrystalline Silicon Wafer (Ferdiansjah)
14:15-14:30	#27 The Effect of Sulfuric Acid Hydrolysis of Bagasse on Glucose Concentration (Nunung Prabaningrum)	#25 Performance Test of VHF-HF-based Emergency Data Communication System between the Sardjito Hospital and District Hospital in the Isolated and Disaster-prone	#31 Modeling of Fluid Flow Rate on Hybrid Solar PV-Thermal Panel (Yakob TH, et.al.)	#17 Research and Development Green School which is located in Jalan Raya Sibang Kaja, Banjar Saren, of Instruments for Measurement, Recording, and Analysis	

Department of Nuclear Engineering and Engineering Physics, Universitas Gadjah Mada

Phone/Fax : +62 274 580882 - Email : astechnova@ugm.ac.id
www.tf.ugm.ac.id/astechnova

Time	Orchid Room	Tulip Room	Magnolia Room	Carnation Room	Dahlia Room
	New and Renewable Energy	**Oil and Gas, Plant Engineering and Instrumentation & Control**	**Green Technologies, Energy Efficiency and Other**	**Nuclear Engineering and Other**	**New and Renewable Energy**
	Chair : Sihana	**Chair : Awang Wardhana**	**Chair : Sentagi Sesotya Utami**	**Chair : Alexander Agung**	**Chair : Nazrul Effendy**
		Areas (Singgih Hawibowo, et.al.)		of The Power Quality toward Sustainable Energy Management System at Green School, Bali, Indonesia (Dita Anggraini, et.al.)	
14:30-14:45	#18 Characterization of Natural Zeolite and Chicken Manure Derived Biochar for Carbon Dioxide Adsorption in Biogas (Margaretha Wuri, et.al.)	#29 Neural Network Model For Smart Lighting Control In Energy Efficient Building Based on Occupants' Activities (Ivan Hidaya, et.al.)	#24 The Wireless Energy Transfer (WET) Using Ultra High Frequency (UHF) For Human Body Implant Recharging (Fahmy Rinanda Saputri, et.al.)		
14:45-15:15	**Closing ceremony (The Grand Ballroom)**				

Department of Nuclear Engineering and Engineering Physics, Universitas Gadjah Mada

Phone/Fax : +62 274 580882 - Email : astechnova@ugm.ac.id
www.tf.ugm.ac.id/astechnova

International Scientific Committee

Prof. Sunil Chirayath (Texas A&M University, USA)

Dr. Liem Peng Hong (NAIS Co., Inc, Japan)

Prof. Agus Pulung Sasmito (McGill University, Canada)

Prof. Christophe Demaziere (Chalmers University of Technology, Sweden)

Prof. Hiroshige Kikura (Tokyo Institute of Technology, Japan)

Prof. Sunarno (Universitas Gadjah Mada, Indonesia)

Prof. Phillip Beeley (Khalifa University, United Arab Emirates)

Prof. Yonghee Kim (KAIST, South Korea)

Dr. Jim Kuijper (NUCLIC, Netherlands)

Dr.-Ing. Sihana (Universitas Gadjah Mada, Indonesia)

Dr. Gea O.F. Parikesit (Universitas Gadjah Mada, Indonesia)

Dr. Totok Soehartanto (Institut Teknologi Sepuluh Nopember, Indonesia)

Dr. Jarot Setyowiyoto (Universitas Gadjah Mada, Indonesia)

Prof. Dr. Ing. Ir. Misri Gozan, M.Tech (Universitas Indonesia, Indonesia)

Prof. John Z. Zhai (University of Colorado Boulder)

Local Organizing Committee (Universitas Gadjah Mada, Indonesia)

General Chair	: Rachmawan Budiarto
Publicity & PR	: Fitrotun Aliyah
Financial Chair	: Faridah
Treasurer	: Amrih Lestari, Safrudin
Publication Chair	: Balza Achmad, Agus Arif, Alexander Agung
TPC Chair	: Awang Wardhana, Anung Muharini, Singgih Hawibowo
Program Chair	: Susetyo Hario Putero
Logistic Chair	: Widya Rosita
Sponsorship & PR	: Sentagi Sesotya Utami Haryono Budi Santosa, Ferdiansjah

edp sciences

Statement of Peer review

In submitting conference proceedings to E3S Web of Conferences, I certify to the Publisher that I adhere to the **Policy on Publishing Integrity** of the journal in order to safeguard good scientific practice in publishing.

1. All articles have been subjected to peer review administered by the proceedings editors.
2. Reviews have been conducted by expert referees, who have been requested to provide unbiased and constructive comments aimed, whenever possible, at improving the work.
3. Proceedings editors have taken all reasonable steps to ensure the quality of the materials they publish and their decision to accept or reject a paper for publication has been based only on the merits of the work and the relevance to the journal.

Title, date and place of the conference

Astechnova 2017
International Energy Conference
1st November 2017
Eastparc Hotel, Yogyakarta, Indonesia

Proceedings editor(s):

Prof. Ir. Sunarno, M.Eng., Ph.D (Universitas Gadjah Mada, Indonesia)
Prof. Agus Pulung Sasmito (McGill University, Canada)
Dr. Liem Peng Hong (NAIS Co., Inc. Japan)

Date and editor's signature

13rd April 2018

Adsorption of CO2 and Methane on Covalent Organic Polymer

Siew-Pei Lee[1], *N.* Mellon[1,*], *Azmi* M. Shariff[1] *and Jean-Marc Leveque*[2]

[1]Chemical Engineering Department, Universiti Teknologi PETRONAS, 32610 Bandar Seri Iskandar, Malaysia
[2]LCME, Universite de Savoie Mont-Blanc, 73376 Le Bourget du Lac cedex, France

Abstract. Development of covalent organic polymer (COP) is a potential new class of adsorbent for CO2 separation from natural gas mainly due to their good hydrothermal stability, chemical tuning flexibility and low cost. CO2 and methane adsorption on COP-1 was studied under atmospheric condition (101.3 kPa, 298 K). COP-1 was synthesized via catalyst-free polycondensation of cyanuric chloride and piperazine. The properties of COP-1 were characterized using several analytical methods such as Fourier Transform Infra-Red (FTIR), N2 adsorption and desorption measurement and Field Transmission Electron Microscopy in coupled of Energy Dispersive X-ray Spectroscopy (FESEM-EDS). Reversible CO2 adsorption isotherm on COP-1 reflects low heat of adsorption which is beneficial to energy minimization in adsorbent regeneration process. Furthermore, moderate specific surface area COP-1 (88.5 m2/g) shows about nine times CO2 uptake higher than methane. The highly selective adsorption performance provides a promising insight in application of COP adsorbent for CO2 removal in natural gas field.

1 Introduction

Removal of CO_2 from gas reservoirs is highly important since the presence of CO_2 can cause operational problem such as solid formation and top-line corrosion in wet-gas pipelines (*1*) . CO_2 is also an inert gas that reduces the heating value of natural gas and the amount of gas available for sales. Thus, CO_2-content must be minimized to within 2% - 3% by volume in order to meet the standard for compositional specification on pipeline gas (*2*).

Currently, the common method for CO_2 removal from natural gas is via chemical absorption process using alkanoamine. However, this technology is subjected to operational drawbacks such as high rate of corrosion on equipment, degradation of solvent through oxidation, high energy consumption for regeneration and large absorber volume (*3, 4*). This drives the research for alternative gas treating technologies for CO_2 removal from natural gas reservoirs that is robust with higher removal efficiency and low operational problems. Among the possible emerging technologies, selective adsorption process is one of the most promising approaches because of the intrinsic eco-compatibility and flexibility such as zero regeneration of chemical solvents and applicability in relatively wide range of temperatures and pressures (*5*). In all adsorption technologies, the separation efficiency and adsorption capacity drive the selection of adsorbing material. Discovery of metal organic framework (MOF) (*6, 7*) and microporous organic polymer (MOP) such as covalent organic framework (COF) (*8*) and covalent organic polymer (COP) (*9-11*) are among the breakthroughs in alternative adsorbent development. These materials are very flexible in terms of structure tuning and properties through rational chemical design and synthesis. For example, the recently generated COP-1 (*10*) has a great potential in CO_2 removal from natural gas stream with high hydrothermal stability, promising CO_2 adsorption capacity and good adsorbent recyclability. COP-1 was obtained by the polymerization of cyanuric chloride (CC) and piperazine as shown in Fig. 1. Free of metal catalyst in the synthesis offers some of the advantages in material development such as low cost production and environmental friendly.

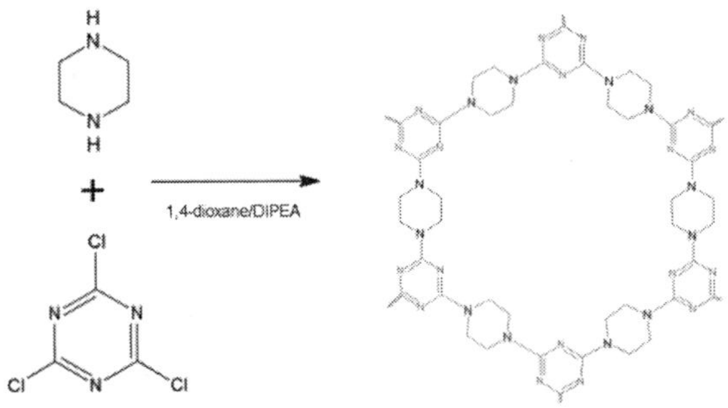

Fig. 1. Synthesis pathway of COP-1

Presently, available information for COP-1 is mainly concentrated on CO_2 removal in post-combustion process. However, scarce information on CO_2 removal from natural gas using COP adsorbent is reported in literature. Thus, the focus of this paper is to report the initial findings of the investigation on CO_2 removal from methane (CH_4), a representative of natural gas, using COP-1 at atmospheric condition

* Corresponding author: norhaye@utp.edu.my

https://doi.org/10.1051/e3sconf/20184301001

2 Methodology

Anhydrous cyanuric chloride (CC), piperazine and N,N-diisopropylethylamine (DIPEA) were purchased from Sigma Aldrich, USA. Dry 1,4-dioxane solvent with ⩽ 0.005% water content and ethanol were purchased from Merck Milipore.

CC (5.00 g, 27.1 mmol) was dissolved in dry 1,4 dioxane (50 mL). The solution was then added dropwise to a solution of piperazine (3.73 g, 43.3 mmol) and DIPEA (18.9 mL) in 1,4-dioxane (150 mL) with continuous stirring at 288 K under N_2 atmosphere. Resulting white mixture was stirred at 288 K for 1 h and then for a further with 2 h at 298 K and then 21 h at 363 K. The off-white precipitate formed was washed with 1,4-dioxane and soaked with ethanol three times over a period of 12 h. COP-1 was obtained after the off-white slurry was dried at room temperature under vacuum for 2 h.

Fourier transform infrared spectroscopy (FT-IR) spectra were recorded Thermo Scientific Nicolet iS 10 FT-IR Spectrometer using ATR technique for CC, piperazine and COP-1. The morphology and Energy-dispersive X-ray (EDX) analysis was performed by FESEM on a Carl Zeiss AG SUPRA 55 VP. N_2 adsorption isotherm was obtained using Micromeritics ASAP 2020 Surface Area and Porosity Analyser at 77 K after the samples had been degassed at 428 K for 5 h. Low pressure CO_2 and CH_4 adsorption (up to 101.3 kPa) at 298 K were measured by Belsorp-mini II, Japan. Sample was degassed under vacuum at 423 K for 5 h prior to analysis.

3 Results and Discussion

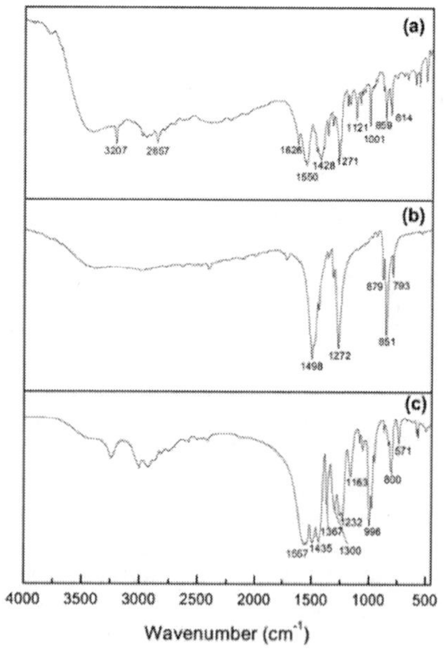

Fig. 2. FTIR spectra of (a) Piperazine (b) Cyanuric Chloride (c) COP-1

FTIR spectra, present the functional groups of CC, piperazine and COP-1. The characteristic stretching band for C-Cl functionality in the FTIR spectrum for CC is clearly visible at 850 cm^{-1} [Fig 2(b)]. The disappearance of C-Cl stretching band in COP-1 structure [Fig 2(c)] suggested that all three chlorine atoms on CC were fully substituted. As a result, it is confirmed that the polycondensation reaction for COP-1 synthesis was fully completed in this study. Additionally, several strong peaks in the 1200-1600 cm^{-1} region appeared in all three spectra, representing the stretching modes of CN heterocycles.

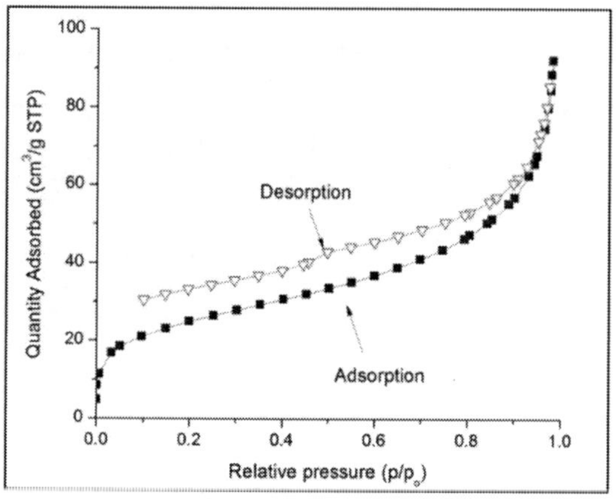

Fig. 3 N_2 adsorption desorption isotherm

Fig. 4. FESEM micrograph of COP-1 (inserted: EDX analysis pattern)

Isotherm of COP-1 is plotted in Fig. 3 as the quantity adsorbed against the relative pressure. The isotherm is slightly convex at very low p/p_0 and no identification of a plateau as the relative p/p_0 increasing (*12*). Hysteresis Type H3 loop was observed when desorption branch follows a different path and its appearance is due to the capillary condensation in mesopores. This isotherm is classified as Type IIb isotherm where the material forms aggregates of plate-like particles with non-rigid slit-shaped pores (*13*). Moderate specific surface area is observed in COP-1 i.e. 88.5 m^2/g, calculated from the

https://doi.org/10.1051/e3sconf/20184301001

linear part of the Brunauer-Emmett-Teller (BET) plot. Besides, about 9.7 nm of Barrett-Joyner-Halenda (BJH) desorption average pore diameter was found on COP-1.

FESEM micrograph Fig. 4 revealed the flaky and flower petals structure of COP-1. Bulk sphere structure is found to be due to the agglomeration of plate shape of the particles which is correlated to Type H3 hysteresis loop found in nitrogen isotherm. However, the bulk agglomerated spherical particle has limited the diffusion of gas molecules which highly attributes to low surface area of COP-1. The EDX analysis performed highlighted that no impurity was found on COP-1 other than C and N elements, as shown in inserted Fig. 4.

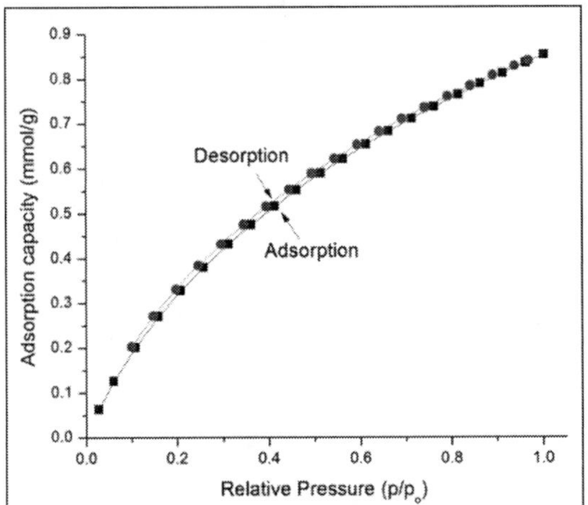

Fig. 5. CO_2 adsorption desorption isotherm at 298 K

CO_2 isotherm on COP-1 for low pressure at 298 K is shown in Fig. 5. It is apparent that CO_2 molecules begin to fill in the mesopores at $p/p_0 \approx 0.03$. The increase of CO_2 adsorption on COP-1 as p/p_0 increases is because of the growth of clusters of molecules on the surface. Mesopores are filling with CO_2 adsorbates when the clusters merge together (*14*). Reversible isotherm without hysteresis in Fig. 5 reflects low heat of adsorption of the host for CO_2 adsorbates due to the affinity of physisorption for CO_2 is dominant on surface of COP-1 instead of chemisorption (*15*). This suggests that COP-1 with unique CO_2 adsorption isotherm competitive for energy minimization in adsorbent regeneration process compared to zeolites(*16*).

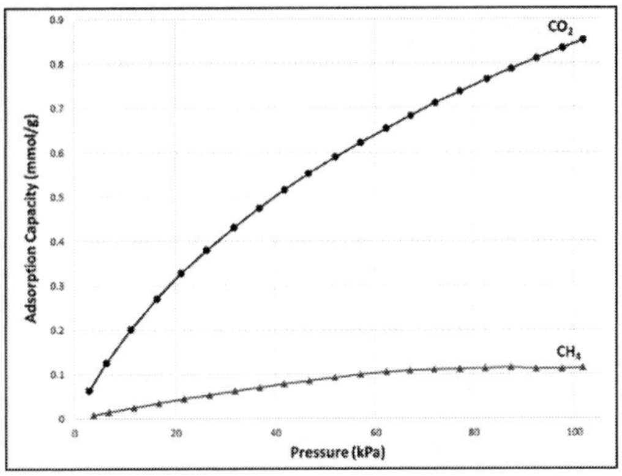

Fig. 6. CO_2 and CH_4 adsorption isotherms at 298 K

The CO_2 and CH_4 adsorption isotherms at 298 K of COP-1 are shown in Fig. 6. In this study, the CO_2 uptake i.e about 0.9 mmol/g is significantly higher than CH_4 uptake (0.1 mmol/g) although both molecules are non-polar and display similar kinetic diameter (0.33 nm and 0.38 nm, respectively). It is interesting to note that the quadrupole moment of CO_2 moiety (4.30) is larger than CH_4 (0.02). Therefore, the interaction between the higher quadrupole moment of CO_2 molecules and dense electron cloud of nitrogen-rich network of COP-1 may facilitate CO_2 adsorption through the surface field gradient-molecular quadrupole interactions (*5*).

4 Conclusion

In conclusion, successfully synthesized COP-1 shows low adsorption energy for regeneration. At atmospheric condition, the dense electron cloud of polymeric network provides nine times of CO_2 uptake as compared to CH_4. It indicates that COP-1 possesses a great potential for CO_2 removal in natural gas field. Therefore, adsorption studies in wider range of temperature and pressure are necessary to evaluate its thermodynamic behaviour and CO_2 separation performance under natural gas drastic conditions.

Acknowledgement

This work is financially supported by Ministry of Higher Education of Malaysia under MyRA Incentive Grant for CO_2 Rich Natural Gas Value Chain program.

References

1. M. Safari, A. Ghanizadeh, M. M. Montazer-Rahmati, Int. J. Greenhouse Gas Control **3**, 3 (2009).
2. A. J. Kidnay, W. R. Parrish, *Fundamentals of Natural Gas Processing*. (CRC Press, 2006).
3. T. E. Rufford, S. Smart, G. C. Y. Watson, B. F. Graham, J. Boxall, J. C. Diniz da Costa, E. F. May, J. Pet. Sci. Technol. **94–95**, 123 (2012).
4. W. Nowak, D. Wawrzynczak, I. Majchzrzak-Kuceba, J. Pacyna.

https://doi.org/10.1051/e3sconf/20184301001

5. M. Tagliabue, D. Farrusseng, S. Valencia, S. Aguado, U. Ravon, C. Rizzo, A. Corma, C. Mirodatos, Chem. Eng. J. **155**, 553 (12/15/, 2009).
6. A. Demessence, D. M. D'Alessandro, M. L. Foo, J. R. Long, J. Am. Chem. Soc. **131**, 8784 (2009/07/01, 2009).
7. H. Furukawa, N. Ko, Y. B. Go, N. Aratani, S. B. Choi, E. Choi, A. Ö. Yazaydin, R. Q. Snurr, M. O'Keeffe, J. Kim, O. M. Yaghi, Sci. **329**, 424 (July 23, 2010, 2010).
8. R. Babarao, J. Jiang, Energy Environ. Sci. **1**, 139 (2008).
9. H. A. Patel, S. Hyun Je, J. Park, D. P. Chen, Y. Jung, C. T. Yavuz, A. Coskun, Nat Commun **4**, 1357 (2013).
10. H. A. Patel, F. Karadas, A. Canlier, J. Park, E. Deniz, Y. Jung, M. Atilhan, C. T. Yavuz, J. Mater. Chem. **22**, 8431 (2012).
11. Z. Xiang, X. Zhou, C. Zhou, S. Zhong, X. He, C. Qin, D. Cao, J. Mater. Chem. **22**, 22663 (2012).
12. F. Rouquerol, J. Rouquerol, K. Sing, in *Adsorption by Powders and Porous Solids,* F. Rouquerol, J. Rouquerol, K. Sing, Eds. (Academic Press, London, 1999), pp. 191-217.
13. F. Rouquerol, J. Rouquerol, K. Sing, in *Adsorption by Powders and Porous Solids,* F. Rouquerol, J. Rouquerol, K. Sing, Eds. (Academic Press, London, 1999), pp. 237-285.
14. F. Rouquerol, J. Rouquerol, K. Sing, in *Adsorption by Powders and Porous Solids,* F. Rouquerol, J. Rouquerol, K. Sing, Eds. (Academic Press, London, 1999), pp. 117-163.
15. S. Choi, J. H. Drese, C. W. Jones, ChemSusChem **2**, 796 (2009).
16. A. Majchrzak, W. Nowak, Journal of CO2 Utilization **17**, 69 (2017).

https://doi.org/10.1051/e3sconf/20184301002

Al-alginate as acid catalyst for FAME synthesis using electrolysis process

Rudy Syah Putra[1,2*], and *Tasyrifatur* Rahma[1]

[1]Department of Chemistry and [2]New and Renewable Energy Research Group, Universitas Islam Indonesia, Jl. Kaliurang km. 14 Yogyakarta 55584, Indonesia

Abstract. A novel solid acid catalyst, which was prepared from sodium alginate and aluminum chlorides and characterized by SEM-EDS, XRD and FT-IR spectrometry. The catalyst was used for the synthesis of fatty acid methyl ester (FAME) via (trans)esterification reaction using electrolysis process. Purity and FAME components of biodiesel were measured by GC-MS. The results showed that the aluminum-alginate prepared in a cheap and easy way exhibited high catalytic activity. FAME conversion from used-cooking oil was obtained by 93.82% in the presence of 10 wt.% of catalyst dosage upon refluxing for 4 h of methanol and oil in a molar ratio of 6:1.

1 Introduction

Fossil fuel is now widely recognized as unsustainable resource. The limitations of fossil fuel had demanded to look for other energy sources that can be renewable. A candidate as renewable energy to substitute petroleum is biodiesel. Biodiesel is a mixture of fatty acid methyl esters (FAME) produced from free fatty acids via esterification and/or (trans)esterification with methanolisis process. Various biodiesel standard namely flash point, cold filter plugging point, sulphur content, and oxygen consumption of biodiesel are improved as compared with that of fossil-based diesel. Moreover, it was found out that biodiesel used as engine fuel can effectively reduce the emission of poisonous gases [1-2].

Typically, conventional biodiesel produces straight forwardly by means of homogeneous alkaline catalyst, such as NaOH and KOH. For high proportion of free fatty acid (FFA) and water containing in a low quality of used-cooking oil, homogeneous alkaline catalyst are markedly sensitive to FFA, leading to catalyst losses, saponification problem, lowering the efficiency and economics of the overall process [3]. The synthesis of biodiesel from used-cooking oil (UCO) by using electrolysis process had been reported by Guan and Kusakabe with the yield > 97% [4]. These methods have some advantages to produce biodiesel from used-cooking oil. For example, the high water content in the oil can be directly used in electrolysis process (i.e. the electron transfer emerged by the formation of $H^{\cdot}$ and OH^{-} ions) is derived indirectly as a catalyst in esterification and transesterification reaction at room temperature [5]. In this study, the enhancement of FAME synthesis from used-cooking oil using electrolysis process by Al-alginate as heterogeneous acid catalyst.

2 Methodology

2.1 Chemical and equipment

Used-cooking oil was provided from home industry activities. High grade of methanol, NaCl, AlCl.6H2O and co-solvent THF were purchased from Merck, Germany and food grade of sodium alginate was purchased from local food industry in Yogyakarta province of Indonesia. Fig. 1 shows an electro-catalytic cell (250 ml) contains two graphite electrodes (Ø 0.8 mm x 130 mm) that were separated by a distance of 2.5 cm [5]. The qualities of used-cooking oil were determined by physicochemical analysis and the results were shown in the Tab. 1

Table 1. Physicochemical properties of used-cooking oil

Oil	FFA (%)	Refractive index	ρ, Kg/L	Moisture, %
UCO	1.89	1.462	0.956	0.099

2.2 Preparation and characterization of catalyst

A novel solid acid catalyst of Al-alginate, which was prepared from sodium alginate and metal chlorides according to publish paper by Qiuyun et al. [6]. Briefly, 2 g of sodium alginate was added to 100 mL of 100 °C distilled water, followed by stirring at 700 rpm until a clear viscous solution was obtained. Then, the viscous solution was added drop wise into 100 mL of 0.1 M

* Corresponding author: rudy.syahputra@uii.ac.id

https://doi.org/10.1051/e3sconf/20184301002

$AlCl_3$ solution. The resulting solution system was stirred vigorously for 13 h to form a white of Al-alginate complex. Finally, the Al-alginate granules were washed until a neutral pH and then dried at room temperature for 48 h. The X-ray diffraction (XRD) patterns of the catalyst were measured using a Bruker D2 Phaser diffractometer system (equipped with a monochromatic Ni filter using CuKa radiation). Fourier transform infrared (FTIR) spectra was recorded on a Perkin Elmer Spectrum Two equipped with the universal ATR. FTIR spectrophotometer without using the KBr disc technique (4000-400 cm-1) and Scanning electron microscope (SEM) was obtained from a Phenom ProX desktop equipped with an energy dispersive spectroscope (EDS) detector to characterize the microstructures and chemical properties of as-prepared Al-alginate material. In addition, a PinAAcle 900 Perkin Elmer measured the sodium and aluminum concentration in the prepared alginate.

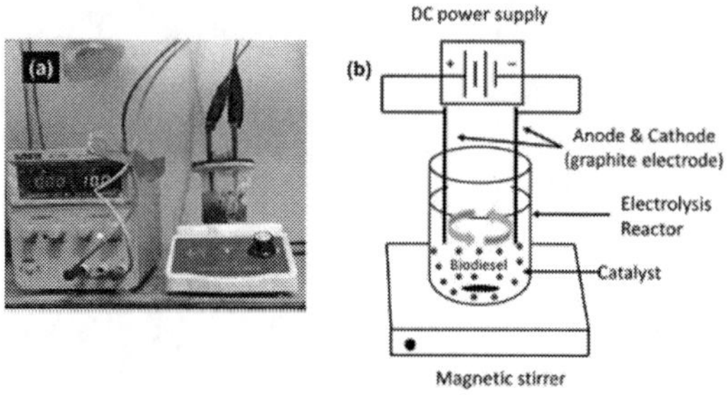

Fig. 1. Electrolysis reactor set up (a) and illustrated functional apparatus in the reactor (b).

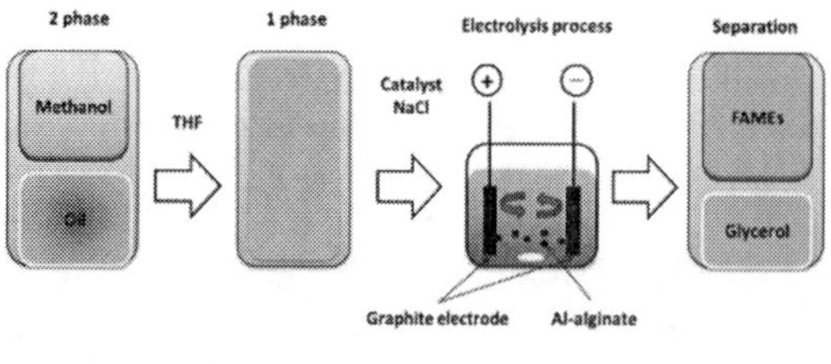

Fig. 2. Steps on FAME synthesis using electrolysis process and Al-alginate as catalyst

2.3 Catalytic testing

Effect of catalyst amount (5, 10 and 15 wt.%) was evaluated on the conversion of FAME from used-cooking oil. The reaction mixture was agitated using magnetic stirrer. The electrolytic cell was filled by 70 ml of reaction mixture containing used-cooking oil, methanol, THF as a co-solvent [5]. The methanol-to-oil molar ratio was 24:1, co-solvent-to-methanol molar ratio was 1:0.25 and NaCl concentration was added on the basis of oil weight (0.56 wt.%) of the entire reaction mixture, respectively. The electro-catalytic was carried out at room temperature using a constant voltage of 18.6 V for 2, 4, 6 h. Fig. 2 shows a step of optimization process for biodiesel production in this study. The FAMEs phase were separated from glycerol and the product was washed with deionized water to remove the residual inorganic components. FAME conversion and unreacted oils remaining in the product was measured by GC-MS.

3 Results and Discussion

3.1 Catalytic characterization

Fig. 3 shows the XRD patterns of Na-alginate and the Al-alginate catalyst. Crystalline of sodium alginate was formed by strong interaction between the alginate chains through intermolecular hydrogen bonding [7]. The degree of crystallinity of the alginate was estimated by using the software DIFFRAC.EVA version 4.0. The Na-alginate presents a 18.6% crystallinity degree; and the Al-alginate, a 11.0%. Three diffraction peaks at 2θ values 13.5°, 22° and 39° were observed for sodium alginate due to the reflection of their (110) plane from polyguluronate unit, (200) plane from polymannuronate and the other from amorphous halo [8]. In case of Al-alginate sample, the intensity of diffraction peaks of alginate decreased notably which indicates that the formation of aluminum oxyhydroxide particles on the grid of alginate affected the crystalline nature of sodium alginate. In other words, this procedure is successful in the preparation of the aluminum-alginate catalyst. The broadening of the diffraction peaks in all prepared alginate indicates nature of the sample.

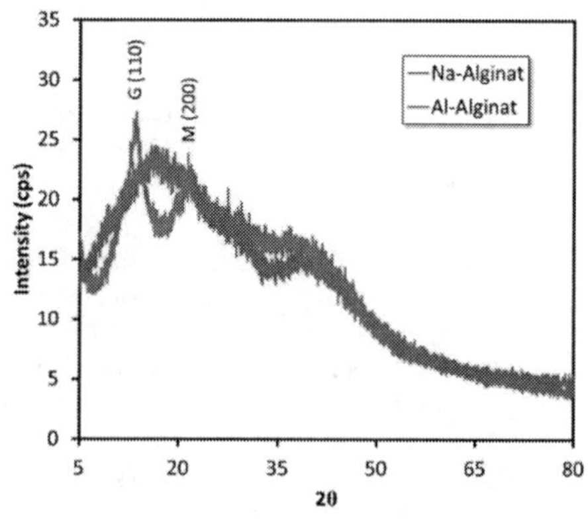

Fig. 3. XRD diffractogram of Na-alginate (a) and Al-alginate (b)

Fig. 4 shows the FTIR spectra of Na-Alginate and Al-Alginate. The FTIR spectrum of Na-alginate shows the characteristic peaks at 3242 cm^{-1} (OH^- stretching) [9], peaks at 1591 and 1407 cm^{-1} (COO^- asymmetric and symmetric stretching), 1081-1024 cm^{-1} (C-O-C antisymmetric stretching), and carboxyl and carboxylate at peaks about 1000 to 1400 cm^{-1} [10]. In FTIR spectra of

Al-alginate, the asymmetric band of carboxylate ion has shifted to lower frequencies from 1591 cm^{-1} to 1593 cm^{-1}, and the hydroxy band of sodium alginate has shifted from 3242 cm^{-1} to 3253 cm^{-1}, because of the interaction between sodium alginate and $AlCl_3$ [9]. The band at 1025 cm^{-1} was given by the guluronic units [11] in all the spectra.

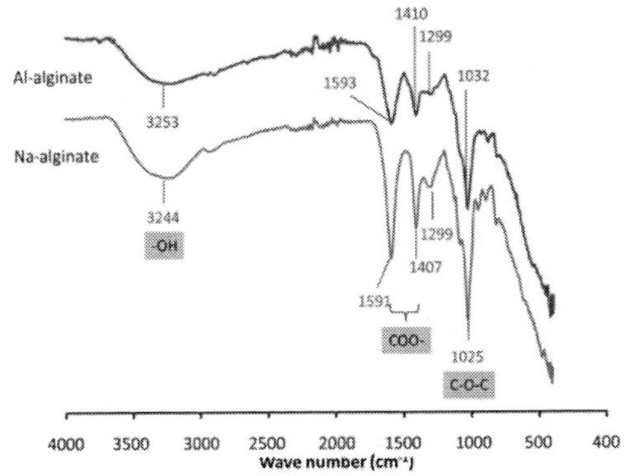

Fig 4. FTIR spectra of Na-alginate and Al-alginate

The results from SEM revealed that bigger micro architecture of Al-alginate formed than Na-alginate in terms of their pore sizes (Fig. 5). A wide and porous surfaces layer of the Al-alginate give a large catalytic activity so that increased the performance of esterification in electrolysis process. EDS spectra of the composite films clearly show the characteristic peaks of aluminum ($AlCl_3$) used for the synthesis along with notable amounts of carbon and oxygen from alginate backbone.

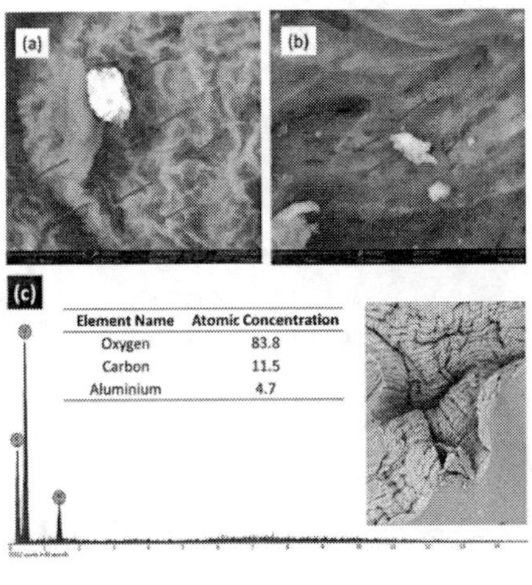

Fig. 5. SEM images showing the topology of Al-alginate (a) and Na-alginate (b) in x10.000 magnificent. The red arrow indicated the porous surface in the material. Al-alginate images and its corresponding EDS analysis (c).

The quantify of ion exchange process between sodium alginate and the contacting $AlCl_3$ solution was measured by an atomic absorption spectrometer. Initial sodium concentration in the Na-alginate was 0.00247 mg/Kg. While, aluminum concentration was 0.0648 mg/Kg in the Al-alginate. Therefore, the ratio of ion exchange between aluminum to sodium was 26.23 Al : Na.

3.2 FAME synthesis with electrolysis

3.2.1 Influence of electrolysis time

Tab. 2 shows the FAME yield increased by the increase of electrolysis time. It can be seen that the FAME yield rapidly reached up to 95.94% within 2 h. After 6 h of reaction, the conversion rate leveled off as the products attained a near-equilibrium composition as high as 96.51%. However, the optimum reaction time for electrolysis was 4 h since total FAME yield was the highest (see Fig. 6) and it was more efficient for electrolysis time. Fig. 6 shows typical of FAME conversion from used-cooking oil in the electrolysis process. When the FFA present in the used-cooking oil, the electrolysis process were successively subjected to esterification reaction with methanol, the high to low FFA conversion (%) from used-cooking oil was as follow oleic acid > palmitic acid > stearic acid > myristic acid, respectively.

Table 2. Effect of electrolysis time on the FAME yield without catalyst

Electrolysis time (hours)	FAME yield (%)
2	95,54
4	95,94
6	96,51

In this regards, the electrolysis was conducted at room temperature (25 °C) without catalyst.

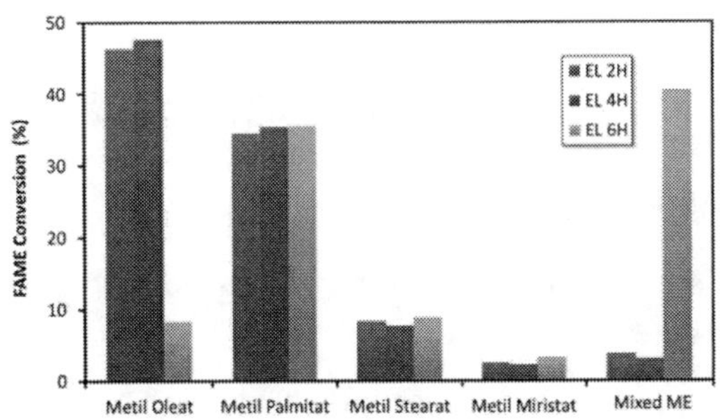

Fig. 6. FAME conversion from used-cooking oil in the electrolysis process without catalyst.

3.2.2 Influence of catalytic loading

Tab. 3 shows the effects of wt.% catalyst to oil ratio on the FAME yield using electrolysis. The biodiesel yield increased by the increase of catalyst loading. It was discovered that excellent conversion rates (as high as 93.76% were obtained in 10 wt.% of catalyst loading since total FAME yield was the highest (see Fig. 7). This fact shows that aluminum-alginate was an appropriate solid catalyst that has a very high reactivity to catalyze esterification reaction. This study showed that the Al-alginate catalyst could be a good candidate to enhance the electrolysis conversion of cheap feed-stocks containing high amount of FFAs (such as inedible oils [12], waste vegetable oils [13], waste cooking oils [14], etc.) into biodiesel.

Table 3. Effect of catalyst loading on the FAME yield in electrolysis process

Catalyst loading (wt.%)	FAME yield (%)
5	87,55
10	93,76
15	93,82

In this regards, the electrolysis was conducted at room temperature (25 °C) in 4 hours process.

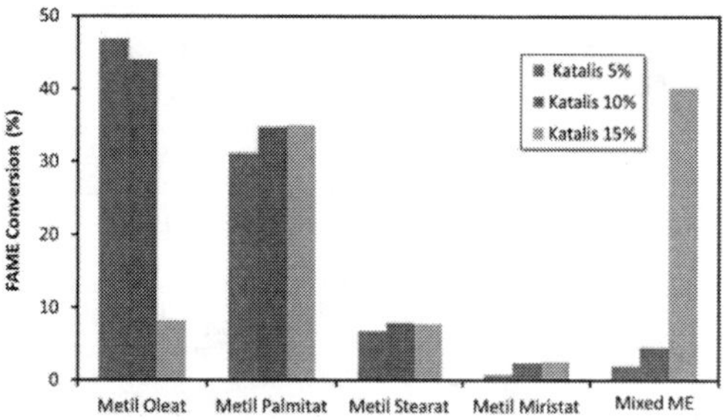

Fig. 7 FAME conversion from used-cooking oil in the electrolysis process enhanced by catalyst

4 Conclusions

The enhancement of FAME yield from the (trans)esterification of used-cooking oil using electrolysis process with Al-Alginate as acid catalyst had already evaluated. High catalyst concentration (10 wt. %) with 4 hours of electrolysis time at room temperature was required in order to get a high FAME yield as much as 93.82%. Future works will need to explore high specific catalytic surface of Al-Alginate aiming to enhance the catalyst in the electrolytic biodiesel production.

Acknowledgement(s)

Authors would like thank to Ministry of RISTEK-DIKTI, the Republic of Indonesia for financial support through Penelitian Produk Terapan 2017. Furthermore, additional support from AIPT program 2016 of Universitas Islam Indonesia is gratefully acknowledged.

References

1. Ng J H. Ng H K. Gan S. J. Applied Energy. 90(1): 58-67 (2012)
2. Ng H K, Gan S. J. Applied Thermal Engineering. 30(16): 2476-2484 (2010)
3. Kayser. H. Pienko, F. and Maria P.D. Fuel. 116:267-272 (2014)
4. G. Guan and K. Kusabe. Chem. Eng. 153:159–163 (2009)
5. R.S. Putra. P. Hartono and T. Julianto. Energy Procedia. 65:309–316 (2015)
6. Z. Qiuyun. L.H.Q. Wenting. L. Xiaofang. Z. Yuping. X.Wei. Y. Song. China Petrol. Process. Petrochem. Tech..14: 2033–2043 (2013)
7. D. Fang. Y. Liu. S. Jiang, J. Nie. and G. Ma. Carbohydrate Polymers, 85 (1): 276–279 (2011)
8. Fabia J. Ślusarczyk C. Gawłowski A. Fibres Text. East. Eur. 13(5): 114-117 (2005)
9. G. Lawrie, I. Keen, B. Drew, A. Chandler-Temple. L. Rintoul, P. Fredericks. and L. Grøndahl. Biomacromolecules. 8 (8), 2533–2541 (2007)
10. M.G. Sankalia, R.C Mashru. J.M. Sankalia. V.B. Sutariya. AAPS PharmSciTech (6) 2: E209-E222 (2005)
11. Pereira. L. Sousa. A. Coelho. H. Amado. A.M. and Ribeiro-Claro. P.J.A. Biomolecular Engineering. 20: 223-228 (2003)
12. R. Wang, M.A. Hanna, W.W. Zhou, P.S. Bhadury. Q.Chen, B.A Song. S. Yang. Bioresource Technology.102(2):1194-1199 (2011)
13. Ceclan R E, Pop A, Ceclan M. Chemical Engineering Transactions, 29: 1177-1182 (2012)
14. Gan S Y, Ng H K, Chan P H, et al. Fuel Processing Technology, 102, 67-72 (2012)

E3S Web of Conferences 43, 01003 (2018)
ASTECHNOVA 2017
https://doi.org/10.1051/e3sconf/20184301003

An application of statistical modelling on power generating asset performance analysis at the Asam-asam steam power plant of South Kalimantan

Aqli Mursadin[1,*]

[1]Faculty of Engineering, Lambung Mangkurat University, Banjarmasin 70123

Abstract. A power generating system in a steam power plant is a complex one. This system involves a large number of variables containing information concerning the operation conditions. It can also be seen as an important asset in both power generation and energy portfolios. At the Asam-asam power plant of South Kalimantan, the complexity of the system has led to difficulties in explaining the apparently steady decrease in the output power. The amount of data collected is simply too big and the dimension too high for analysis purposes based on conventional thermodynamics. This study was performed to tackle the problem using statistical modelling. This approach can accommodate empirical behaviors of the variables and the probabilistic nature of the system. Information obtained from this modelling can be valuable for various purposes. The method consists of literature review, model development, data collection and analysis, and model fitting. Generalized additive models were chosen. Data were available from the company as observed from more than 140 variables. The resulting model identifies variables significantly related to the output power and locate subsystems whose fluctuating behaviors are usually ignored in a conventional thermodynamic analysis. A direction for future research is recommended.

1 Introduction

A power generating system in a steam power plant is a complex one. It converts chemical energy in the fuel supplied to the system into heat. The circulating water fed to the boiler as liquid captures this heat. This surge in energy is such that it turns the liquid water into superheated vapor, which then turns around the blades of the turbine. The resulting mechanical work becomes electricity through the coupling of the rotor with the shaft of an electrical generator. This complex mechanism works in a cycle within a configuration of equipment [1, 2].

Such a complex system involves a large number of variables. These variables carry with them information concerning the operation conditions of the plant. Consequently, analysis of these variables and the relationships between them is essential for explaining the behavior and the performance of the system. These variables can be considered as random variables whose observed values are subject to random chances, and their analysis requires an approach that accommodates their probabilistic behaviors.

Further, such a system can be seen as an important asset in both power generation and energy asset portfolios. The performance of this system requires the use of statistical techniques as pointed out by [3] in the context of equipment and process monitoring in engineering asset management. A promising approach is based on modelling such behaviors.

The Asam-asam steam power plant has four separate generating units located in the Asam-asam Village of the Tanah Laut Regency 122 km northeast of Banjarmasin. Of these generating units, Unit 1 was initially synchronized into the grid on 28 June 2000, Unit 2 on 25 October 2000, Unit 3 on 24 March 2012, and Unit 4 on 7 November 2012. Each of these units was originally designed to have a maximum output power of 65 MW.

Data collected until December 2016 indicate that there has been a steady decrease in the output of Unit 2 with the lowest value of around 85 percent of the maximum. A study by [4] shows that the vacuum pressure at the main steam condenser has an increase of 20 percent. This may suggest that the main steam leaving the turbine still has a sufficiently high enthalpy (total energy), or the fouling growth inside the condenser is unusually rapid, or else. Any one of these may contribute to the above-mentioned decrease in the output. However, this study only considered variables observed on equipment and systems directly connected to the turbine. For examples, several devices like the deaerator (used to remove gas from liquid feed water) and the pulverizer (which supplies the coal powder) may perform processes

* Corresponding author: a.mursadin@gmail.com

whose variables fluctuate in values in such a pattern consistent with that of output power.

It is difficult to explain this steady decrease due to the high dimension of variables and the big size of data. It is also worsen by the fact that the available measurement data have never been properly used for studying the system.

An ability to model the relationships between these process variables can be useful in uncovering information concerning the operation behaviors of the system. This, in turn, can provide knowledge and understanding beneficial to the maintenance of the power plant and the improvement of its performance in the future.

For this reason, it is important to study how it is possible to model the power generating behavior of the system in such a way that completely explains the effect of each of the system variables. This study should give insight into the empirical behavior of the system. It can provide a set of models with which the relationship between system variables can be represented and analyzed. The resulting information will be useful for operation and maintenance purposes as well as for future improvement of the system.

2 A problem of decreasing output power

A steam power plant utilizes total energy or enthalpy of circulating water to generate electricity. It basically consists of several main components including a steam turbine, a boiler, a heat exchanger known as a condenser, a water feed pump that supplies condensed water to the boiler, and a complex piping system along with several other auxiliary devices and pieces of equipment. The rotating shaft of the turbine is then coupled with its counterpart in the electrical generator to produce electricity. This generating system works in a cycle. A theoretical introduction to this type of power plant from the thermodynamic viewpoint can be found in [1].

A modern modification of the above cycle is the one currently installed in the Asam-asam power plant, especially the generating unit 2. Here, several other devices are added such as a low-pressure heater, a high-pressure heater, a deaerator, and an economizer. Fig. 1 shows the diagram of the cycle at this generating unit.

Processes in such a cycle usually involve a large number of variables. For instance, at the generating unit 2 as mentioned above, more than 140 variables are monitored and their values logged every two hours on a daily basis.

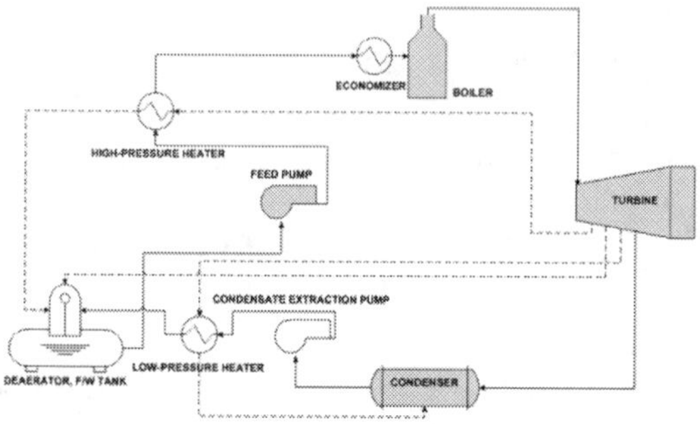

Fig. 1. A diagram of the generating cycle at the generating unit 2 of Asam-asam.

As previously mentioned, there has been a problem of decreasing output power at the Asam-asam power plant. The worst case can be found at the generating unit 2 where the lowest output power is around 85 percent of its maximum. Fig. 2 shows the empirical distribution of the output power at the generating unit 2 of the plant during the period between 01/12/2016 and 14/12/2016 (in MW). Apparently, the average output is far below its maximum capacity.

Moreover, the above mentioned period saw several processes involved in the generating system of that particular unit being in their unstable conditions. This means that the data simply represent those processes' states of instability. This is, however, a short period before a general overhaul was carried out in the beginning of the following year.

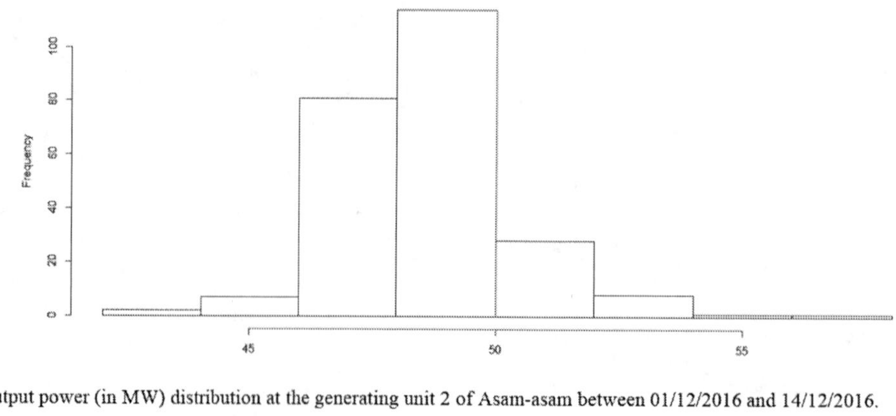

Fig. 2. Output power (in MW) distribution at the generating unit 2 of Asam-asam between 01/12/2016 and 14/12/2016.

https://doi.org/10.1051/e3sconf/20184301003

As reported in [4], the vacuum pressure at the main steam condenser has an increase of 20 percent. This may suggest that the main steam leaving the turbine still has a sufficiently high enthalpy (total energy), or the fouling growth inside the condenser is unusually rapid, or else. Any one of these may contribute to the above-mentioned decrease in the output. However, this study only considered variables observed on equipment and systems directly connected to the turbine. For examples, several devices like the deaerator (used to remove gas from liquid feed water) and the pulverizer (which supplies the coal powder) may perform processes whose variables fluctuate in values in a pattern consistent with that of output power.

3 Modelling engineering systems

Complex engineering systems involve a large number of variables. These variables carry with them information concerning the operation conditions of the systems. Consequently, analysis of these variables and the relationships between them is essential for explaining the behavior and the performance of the systems. These variables can be considered as random variables whose observed values are subject to random chances, and their analysis requires an approach that accommodates their probabilistic behaviors.

According to [5], especially under unstable conditions, engineering processes are better monitored using methods based on automatic process control. This approach, however, does not necessary provide opportunities for studying the root causes of the problems. In [6] it is clear that statistical process control does provide such opportunities. Unfortunately, the latter requires that the process be stabilized first before observations can result in any meaningful information about the behaviors.

A promising approach is based on modelling such behaviors. An approach centered on system assets in information modelling has been proposed by [7]. The use of statistical modelling techniques such as those with regression and multivariate models is required when the number of variables involved is large and they have to be simultaneously observed as multivariate variables. Several applications have been reported for monitoring and controlling engineering processes as in, *e.g.*, [8, 9].

Generalized additive models (GAMs) can be used to fit engineering data using smoothing terms through addition. The general form is given, *e.g.*, in [10] as follows:

$$g(\mu_i) = \beta_0 + \sum_{j=1}^{k} f_j(x_{ij}) \qquad (1)$$

where the link function $g(\mu_i)$, with μ_i being the expected value of the ith observation, is given as a linear combination of k smoothing functions or smoothing terms $f_j(x_{ij})$ and the intercept β_0. The function $g(\mu_i)$ is chosen to be associated with a certain exponential family of probability functions such as the normal family or the gamma family. For instance, the empirical distribution in Fig. 2 may be modelled using (1) where an f_j would be a smoothing function of one of the other system variables. Computing packages are now available for this purpose in a statistical programming language like R.

4 Methods

The modelling approach chosen in this study was based on generalized additive models. This allows relationships to be inferred between quantitative values in data in an additive fashion, thoroughly tested, and easily verified and understood. It also allows function smoothing to be performed on the relationship.

The object was chosen to be the power generating system of generating unit 2 at the Asam-asam steam power plant in South Kalimantan. Data were taken from the already available log sheets at the plant with permission, that is, from 01/12/2016 to 14/12/2016. The total number of variables observed was 144. There were 151 observations for each of these variables. Part of a log sheet used at the plant is shown in Fig. 3.

This is a period of time that saw several processes involved in the generating system of that particular unit being in their unstable conditions. This means that the data represent those processes' states of instability. A general overhaul was carried out in the beginning of the following year.

The resulting matrix of data was then fed to the model as in (1) using an interface of the programming language R (available from http://cran.r-project.org). The mgcv package in R was used to fit the data.

5 Data analysis and results

Before applying model (1) it would be useful for the modelling process to reduce the number of system variables. This was done by screening for system variables having a sufficiently high coefficient of determination (R^2) with output power. Such variables and their location in the cycle are given in Table 1.

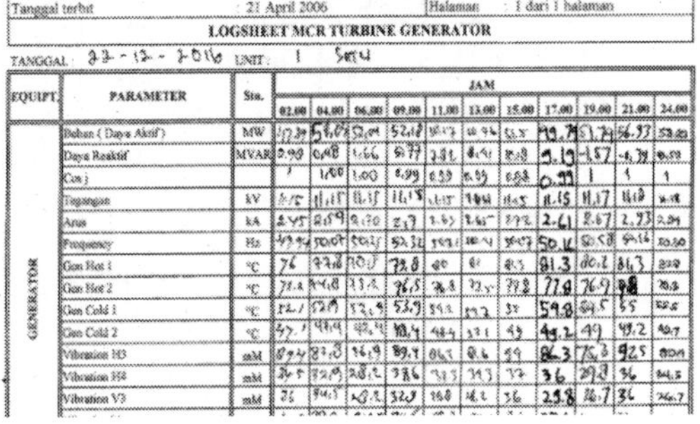

No. Dokumen : SMM-SASM 04.OPR.23 Revisi : 2
Tanggal terbit : 21 April 2006 Halaman : 1 dari 1 halaman

LOGSHEET MCR TURBINE GENERATOR

TANGGAL: [illegible] UNIT: 1

EQUIPT.	PARAMETER	Sat.	02.00	04.00	06.00	09.00	11.00	13.00	15.00	17.00	19.00	21.00	24.00
GENERATOR	Beban (Daya Aktif)	MW	[illegible]	[illegible]	[illegible]	52.18	[illegible]	[illegible]	[illegible]	[illegible]	[illegible]	[illegible]	[illegible]
	Daya Reaktif	MVAR	[illegible]	[illegible]	[illegible]	[illegible]	[illegible]	[illegible]	[illegible]	[illegible]	[illegible]	[illegible]	[illegible]
	Cos φ		[illegible]	1.00	1.00	0.99	[illegible]	[illegible]	[illegible]	0.99	1	1	1
	Tegangan	kV	[illegible]	[illegible]	[illegible]	[illegible]	[illegible]	[illegible]	[illegible]	11.15	11.17	[illegible]	[illegible]
	Arus	kA	[illegible]	[illegible]	[illegible]	[illegible]	[illegible]	[illegible]	[illegible]	2.61	2.67	2.73	[illegible]
	Frequency	Hz	[illegible]	[illegible]	[illegible]	[illegible]	[illegible]	[illegible]	[illegible]	[illegible]	[illegible]	[illegible]	[illegible]
	Gen Hot 1	°C	76	[illegible]	[illegible]	[illegible]	[illegible]	[illegible]	[illegible]	81.3	[illegible]	[illegible]	[illegible]
	Gen Hot 2	°C	[illegible]	[illegible]	[illegible]	76.5	[illegible]	[illegible]	[illegible]	77.8	76.9	[illegible]	[illegible]
	Gen Cold 1	°C	[illegible]	[illegible]	[illegible]	53.9	[illegible]	[illegible]	[illegible]	59.8	[illegible]	55	[illegible]
	Gen Cold 2	°C	[illegible]	[illegible]	[illegible]	[illegible]	[illegible]	[illegible]	[illegible]	[illegible]	49	49.2	[illegible]
	Vibration H3	mM	[illegible]	[illegible]	[illegible]	[illegible]	[illegible]	[illegible]	[illegible]	86.3	75.3	92.5	[illegible]
	Vibration H4	mM	[illegible]	[illegible]	[illegible]	[illegible]	[illegible]	[illegible]	[illegible]	36	[illegible]	36	[illegible]
	Vibration V3	mM	[illegible]	[illegible]	[illegible]	[illegible]	[illegible]	[illegible]	[illegible]	[illegible]	[illegible]	36	[illegible]

Fig. 3. Part of a log sheet used at the plant.

Generated current is the electrical current generated in the generator. First stage pressure is the pressure of the first stage turbine. Condensate outlet pressure is the pressure of the condensed water leaving the low-pressure heater. Extracted steam pressure is the pressure of steam extracted from water by the deaerator. Feed water tank pressure is the pressure inside the feed water tank. Generated current and first stage pressure are directly related to output power at the generator. Hence, their relationship with output power is obvious. The other three variables, however, are not directly related to output power.

Table 1. Variables screened for the model.

Variable	Location	Symbol
Generated current	Generator	x_1
First stage pressure	Turbine	x_2
Condensate outlet pressure	Low-pressure heater	x_3
Extracted steam pressure	Deaerator	x_4
Feed water tank pressure	Feed water tank	x_5

Model fitting was performed using R programming language. The distribution family chosen for the expected value of output power is a gamma distribution. The shape of the histogram in Fig. 2 has motivated this choice. This requires the use of an inverse link function [10]. The model is given as follows:

$$1/\log(\mu_i) = \sum_{j=1}^{5} f_j(x_{ij}) \quad (2)$$

In general, the five system variables can explain more than 83 percent of variation in output power value. The other 138 variables are responsible for less than 17 percent of the variation. The significance of the effect of the five variables can be seen from the resulting probability value (p-Value) of their corresponding smoothing term. The smaller this value is, the more significant the effect will be. This is given in Table 2.

Table 2. Significance of system variables.

Variable	p-Value
Generated current	0.001428
First stage pressure	0.012185
Condensate outlet pressure	0.000734
Extracted steam pressure	0.001069
Feed water tank pressure	0.000776

The overall fit of the model can also be seen from its adjusted R^2 value (0.805) and its generalized cross validation score (3.26×10^{-6}). A higher R^2 value and a smaller generalized cross validation score mean a better fit [10].

It is obvious that there is a strong relationship between generated current and the output power. However, it is quite surprising that variables condensate outlet pressure and feed water tank pressure contribute to variation in output power more significantly than the rest of the system variables. From an engineering viewpoint, the two variables are not directly related to output power at the generator, and this relationship might not be apparent from the measured values of those variables. The generalized additive model used for this purpose was based on a gamma distribution family. The link function is an inverse of a logarithmic function which makes the function more sensitive to changes in the two variables.

This result shows how it is possible discover empirical relationships between variables previously ignored or considered trivial. The model provides opportunities to study an engineering system even when it is at its unstable state.

Information from such a study can be very useful for operation and maintenance purposes. For instance, it might be recommended that the engineer in-charge of the operation look at the processes involving the corresponding variables. Although the model can also be used to predict the values of output power, its main purpose is to explain the variation of the variable.

6 Conclusions and recommendations

A steady decrease in output power of a steam power plant generating unit is a problem faced at the Asam-asam power plant. It is important to explain this decrease. In doing so, it is necessary to be able to explain the effects of hundreds of variables on output power. The approach based on conventional thermodynamic analysis is not sufficient for this purpose since several assumptions have been violated by actual operation conditions.

The study has demonstrated how it is possible to model a complex engineering system, *i.e.* a power generating system using statistical modelling even when it is at its unstable state. The complexity comes from the fact that the number of variables involved is large and the empirical relationships between variables are not necessarily known.

A generalized additive model has been used to model the relationship between output power and five other variables. By fitting this model, it is discovered that condensate outlet pressure and feed water tank pressure contribute very significantly to changes in output power. It is recommended that this study be taken further to examine the implementation of the modelling results on the actual operation of the system.

https://doi.org/10.1051/e3sconf/20184301003

The author would like to thank Mr. Andi Y. Iriyanto from PT. PLN at the Asam-asam power plant for his assistance during the data collecting process, and students under his supervision at the Mechanical Eng. Dept., Lambung Mangkurat Univ. for preparing a digital version of the data.

References

1. Y. A. Çengel and M. A. Boles, *Thermodynamics: An Engineering Approach 8th Ed.* (McGraw-Hill Education, New York, NY, 2015).
2. H. F. Martin, "Steam Turbines for Power Generation", in K. R. Rao (ed.), *Energy and Power Generation Handbook: Established and Emerging Technologies* (ASME Press, New York, NY, 2011).
3. C. Emmanouilidis and P. Pistofidis, "Wireless Condition Monitoring and Embedded Novelty Detection", in J. E. Amadi-Echendu, K. Brown, R. Willett, and J. Mathew (eds.), *Definitions, Concepts and Scope of Engineering Asset Management* (Springer-Verlag, London, 2010).
4. A. Y. Iriyanto, *Analysis of Steam Turbine Performance at the Generating Unit 2 of the Asam-asam Power Generating Plant* (transl. unpublished Final Project Report) (Department of Mechanical Eng., Lambung Mangkurat University, 2017).
5. G. E. P. Box, D. E. Coleman, and R. V. Baxley, Jr. "A comparison of statistical process control and engineering process control" Journal of Quality Technology, **29** (1997): 128-130.
6. H. S. Gitlow, A. J. Oppenheim, R. Oppenheim, and D. M. Levine, *Quality Management 3rd Ed.* (McGraw-Hill, New York, NY, 2005).
7. N. Fujisawa, S. Matsuo, Y. Yamamoto, and K. Nonaka, "Thermal power plant asset management with asset-centric data model", Fujitsu Sci. Tech. J. **50** (2014): 76-84.
8. J. F. MacGregor, H. Yu, S. G. Munoz, and J. Flores-Cerrilo, "Data-based latent variable methods for process analysis, monitoring and control", Computers and Chemical Eng. **29** (2005): 1217-1223.
9. L. Pan, D. Flynn, and M. Cregan, "Statistical model for power plant performance monitoring and analysis", Universities Power Eng. Conf. (2007): 121-126.
10. A. Agresti, *Categorical Data Analysis 2nd Ed.* (John Wiley & Sons, Inc., Hoboken, NJ, 2002).

 https://doi.org/10.1051/e3sconf/20184301004

Analog tachometer as an indicator motorcycle machine wearing based on moving coil

Dhanis Woro Fittrin Selo Nur Giyatno[1,*], *Tommy* Richard Orlando[1], and *Nining* Supriatin[1]

[1]Department of Electrical Engineering and Informatics, Vocational College, Universitas Gadjah Mada

Abstract. As an increasing highly mobility and high traffic, the necessary of motorcycle´is highly increasing. The condition makes user ride motorcycle with highly speed in highly frequency. Then, these conditions make motorcycle machine reliability is highly decreasing. Finally, it made machine is run to damage and maintenance cost to be high. Analog tachometer is an electronic instrumentation that proposed to solve these problems. Actually, instrumentation system of tachometer is an electromechanical system. A wire in a control unit is embedded into shaft of crank. Then, magnet in control unit will convert rotary machine energy into electrical energy with d'Arsonval meter. Current sensor and small variable resistor are the kind of sensor that are used in tachometer. Small variable resistor is used for tuning and recalibration. Utilization small variable resistor in tachometer circuit is make calibration and recalibration current sensing of electrical current that rectified by diode. Tachometer for counting rotation per minute (RPM) motorcycle machine is built up. The tachometer has capability to count RPM motorcycle machine 1,000 – 13.000 RPM. The range is reliable as an indicator for user to minimize motorcycle machine wearing.

1 Introduction

Motorcycle is a popular modern transportation in Indonesia because cheap and high mobility. Therefore, almost people ride with high speed without consider safety and machine wearing. High working intensity make machine reliability decreases. Then, motorcycle machine need high cost to maintenance.

This paper a electronic instrumentation to count the number of motorcycle machine revolution with analog display. Analog tachometer is an electronic instrumentation to count machine revolution based on time period – revolution per minute (RPM). Selection analog display as an RPM measured because for user friendly purpose as utilization analog tachometer for indicator so detailing value is not priority; not for measurement. RPM is a parameter to look process of machine performance.

Reference [1] explains designing and assembling digital tachometer. Reference [2] mentioned that tachometer is used to get information about machine performance. Then, reference [3] explains utilization inertia sensor with generator tachometer. Then, in the reference [4], differential tachometer is designed with arduino as control center to reduce noise. Furthermore, the result of machine revolution is transferred into personal computer. In the reference [5] [6] [7], digital tachometer is used in brushless dc motor testing. Reference [8] [9] use contactless tachometer.

Instrument based electronics circuit such as analog tachometer is used to give weathering or reliability indicator through showing rpm value of motorcycle machine for user. Furthermore, other researchers are used different method to show machine reliability. Reference [10] used certain algorithm to shown machine reliability, namely machine vibration. Some selection indicators are used to increased diagnosis performance. This diagnose model is conform to implemented for system with profesional user and complex machine that its reliability is affected by many factors. Then, reference [11] used machine vibration as parameter to determine machine fault or the reliability of synchrone rotating machine with permanent magnet. Vibration model is validated by finite element analysis (FEA) conforms to assess machine reliability so the machine fault and isolation fault can be detected and then isolated. Reference [12] explains fault diagnosis for sensor and actuator in the HVAC (heating, ventilation, air conditioner, and cooling) system with temperature as based parameter. Diagnosis process is consisted of 3 steps, namely detecting with distributed linear estimation. Then, local detecting that is consisted of adaptive estimation scheme and decision determining with combinatorial logical decision. Finally, differentiation between actuator fault and sensor fault is done then fault isolation is implemented with neighborhood decision agent. This method is used to diagnose fault which is caused by sensor and actuator or

* Corresponding author: dhanisworo@ugm.ac.id

front end and back end process. Reference [13] explain utilization of active-tolerant control strategy (AFTCS) for octorotor unmanned aerial vehicle (UAV) with considering some motor faults. AFTC have facility to detect error, fault isolated and recovery system. While, parameter for diagnosis are motor velocity and current measurement. When energy losses is detected, algorithm would activated recovery system facility. Reference [14] explains algorithm model or software which can be implemented in industrial manufacture to detect fault process. Alarm in automacy industry is validated as the best solution to be actuator in the method. Reference [15] explain fault detection for gearbox in wind turbine generator. With fault tree analysis, intelligent system is made to detect the machine fault. Qualitative and quantitative analysis are made with C++ in the platform software .Net. Reference [16] explains complexity fault detection process to detect fault of rotating machine because its non-stationery character and non-linear vibration since its operated. Then, machine fault detection which is can extract and classified fault character is proposed. Machine fault detection can detect operation fault and forecasting fault. Reference [17] explains utilization of wavelet analysis to detect machine fault. In the wavelet analysis, every fault can be interpreted as energy losses so fault detection can be done fastly. Reference [18] explains spectrum utilization in the machine fault detection. Differentiation between signal and noise can be done with dividing signal into small signal. Then, power parameter is better to be based parameter to fault detection than voltage or current signal only. Reference [19] explain about complex signal identification in the machine fault detection. Signal sampling toward measurement signal is done correctly. Reference [20] explains utilization of locality constrained linear coding with support vector machine (LLC-SVM) to detect machine fault with vibration as based parameter. LLC-SVM is used to identificate vibration signal and classfified into the kind of fault in the rotating machine.

Based on these researches, there are many information to support our proposed design, namely instrument facility design should be conformed to instrument's function. Detail result is needed if tachometer is used to be instrument measurement. Furthermore detailing and complexity should be reduced when tachometer is used to be an indicator because fast response is priority. Digital output is priority to fulfill tachometer function as an instrument measurement and analog output is used when tachometer's function is an indicator. Arduino as microcontroller is used to control signal output for tachometer

2 Analog Tachometer

2.1 Tachometer

Tachometer is an instrument to count the number machine revolution based on certain period. Then, tachometer will display the output in the analog or digital number. Motorcycle use tachometer as an indicator machine wearing based on machine revolution.

On the motorcycle machine, tachometer is used to measured direct or indirectly machine revolution based on axis and typically shown the safety range for user. Then, tachometer is controlled by cable moving from control unit which is inserted into machine by electrical method or non-electrical method

2.2 Moving Coil

Instrument that work based on moving coil meter is known with d'Arsonval meter. This coil is consisted permanent magnet and coil so an electromagnet phenomenon is happened. Therefore, because of the sensitivity from magnet effect, electromagnet field effect must be avoided from moving coil area. Figure 1 shows physics of moving coil in tachometer.

Fig. 1. Moving coil.

2.3 Electromechanics Concept

Torque is machine capability to work. Torque parameter is a differential parameter to counting produced energy by rotate machine based on its centered. Therefore, torque is formulated as multiplication between centrifugal force and distance between object and centered or axis. On the combustion motor, axis power is known by measure torque with dynamometer. The principal work of the instrument is giving opposite load with rotation direction until machine revolution 0.00 RPM.

2.4 Fuel – Mechanics and Mechanics – Thermal Energy Conversion

There are many component in the motorcycle system that support motorcycle performance. There are needed primary and secondary gear that have connection with centered and gearbox. Fuel oil in motorcycle will converted into thermal and mechanics energy.

https://doi.org/10.1051/e3sconf/20184301004

1. Thermal energy is produced by fuel burning and machine compression.

2. Mechanical energy is produced by thermal chamber with caused by machine compression. The kind of energy is used to make primary gear to be move.

3. Primary gear is used to induction process, especially in mechanical energy conversion into thermal energy conversion.

4. Secondary gear is used to reduce gearbox level, especially in thermal energy conversion into mechanical energy.

3 Design Method

3.1 Principal Work of Tachometer

Voltage input in analog tachometer indicated tachometer ready to count machine revolution. When transducer detects machine revolution, sensor will detect the electrical current that flowed in the circuit. Electrical current in sensor will be rectified by diode then used to make coil to be moved. Furthermore, the coil will make pointer to be move based on the number of machine revolution.

3.2 Designing Electronic Circuit for Analog Tachometer

Voltage input in analog tachometer indicated that tachometer is ready to count the machine revolution. Current detected by current sensor will rectify by diode and be calibrated with electronic component. Then, electrical current will make coil to be move and the voltage produced will measured by IC.

Fig. 2. Electronics circuit of analog tachometer

Furthermore, Figure 2 shows electronic instrumentation circuit of analog tachometer with Eagle software. In this circuit, analog tachometer is designed with 3 cables by different function, namely yellow cable for connecting pulser with diode, black cable for connecting power supply with analog tachometer, and green cable for connecting analog tachometer with ground. Analog tachometer wiring is shown by Figure 3.

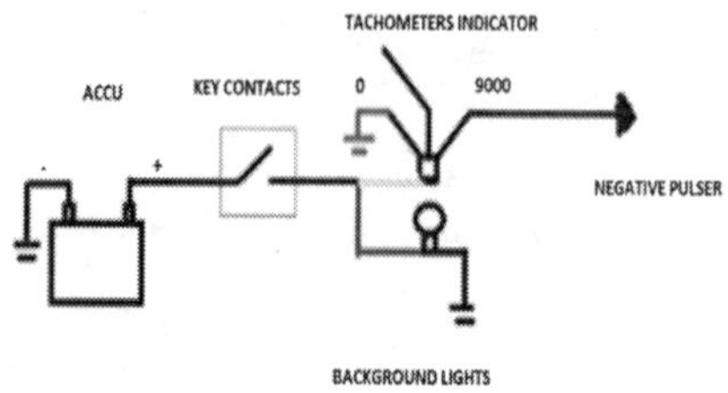

Fig. 3. Analog tachometer wiring

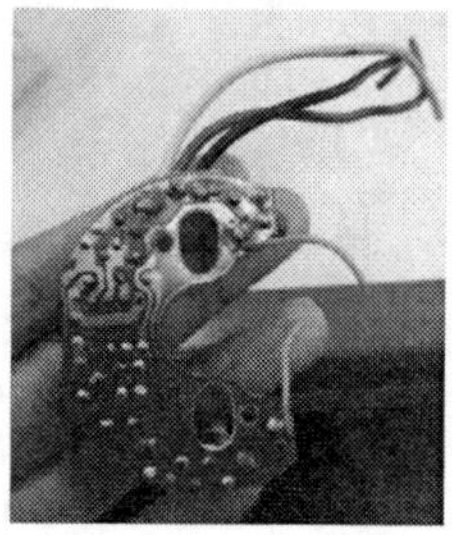

Fig. 4. Electronics circuit board of analog tachometer

3.3 Designing Analog Tachometer

Mechanical design for analog tachometer is made after electronic instrumentation circuit is designed. Figure 5, Figure 6, and Figure 7 show prototype planning, analog tachometer without package, and the final product of analog tachometer.

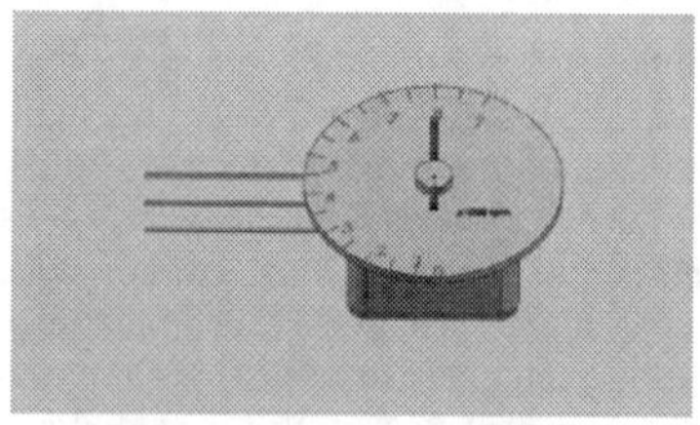

Fig. 5. Proposed design of analog tachometer

Fig. 6. Analog tachometer without packaging

Figure 6 shows the result of measurement of analog tachometer without package. The frontend is used to display the result of motorcycle machine revolution. The subsystem is made by acrylic then connected with machine revolution parameter. The parameter have 13 as highend with 1,000 RPM as lowest machine revolution and 13,000 RPM as highest machine revolution.

Fig. 7. Analog tachometer witH packaging

Then, pointer is designed moved clockwisely. As an increasing of pointer deviation, the electrical current that readable is higher. Figure 7 shows the product of measurement by analog tachometer. The final step of design method is testing, namely connecting the three cable into motorcycle.

4 Result and Discussion

Tachometer is tested with install these electronics circuit into motorcycle. Then, machine revolution is measured when motorcycle was operated. The result shows as follow.

4.1 Tachometer testing

Tachometer is designed have machine revolution with range 0 – 13,000 per minute – RPM. Though motorcycle machine is still started to rotate, machine have rotate with 1,000 RPM. It is caused machine must be against machine torque firstly to rotate. Then, when motorcycle machine have machine revolution 1,000 – 3,000 RPM, machine on the stationery or ready to rotate. Therefore, this range indicates transient condition of motorcycle machine. In this range, machine still works normally. Furthermore, when motorcycle machine have revolution speed 3,000 – 10,000 RPM, machine have optimal condition. At this condition, machine has high performance condition. Therefore, user should have ride motorcycle until 10,000 RPM.

When motorcycle machine have revolution more than 10,000 RPM until 13,000 RPM, machine works inefficient. Therefore, when tachometer shows machine revolution more than 10,000 RPM, rotation speed of motorcycle must be decreased. Though the maximum of motorcycle machine revolution is 10,000 RPM, tolerance about 3,000 RPM is given because the difference is needed machine to change its position, namely starting from OFF – ON and ON – OFF. When tachometer have machine revolution more than 13,000 RPM, analog display will move to 0 RPM that indicate user must decrease its rotation speed instantly.

4.2 Selection Consideration of Analog Display for Tachometer

Though many indicator panel on the modern vehicle with digital display, tachometer for motorcycle machine performance is designed with analog display. It is caused tachometer for motorcycle has function to give indicator and not to measure so fast response is priority than readability detailing.

Besides that, the trend of machine revolution is dynamically. For example, when motorcycle machine is started, machine will have revolution 1,000 – 3,000 RPM at less than 1 second. Then, when tachometer has digital display, the change of measurement result cannot detect effective.

4.3 Relationship Between Machine Revolution, Motorcycle Speed, and Fuel Efficiency.

Motorcycle machine revolution is not connecting with motorcycle speed directly. Motorcycle machine revolution is comparable with work level for same certain machine type and character. Therefore, work level is comparable with fuel consumption.

Motorcycle machine have several gear that effective to use on the certain machine revolution. Because of each gear have different size of gear, the gear torque in these range is different. In the motorcycle machine with manual transmission, utilization of gear is maintained manually with itself. For example, on the motorcycle machine with 2nd gear, machine revolution to get rotation speed 40 km/hour is needed 3,000 RPM. On the other hand, utilization 3rd gear to get same rotation speed, revolution speed that needed 2,000 RPM. With the example, it is known that utilization 2nd gear to get rotation speed 40 km/hour is needed more fuel. Then, in the motorcycle machine with manual transmission, it will maintain its gear with itself to get every same rotation speed. Though automatic gear selection uses machine revolution parameter as control parameter.

4.4 Ideal Machine Revolution

The number of ideal machine revolution for motorcycle machine is depending on motorcycle specification. Usually, motorcycle machine has ideal revolution with range 6,000 – 10,000 RPM. In this range, user is suggested use higher gear level. On the other hand, the regulation is not applied for motorcycle machine with automatic transmission.

Table 1. Measurement and validation revolution per minute of motorcycle machine.

No.	Revolution per minute (RPM)	Motorcycle speed (km/hour)	Safety level
1.	4,000	30.00	Safe
2.	6,000	50.00	Safe
3.	8,000	70.00	Prone
4.	9,000	80.00	Prone
5.	11,000	100.00	Damage
6.	13,000	120.00	Damage

Based on Table 1, it is known that as an increasing machine revolution (indicated with RPM), the speed of motorcycle is more than before. Therefore, as an increasing the number of RPM, the need of fuel is higher.

5 Conclusion

Tachometer is an electronic instrumentation used to motorcycle user to keep the machine reliability. Knowing revolution per minute (RPM) of machine, user can anticipate gear level and speed of motorcycle. It is because an increasing the number of RPM of motorcycle machine make an increasing of machine wearing due to machine component friction. An increasing torque of motorcycle, it is needed high level of fuel.

References

1. M. Rana, S. Sahabuddin, and S. Mondol, "Design and Implementation of a Digital Tachometer," *Int. J. Sci. Eng. Technol.*, vol. 5, no. 1, pp. 85–87, (2016).
2. G. Sachdev, A. Goyal, A. Jain, and J. Surana, "A Study on Vehicle and Counter Measure," *Int. J. Eng. Dev. Res.*, vol. 5, no. 2, pp. 909–921, (2017).
3. K. Iwata, T. Yamamichi, Y. Sugiyama, and T. Kitano, "Train Position Detection System by Means of Inertial Sensors together with a Tachometer Generator," *QR RTRI*, vol. 57, no. 4, pp. 293–299, (2016).
4. R. Jia, J. O. Yang, H. Lin, and J. Chen, "A Real-Time Differential Anti Strong Interference Tachometer based on Arduino," *Int. Conf. Appl. Mech. Electron. Mechatron. Eng. AMEME*, pp. 1–6, (2016).
5. S. Nashit, S. Adhikari, S. Farhan, S. Avinash, and A. Gambhire, "Design, Fabrication and Testing of Regenerative Braking Test Rig for BLDC Motor," *Int. Res. J. Eng. Technol. IRJET*, vol. 3, no. 5, pp. 1881–1884, (2016).
6. N. Mane, P. Mane, K. Sabale, and M. Mokashi, "Design of Multi-rotor Wind Turbine with Solar Hybrid System," *Int. J. Curr. Eng. Technol.*, no. 6, pp. 1–4, (2016).
7. M. Muchtar, S. Manjang, D. A. Suriamiharja, and M. A. Thaha, "Physical Model Performance of Wave Energy Converter based on Water Mass Gravity Force under Container Shape Variation," *Int. J. Sci. Res. IJSR*, vol. 5, no. 9, pp. 1627–1630, (2016).
8. A. Bhadani, D. Koladiya, J. Devani, and A. Tahiliani, "Modelling and Controlling of BLDC Motor," *Int. J. Adv. Eng. Res. Dev.*, vol. 3, no. 3, pp. 139–144, (2016).
9. S. Das, P. B. Deb, S. Chakraborty, N. Nag, M. Laha, and P. S. Majumdar, "Contactless Speed Monitoring and Displaying," *Int. Res. J. Eng. Technol.*, vol. 3, no. 12, pp. 876–878, (2016).
10. I. Khelf, L. Laouar, A. M. Bouchelaghem, D. Remond, and S. Saad, "Adaptive fault diagnosis in rotating machines using indicators selection," *Mech. Syst. Signal Process.*, vol. 40, pp. 452–468, (2013).
11. K. Alameh, N. Cite, G. Hoblos, and G. Barakat, "Vibration-based Fault Diagnosis Approach for Permanent Magnet Synchronous Motors," *IFAC - Pap. OnLine*, vol. 48–21, pp. 1444–1450, (2015).
12. P. M. Papadopoulos, V. Reppa, M. M. Polycarpou, and C. G. Panayiotou, "Distributed Diagnosis of Actuator and Sensor Faults in HVAC Systems," *IFAC - Pap. OnLine*, vol. 50–1, pp. 4209–4215, (2017).
13. M. Saied, B. Lussier, I. Fantoni, H. Shraim, and C. Francis, "Fault Diagnosis and Fault-Tolerant Control of an Octorotor UAV using motors speeds," *IFAC - Pap. OnLine*, vol. 50–1, pp. 5263–5268, (2017).
14. F. L., A. M., and D. A., "A real-time algorithm for fault identification in machining centres," *18th World Congr. Int. Fed. Autom. Control*, vol. 44, no. 1, pp. 5201–5206, (2011).
15. Y. Zhi-Ling, W. Bin, D. Xing-Hui, and L. Hao, "Expert System of Fault Diagnosis for Gear Box in Wind Turbine," *Syst. Eng. Procedia*, vol. 4, pp. 189–195, (2012).
16. W. Li, S. Zhang, and S. Rakheja, "Feature Denoising and Nearest-Farthest Distance Preserving Projection for Machine Fault Diagnosis," *IEEE Trans. Ind. Inform.*, vol. 12, pp. 393–404, (2016).
17. T. N. Kruglova, "Wavelet analysis for fault diagnosis of electrical machines using current signals," *2nd Int. Conf. Ind. Eng. Appl. Manuf. ICIEAM*, pp. 1–5, (2016).
18. N. Hamad, K. F. Brethee, F. Gu, and A. D. Ball, "An investigation of electrical motor parameters in a sensorless variable speed drive for machine fault diagnosis," *22nd Int. Conf. Autom. Comput. ICAC*, pp. 329–335, (2016).
19. Z. Du, X. Chen, H. Zhang, H. Miao, Y. Guo, and B. Yang, "Feature Identification with Comprehensive Measurements for Machine Fault Diagnosis," *IEEE Trans. Instrum. Meas.*, vol. 65, no. 5, pp. 977–987, (2016).
20. K. T. Sreekumar *et al.*, "Locality constrained linear coding for fault diagnosis of rotating machines using vibration analysis," *Annu. IEEE India Conf. INDICON*, pp. 1–6, (2016).

Analysis Characteristics Of Pendulum Oscillation In PLTGL-SB

I.S. Arif[1], *I.K.A.P* Utama[2,*], *R.* Hartono[3], *J.* Prananda[1], *R.* Isnaini[3], *T.* Rachmattra A[1]

[1]Dept. of Marine Engineering, Institute of Technology Sepuluh Nopember, Indonesia
[2]Dept. of Naval Architecture, Institute of Technology Sepuluh Nopember, Indonesia
[3]Dept. of Physics Engineering, Institute of Technology Sepuluh Nopember, Indonesia

Abstract. Energy is an everlasting demand which sustains humanity and its activity throughout the world. The main problem energy for all country is still using fossil energy. Using fossil energy caused pollution and contaminate environment. Adopting Ocean wave energy we can convert the wave into eco-green energy. Wave energy in Indonesia very abundant, it's coastal line is 95,181 km number 2 in the world after Canada. The device that used to convert ocean wave energy into electrical energy is called WECs (Wave Energy Converters). Latterly, there are many kinds of WECs that already developed by an engineer. Pendulum system is one sample of WECs, its has uncomplicated working principle. Zamrisyaf is The man who first invented the device in 2010. Recently, the researcher still haven't found the parameter of pontoon geometry that can afford a good sea keeping for this WECs. In this research, the authors has been performed using experimental approach to test the pendulum system. The purpose of this study is to analyze the novel geometry of pontoon that identical to trimaran that can produce a large amplitude of pendulum oscillation through an experimental approach. The analyzed aspect is the combination of outrigger length, outrigger height, pendulum rod length, pendulum mass, and wave periods which have a maximum amplitude of pendulum oscillation. The analysis results show that the better solution is pontoon thus has outrigger height of 40 mm, pendulum rod length of 106.7 mm, outrigger length of 413 mm, pendulum mass of 20, and wave periods of 0.8 s have a maximum amplitude of pendulum oscillation as big as 60 degrees.

1 Introduction

As energy demand increases, especially electrical energy in the world, it will result in an increasingly severe greenhouse effect as an effect of dependence on fossil energy sources [1]. Additionally, the dependence on fossil energy as an energy source can lead to other negative effects such as greenhouse gas emissions that lead to faster earth temperature rises, reduce energy supply, and cause geopolitical issues with respect to control of oil reserves that are at the core of some The issue of armed conflict in recent decades [2]. Due to all these problems the world is beginning to realize the importance of clean energy sources or renewable energy sources along with its conversion technology [1,2]. One source of renewable energy that has the potential to be converted into electrical energy is derived from the sea that can be broadly classified into several types, namely tidal energy, wave energy, ocean thermal energy, ocean currents energy, and sea wind energy and Salt gradient [3].

The three most common types of marine energy are tidal energy, wave energy, and ocean thermal energy [1]. Sea waves as one of renewable energy are known as resources that have the potential to be exploited in a sustainable manner to be converted into electrical energy [4].

As in the world, electricity demand in Indonesia is also increasing along with the times. In addition, there are at least three major groups of national energy problems: low electrification rates, dependence on fossil energy sources, and low utilization of renewable energy [5]. To pursue future energy needs the policy to be done is to integrate renewable and nonrenewable energy, with a greater change in the composition of use in renewable energy types. Indonesia itself is a maritime country with great wave energy potential.

Sea wave energy is a renewable energy that is continuous and environmentally friendly [5]. Tool to convert ocean wave energy into electrical energy is called Wave Energy Converter (WECs) [6]. To date, there have been several types of WECs that have been developed in the world, including submerged pressure differentials, oscillating wave surge converters, overtopping devices, oscillating water column, and others [7]. In addition, there is also a tool of ocean wave energy conversion into electrical energy developed by researchers from Indonesia that is using pendulum system or called as a wave powerhouse wave pendulum system (PLTGL-SB). The study of PLTGL of this pendulum system was first performed by Zamrisyaf, one of the PLN planning staffs in West Sumatra. The concept of a wave powerhouse of the pendulum system is simple, that is, by exploiting the fluctuation of the movement of

* Corresponding author: kutama@na.its.ac.id

 https://doi.org/10.1051/e3sconf/20184301005

sea water waves which will make the pendulum move like a bell, the movement is then transmitted into a rotary motion to move the dynamo then into electrical energy [8]. The power generated by the wave powerhouse of the pendulum system is much influenced by the dimensions of the pontoon, the length of the pendulum, and the mass of pendulum [9]. The parameters affecting the magnitude of power are the amount of torque associated with the amplitude and the frequency of the resulting vibration [10].

In previous studies, the observed pontoons were rectangles by Zamrisyaf. After examining the rectangular shape, research on the shape of the pontoon was again carried out with a circular pontoon, a square of eight, then a square of six. The best results for the wave height and the same period are obtained when the pontoon is square. Parameters that affect the motion of the pendulum itself is the stability of the pontoon used. In this final project will be investigated the power generated by PLTGL-SB system with cylindrical pontoon which has symmetrical outrigger on both sides. In addition, it will also examine the relationship of mass, length of the pendulum arm, the length of the outrigger sides, and the outrigger heights from the base to the torque generated by the pendulum. The research will be done through experiments in the laboratory by using a prototype that has a scale of 1:10 to the real dimension

.

2 Methods

2.1 Literature Study

In previous studies, the observed pontoons were rectangles by Zamrisyaf. After examining the rectangular shape, research on the shape of the pontoon was again carried out with a circular pontoon, a square of eight, then a square of six. The best results for the wave height and the same period are obtained when the pontoon is square. Parameters that affect the motion of the pendulum itself is the stability of the pontoon used.

In this final project will be investigated the power generated by PLTGL-SB system with cylindrical pontoon which has symmetrical outrigger on both sides. In addition, it will also examine the relationship of mass, length of the pendulum arm, the length of the outrigger sides, and the outrigger heights from the base to the torque generated by the pendulum. The research will be done through experiments in the laboratory by using a prototype that has a scale of 1:10 to the real dimension.

2.2 Analytical Calculation

Once determined the dimensions of the pontoon prototype will be made, then the analytical calculation to support the data so that it can be known combination of sizes in accordance with the desired specification, the condition where the pontoon continues to shake but still in a stable positive condition. The dimensions of the self-made pontoon prototype are 32 cm long cylinders with 16.5 cm diameter. which has a symmetrical outrigger on both sides as long as 32 cm with a diameter of 4 cm. A pendulum made is a vertical pendulum. While the given wave direction is 90 degrees from the face of the pontoon.

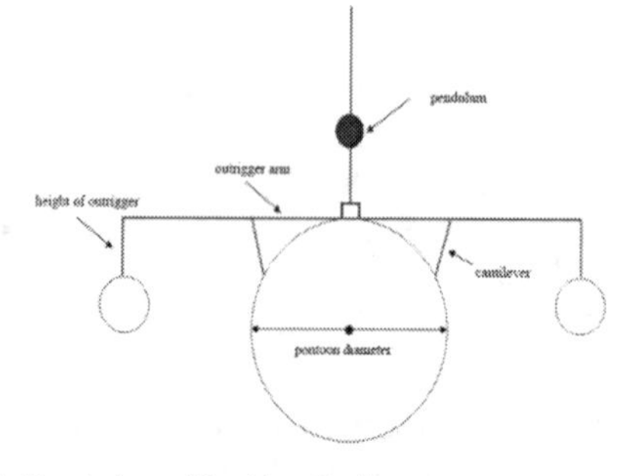

Fig. 1. Front view of Pontoon Design.

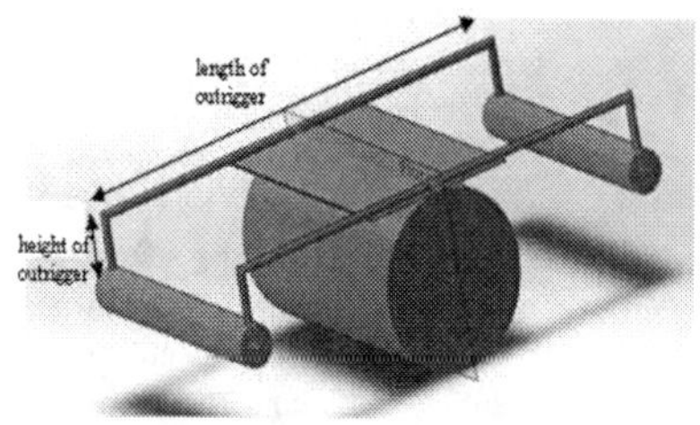

Fig. 2. Pontoon 3D Design.

Specification of the design of the pontoon prototype to be performed calculation analysis is as follows:

Large Cylinder Diameter: 16.5 cm
Large Cylindrical Thickness: 0.7 cm
Small Cylinder Diameter: 4 cm
Thickness of Small Cylinders: 0.2 cm
Large Cylinder Length: 32 cm
Small Cylinder Length: 32 cm
Mass Cylindrical Type: 1.39 gr / cm3
Full: ½ and 2/3 diameter of pontoon
High of Outrigger: 1.5 cm and 4 cm
Plate length: 32 cm
Plate width: 17 cm
Plate Density: 1.2 gr / cm3

Once determined the main size of the pontoon dimension along with the materials used, then calculated parameters that will determine the stability of the prototype made. The parameters are COB, COG, and metacenter. The first thing to be calculated is to determine the mass of each component so that it can be known the total mass of the whole system. The mass calculation can be done with ϱ is the density of the object and V is the volume of the object. After the mass of the object is found, it is necessary to look for the weight of the object as equation 2.

https://doi.org/10.1051/e3sconf/20184301005

$$m_{tot} = \sum \rho_n V_n \quad (1)$$
$$B_{tot} = \sum m_n g \quad (2)$$

When the weight of each component and total weight has been found, the center of gravity (COG) can be searched using equation 3 where B is the weight of the object and l is the center point of each section in terms of the abscissa axis.

$$COG = \frac{B_1 l_1 + B_2 l_2 + B_n l_n}{B_{tot}} \quad (3)$$

Meanwhile, to find the value of the center of buoyancy (COB), will be more easily searched because we have determined from the beginning the size of the submerged pontoon. COB is likened to COG from the submerged parts, so the search will be the same as COG with a record of just the submerged pontoon that is used as a review. Thus, COB can be searched using equation 4.

$$COB = \frac{B_{submerged} \cdot l_{submerged}}{B_{submerged}} \quad (4)$$

The next step is to calculate the metacenter of the pontoon. Metacenter itself can be searched using equation 5 with I_0 is the moment of system inertia, V_t is the volume of the pontoon submerged, and G_0B_0 is the distance between COG and COB.

$$M = \frac{I_0}{V_t} - G_0 B_0 \quad (5)$$

On a floating object, the magnitude of the Archimedes style will be the same as the force of gravity. So from the principle, we can know how big the submerged so that we can find out how big ballast needed for the pontoon can reach laden pre-determined. Then to determine the ballast, can be searched as equation 6 to 8 with ϱf is the fluid type mass, V_t is the volume of the submerged pontoon, ms is the total mass of the system so that the pontoon can be submerged in accordance with the specified load, and g is the acceleration of earth gravity.

$$Fa = m_s g \quad (6)$$
$$\rho_f V_t g = m_s g \quad (7)$$
$$m_s = \rho_f V_t \quad (8)$$

If the mass of the system has been obtained, then the ballast can be found by reducing the mass of the system with the total mass of the pontoon previously calculated using equation 1.

2.3 Determination of Prototype

After the analytical calculation, it can be determined the feasibility of the prototype related with specifications that have been in accordance with the purpose of research. If at the previous calculation the metacenter point is at the top, followed by COG, then COB, then the pontoon has stabilized positively so that a predetermined specification can be made.

In this experiment, there were 48 variations with 3 variations of wavelength periods, 2 variations of the outer sleeve length, 2 variations of the outrigger that can be represented by water-laden, 2 variations of pendulum arm length, and 2 pendulum mass variations. The wave height used in the data retrieval process is 1.5 cm for all variations.

2.4 Prototype Fabrication

After the analytical calculation and determination of the dimensions of the pontoon, the prototype can be fabricated in accordance with the design specifications. Figures 3 and 4 show the pontoon prototype that has been created.

Fig. 3. Pontoon with ballast 2.3 kg.

Fig. 4. Pontoon with ballast 1.2 kg.

2.5 Data Retrieval

In this experimental study, the data were collected in the towing tank of Hydrodynamics Laboratory of Marine Faculty Technology-ITS. Which has a length of 50 m, width 3 m, and a depth is 2m. The data taken is the video movement of the pontoon and pendulum to know the deviation, as well as the amplitude of the roll and heave of the pontoons using sensors that have been available in the laboratory.

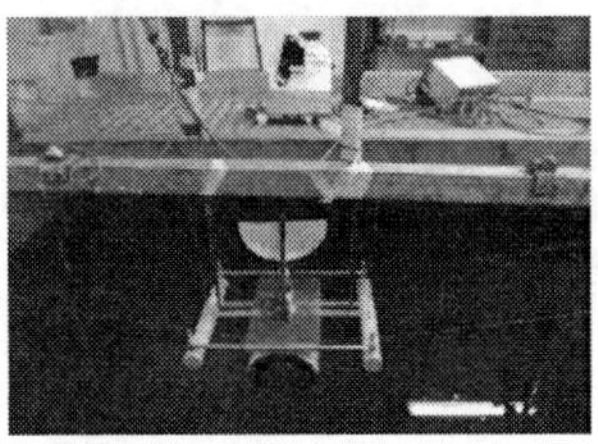

Fig. 5. Setting tool for data retrieval

2.6 Data Analysis and Paper Making

As mentioned in Section 3.5, the data obtained from the experiments are the roll and heave drift of the pontoon and also the videotape of the pendulum data. The first thing to do is to process the video so that the data of pendulum deviation is quantitative. After the data obtained pendulum deviation and also pontoons, then done through data so that obtained graphs and tables that can represent the data. In addition, it is necessary to calculate the power generated by each condition of the experiments performed. Before sought power, the torque value (τ) must be known. The torque can be admitted applying equation 8 where m is the pendulum mass, g is the acceleration of gravity, l is the length of the pendulum arm, and θ is the confluence at the pendulum oscillation.

$$\tau = mgl\sin\theta \tag{8}$$

In addition to torque, the angular velocity of the pendulum oscillations should also be known. The angular velocity of the pendulum itself can be determined using the equation 9 with g is the acceleration of gravity, l is the length of the pendulum arm, and h is the difference of the pendulum height when in the θ junction which can be searched using equation 10. After knowing torque and angular velocity, power can be searched using equation 11.

$$\omega = \frac{\sqrt{2gh}}{l} \tag{9}$$

$$h = h - h\cos\theta \tag{10}$$

$$P = \tau\omega \tag{11}$$

After all the data is processed, then the preparation of the making paper which is the final stage of this research project work and do as delivery of results and analysis of the implementation of research programs. The paper refers to the previous stages and describes the entire process of activity using a particular systematic.

3 Results and Discussion

3.1 Pontoon Dimension

The tested pontoon is a prototype with a ratio of 1:10. Prototype pontoons are made of cylindrical shape with outrigger symmetrical on both sides with a length of 320 mm. The diameter of the pontoon is 165 mm, while the outrigger has a diameter of 40 mm. For the pendulum used consisted of 2 variations of mass, ie 20 grams and 40 grams with the variation of arm length 106.7 mm and 165 mm. For more details, the shape of the prototype and pendulum can be seen in the picture below.

Fig. 6. Dimension of the pontoon with ballast 1.2 kg.

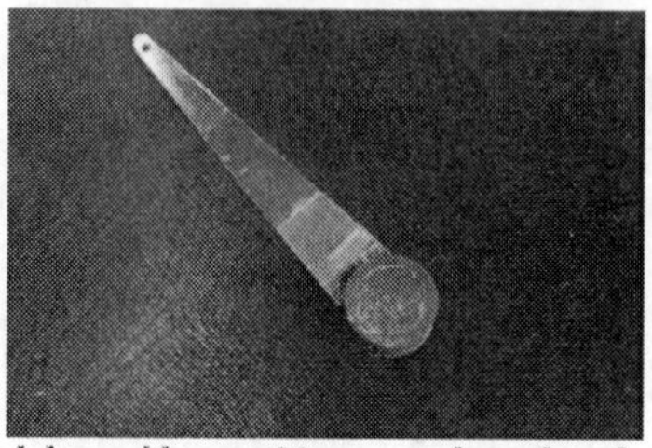

Fig. 7. Pendulum with mass 20 gram and arm length 106.7 mm.

3.2 Mass Calculation, Volume and Stability of Pontoons

Once the geometry of the pontoon is determined, it is necessary to calculate the stability of the pontoon when exposed to the excitation force of the wave. The thing that most affects the stability of an object when immersed in water is the number of submerged parts of the object. The height of the part of the object dyed in the water is called as laden (t). If the load is known, then the center of buoyancy (COB) will also be known. To know all the data, it is necessary to calculate the mass and volume of the pontoon to be investigated.

Table 1. Weight of pontoon and pendulum construction.

No.	Part of Pontoon	Volume (cm^3)	Density (gr/cm^3)	Mass (gr)
1.	Pontoon	1076.58	1.39	1496.45
2.	Outrigger	147.95	1.39	205.66
3.	Cover of the pontoon	213.72	1.20	256.46
4.	Cover of the outrigger	25.12	1.20	30.14
5.	Outrigger arm	7.61	2.71	57
6.	Outrigger arm 2	6.33	2.71	47.8
7.	Height of outrigger	15.2	2.71	70.68
8.	Pendulum construction	-	-	80.00

Based on the calculation, then obtained the total mass of the pontoon along with construction outrigger and pendulum that is equal to 2.12 kg. The selected self is 0.5 and 0.67 diameter of the pontoon. Through analytical

https://doi.org/10.1051/e3sconf/20184301005

calculations, it can be obtained that the location of COB, COG, and metacenter point in accordance with the theory of stability so that the pontoon can return to its original position (stable positive).

Fig. 8. Position points for positive stable objects (K = keel, B = COB, G = COG, and M = metacenter)

According to calculations, the mass of the system that is equal to 2.12 kg. However, with the desired weight of 2.12 kgs is not reached, the ballast is required. Data on point COG, COB, metacenter, and also ballast can be seen in table 2.

Table 2. Formatting sections, subsections and subsubsections.

Draught	COG (cm)	COB (cm)	Metacenter (cm)	Ballast (kg)
0.5 D	8.67	7.49	12.17	1.2
0.67 D	9.26	7.68	11.78	2.3

3.3 Experimental Results of Pontoon and Pendulum Deviation

The number of variations tested was 48, with 3 variations of wavelength period, 2 variations of the outrigger sides, 2 variations of outrigger height, 2 pendulum arm variations, and 2 pendulum mass variations.In general, the oscillation characteristics of the experimental results are divided into 4 major sections. The first characteristic is that there is no oscillation at all. This phenomenon occurs in variations that are given a coming wave with a period of 0.4 seconds.

The second characteristic is when the pendulum oscillates with a smaller amplitude equal to 20 degrees. This phenomenon occurs in several variations of the pontoon given the wave style with a period of 0.6 or 0.8 seconds.The third characteristic is that the pendulum vibrates back and forth, not oscillating as expected by the researcher. This phenomenon occurs in 5 variations given the wave style with a period of 0.6 seconds (variation 17, variation 26, variation 32, variation 38, variation 47).

The fourth characteristic is when the amplitude of the pendulum oscillation is of great value, ie more than 20 degrees. This fourth characteristic occurs in some variations that are given waves with periods of 0.6 and 0.8 seconds.

From the four characteristics mentioned above, the fourth type is eligible to be used as an ocean wave power plant because it produces a large amplitude of oscillation so that it can produce great power as well. From several variations on the incoming of the fourth characteristic, there are three variations that produce the largest amplitude and maximum amplitude, ie variations 12, 36, and 42. Figure 9 to 11 shows the result of pendulum oscillation along with the heaving value and rolling of pontoons on each of these variations.

The largest mean amplitude occurring at the time-tested was a pontoon of variation 12 with an average amplitude value of 41.82 degrees and a maximum amplitude of 60 degrees. While the average amplitude at 36 and 42 variations in a row is 36.52 degrees and 32.54 degrees respectively. Variation 12 produces the largest pendulum shrinkage value because it is the most appropriate combination to obtain large amplitude.

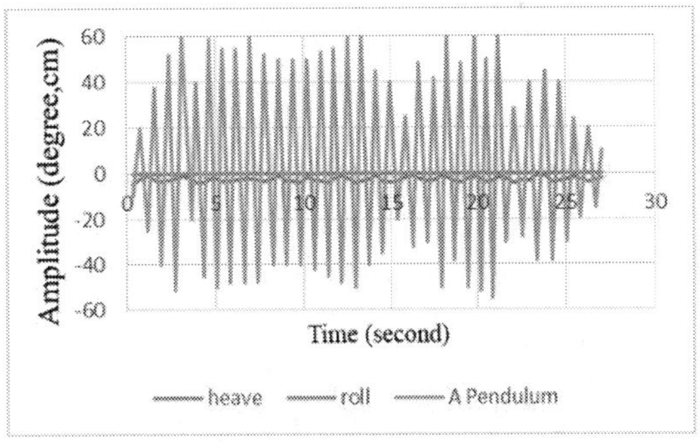

Fig. 9. Motion of pontoon and pendulum for variation 12 (T=0.8 s; Pc=413 mm; Hc=40 mm; Lp=106.7 mm; Mp=20 gr)

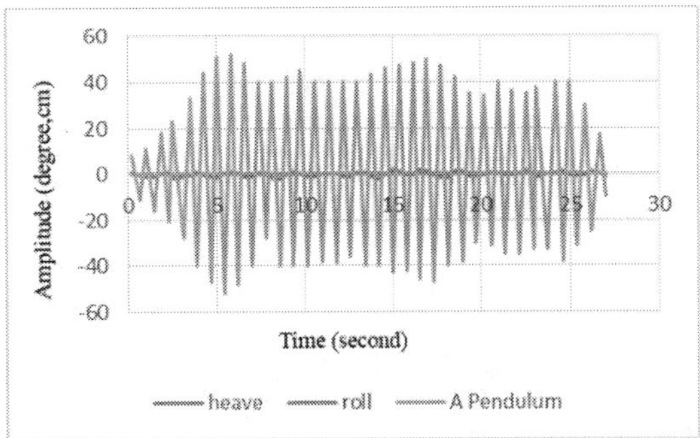

Fig. 10. Motion of pontoon and pendulum for variation 36 (T=0.8 s; Pc=495 mm; Hc=15 mm; Lp=106.7 mm; Mp=20 gr)

https://doi.org/10.1051/e3sconf/20184301005

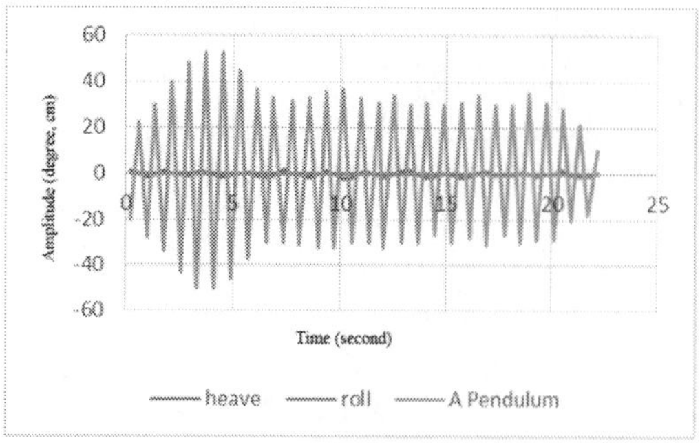

Fig. 11. Motion of pontoon and pendulum for variation 42 (T=0.8 s; Pc=413 mm; Hc=15 mm; Lp=106.7 mm; Mp=20 gr)

If the dimensional analysis is used, then when enlarged to the original scale variation 12 has a pontoon finger size of 1.65 m, outrigger fingers of 0.4 m, pontoon length and outrigger 3.2 m, 4.13m outrigger, outrigger 0.4 m, pendulum arm 1.067 m, and pendulum mass of 20 kg.

3.4 Effect of Variables on Pendulum Oscillation

After 48 experiments, the effect of the related variables on the pendulum oscillation. Data obtained that the greater the given wave period, the amplitude of the pendulum oscillation will also be greater. The greater the wave period does cause the rolling angle to become smaller, but not so with the rolling speed that occurs. If the rolling speed is large, the frequency of the pontoon vibrates also will be faster so that the pendulum will more easily oscillate.

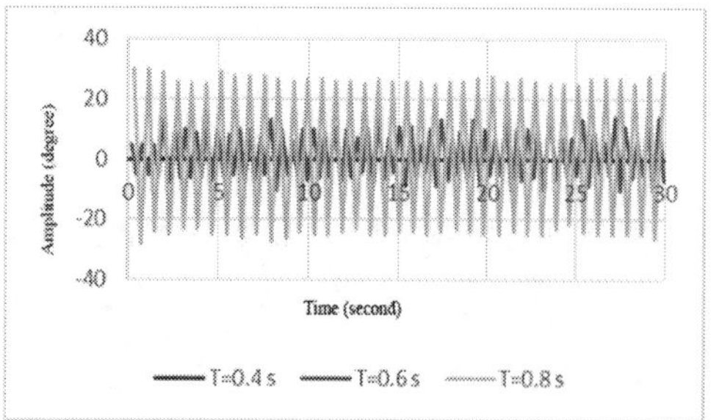

Fig. 12. Pendulum oscillations with different wave periods

Based on Figure 13 the variation of 15 is when the pontoon has an outrigger with a 495 mm sleeve length from end to end, whereas, in variation 12, the length of the outrigger arm from end to end is 413 mm. Based on the experiment, we can see if the length of the outrigger arm is greater, the average amplitude generated by the pendulum oscillation will be smaller. At the time of variation 12 (outrigger sleeve = 413 mm), the average amplitude of pendulum oscillation is 41.82 degrees with a vibration frequency of 1.23 Hz. Whereas when a variation of 15 that has a 495 mm outrigger sleeve, the pendulum will oscillate with a maximum amplitude of 21.44 degrees and a frequency of 1.355 Hz. According to D. R. Berrett and C. B. Barrass (1999) by increasing the width of the ship (the length of the vessel is considered fixed), the vessel will be more stable [11]. This is what happens to the pontoon tested by the author. Seen in Figure 13 that the pontoon variation 15 with the larger outrigger sleeve length has a smaller rolling motion, then as a consequence, the pendulum oscillation will have a smaller amplitude as well.

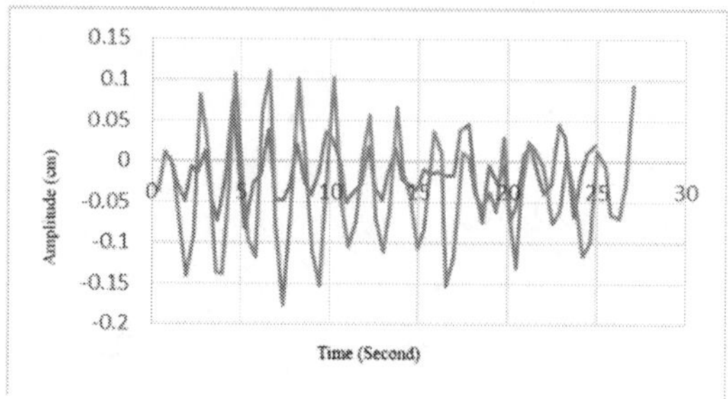

Fig. 13. Rolling motion for variation 12 and 15

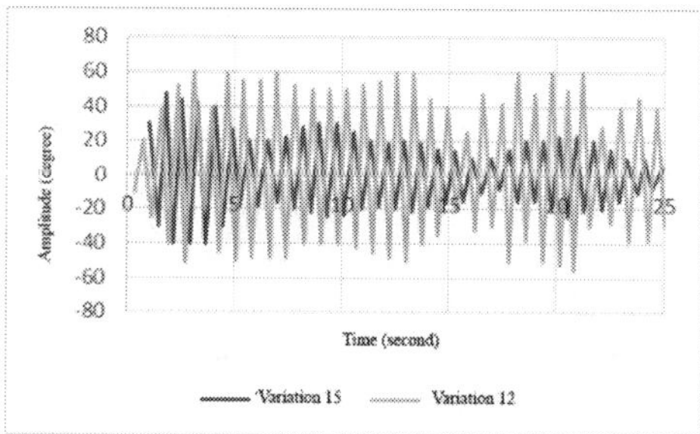

Fig. 14. Pendulum oscillation with changes of the outrigger arm length

In this experiment, changes in the outrigger level will affect the magnitude of the pontoons that are submerged in water. The bigger the outrigger, the value of the pontoon will be smaller. The larger-laden pontoon (variation 27, laden as high as 106.7 mm) has a larger rolling amplitude compared to the 82.5 mm loaded pontoon. Comparison of this rolling motion can be seen as figure 16. The larger rolling amplitude is why the variation 27 has a greater oscillation amplitude compared to the variation 6.

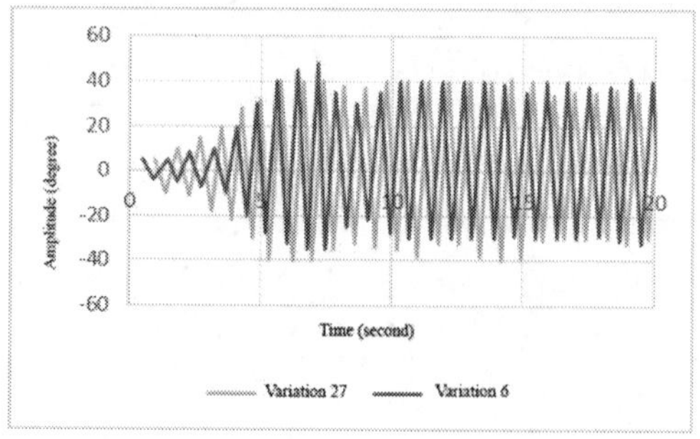

Fig. 15. Pendulum oscillation with outrigger high change

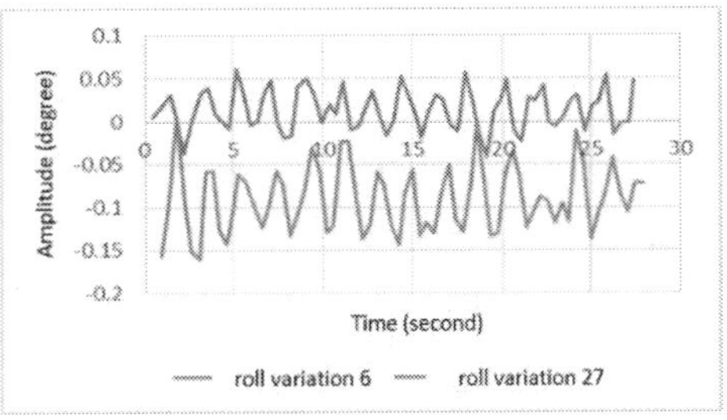

Fig. 16. Motion rolling when variations 6 and 27

If we see again equation 2.9 it can be seen that the mass will be inversely proportional to the size of the pendulum. This also happens to experiments that have been done by the author. It can be seen figure 17 that the pontoon with a variation of 18 having a pendulum mass of 40 grams has a mean deviation smaller than when the pontoon tested is a variation of 21 with a pendulum mass of 20 grams. The pendulum amplitude at variation 18 is 16.61 degrees, while variation 21 has an average pendulum oscillation amplitude of 19.32 degrees. The frequency of the pendulum oscillation itself does not have a significant difference value so that in the calculation of power will not be too much effect.

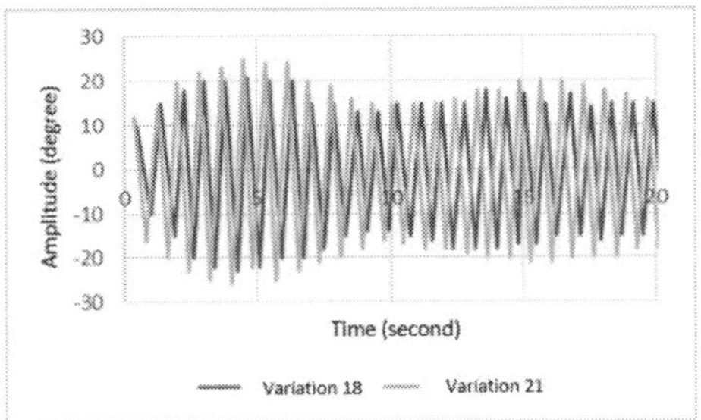

Fig. 17. Pendulum oscillation with mass changes

We can see that the root of the pendulum arm will be proportional to the period. The angular deviation from the pendulum will be inversely proportional to the period. So it can be concluded that the shorter the pendulum arm the larger the deviation. This is why the pendulum pendants with longer pendulum sleeves have smaller deviations compared to pendulums with shorter arms as shown in Figure 18. Variation 30 was pontoon with pendulum sleeve of 106.7 mm with mean pendulum value of 38.01 degrees, while variation 33 has pendulum arm length of 165 mm with average pendulum deviation value of 31.87 degrees.

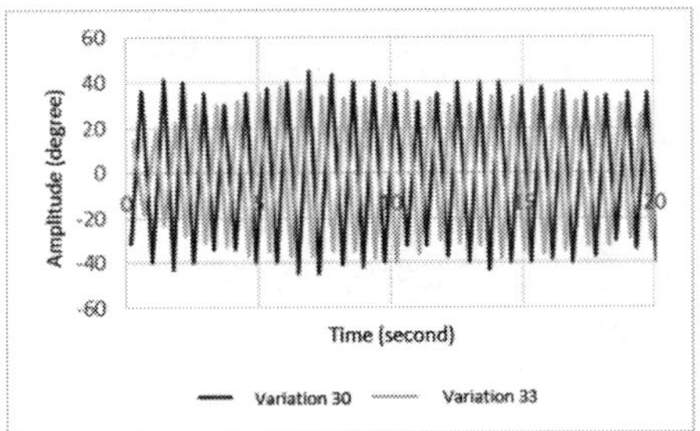

Fig. 18. Pendulum oscillation with pendulum arm length change

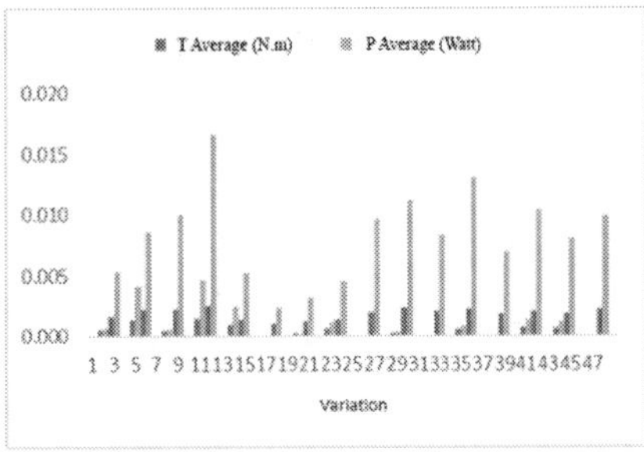

Fig. 19. Torque, angular velocity, and average power on each variation

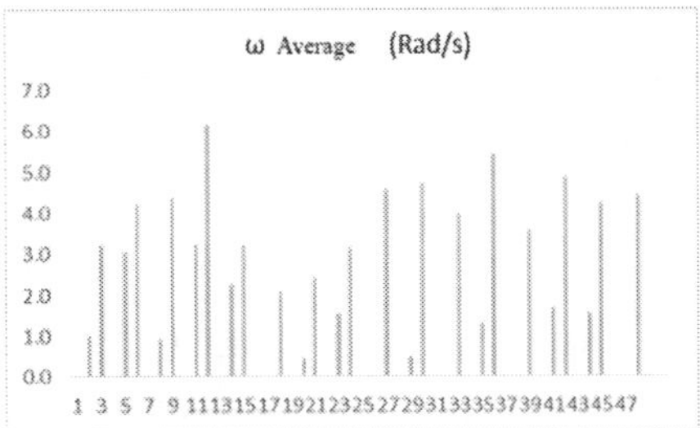

Fig. 20. Average angular velocity on each variation

Based on analytical calculations, it was found that the power obtained for variation 12 is 0.017 watts with an average angular velocity of 6.173 m/s.

4 Conclusion

Referring to the results of data analysis and discussion in the previous chapter, it can be taken some conclusions about this experimental study as follows.The amplitude of the pendulum oscillation decreases as the length of the outrigger arm, the outrigger, the pendulum mass, and the length of the pendulum arm have, but unlike the rise of the wave period, the pendulum oscillation amplitude increases as the parameter increases.

The variations that resulted in the maximum deviation value and the largest mean deviation were variation 12 with the wave period 0.8 s, the outer sleeve length 413 mm, the outrigger height 40 mm, the pendulum arm length of 106.7 mm, and the pendulum mass of 20 grams with a maximum deviation of 60 degrees and an average deviation of 41.82 degrees.

Some things that can be suggested from the results of this research for the continuation of its development. Further research can be a variation of the influence of the direction of the arrival of the wave of pontoon movement which will affect the pendulum oscillation. Data retrieval on subsequent variations is done when the water conditions in the towing tank are completely quiet. In addition, the sensor also needs to be checked each turn of variation so that the data obtained can be really ideal and accurate.

References

1. Indonesia Energy Outlook 2014, Badan Pengkajian dan Pengembangan Teknologi , (2014).
2. Pengenalan Teknologi Pemanfaatan Energi Angin, Tasikmalaya: Tim Lentera Angin Nusantara, (2014).
3. Tutorial Blade 3 Dimensi, Tasikmalaya: Tim Lentera Angin Nusantara, (2015).
4. H. Piggot, Windpower Workshop: Building Your Own Wind Turbine, Centre of Alternative Technology Publications.
5. M. H. O.L., Aerodynamics of Wind Turbine, Second Edition, London, (2008).
6. S. Mehrangiz, Y. Emami, S. H. S. Sadigh and A. Etemadi, " Various Technologies for Producing Energy from Wave: A Review," *International Journal of Smart Grid and Clean Energy,* vol. 2, no. 2, pp. 289-294, (2013).
7. G. Iglesias, M. Alvarez and P. Garcia, "Wave Energy Converters," *EOLLS,* vol. 55, pp. 235-241, (2013).
8. H. T. Benbouzid and M. Benbouzid, "An Up-to-Date Technologies Review and Evaluation," *INTERNATIONAL REVIEW OF ELECTRICAL-IREE,* vol. 10, no. 1, pp. 56-61, (2015).
9. J. Falnes, "A review of wave-energy extraction," *Science Direct Marine Structure,* vol. 20, pp. 185-201, (2007).
10. F. Triatmojo, "Dinamika Kebijakan Diversifikasi Energi di Indonesia : Analisis Kebijakan Pengembangan Energi Terbarukan di Indonesia," *Jurnal Ilmiah Administrasi Publik dan Pembangunan,* vol. 4, pp. 146-159, (2013).
11. B. Bryan and D.R. Derrett, Ship Stability For Masters And Mates, Oxford, UK: Butterworth-Heinemann, (2006).

E3S Web of Conferences **43**, 01006 (2018) https://doi.org/10.1051/e3sconf/20184301006
ASTECHNOVA 2017

Analysis of Back Surface Field (BSF) Performance in P-Type And N-Type Monocrystalline Silicon Wafer

Ferdiansjah[1,*], *Faridah*[1] ,and *Kelvian Tirtakusuma* Mularso[1]

[1] Department of Nuclear Engineering and Engineering Physics, Universitas Gadjah Mada, Indonesia.

Abstract. Back Surface Field (BSF) has been used as one of means to enhance solar cell performance by reducing surface recombination velocity (SRV). One of methods to produce BSF is by introducing highly doped layer on rear surface of the wafer. Depending on the type of the dopant in wafer, the BSF layer could be either p+ or n+. This research aims to compare the performance of BSF layer both in p-type and n-type wafer in order to understand the effect of BSF on both wafer types.

Monocrystalline silicon wafer with thickness of 300 μm, area of 1 cm^2, bulk doping level NB $=1.5\times10^{16}/cm^3$ both for p-type wafer and n-type wafer are used. Both wafer then converted into solar cell by adding emitter layer with concentration NE $=7.5\times10^{18}/cm^3$ both for p-type wafer and n-type wafer. Doping profile that is used for emitter layer is following complementary error function (erfc) distribution profile. BSF concentration is varied from $1\times10^{17}/cm^3$ to $1\times10^{20}/cm^3$ for each of the cell. Solar cell performance is tested under standard condition, with $AM_{1.5G}$ spectrum at 1000 W/m^2. Its output is measured based on its open circuit voltage (V_{OC}), short circuit current density (J_{SC}), efficiency (η), and fill factor (*FF*).

The result shows that the value of V_{OC} is relatively constant along the range of BSF concentration, which is 0.694 V – 0.702 V. The same pattern is also observed in *FF* value which is between 0.828 – 0.831. On the other hand, value of J_{SC} and efficiency will drop against the increase of BSF concentration. Highest J_{SC} which is 0.033 A/cm^2 and highest efficiency which is 18.6% is achieved when BSF concentration is slightly higher than bulk doping level. The best efficiency can be produced when BSF concentration is around $1\times10^{17}cm^{-3}$.. This result confirms that surface recombination velocity has been reduced due to the increase in cell's short circuit current density and its efficiency.

In general both p-type and n-type wafer will produce higher efficiency when BSF is applied. However, the increase is larger in p-type wafer than in n-type wafer. Better performance for solar cell is achieved when BSF concentration is slightly higher that bulk doping level because at very high BSF concentration the cell's efficiency will be decreased.

1 Introduction

Solar cell has grown thinner in past decade due to the desire to reduce material cost. A lot of research has been performed in order to find ways of reducing cell's thickness without sacrificing its efficiency. Current solar cell that widely produced has P-type based monocrystalline wafer and it shows an increase in its effciency. However when the wafer become thinner, effect such as surface recombination will become dominant, hence a way to limit or reduce this effect should be applied. One the method to limit backsurface recombination in a P-type wafer is by applying aluminium layer upon the rear surface of the cell. This layer will function as internal electric field generator that will repel photogenerated minority carrier near the rear surface, and therefore will reduce back surface recombination velocity. The effectiveness of aluminium layer as back surface field generator will depend on Al concentration. However this concentration will be limited to its maximum solid solubility in Si, which is around 1018 cm-3 at the standard firing temperature of around 750oC [1].

On the other hand, N-type silicon wafer exhibits several advantages compared to the P-type wafer. Some of the advantages of N-type solar cells are : no formation of boron-oxygen complexes that will produce degradation in wafer, high tolerance for common impurities, having longer diffusion length at similar doping level with the other dopantts, lower doping level is needed to produce same value of conductivity on the nn+ junction formation [2].

2 Theory

Minority carrier recombination is one of mechanism that reduce quantum efficiency as well as power efficiency [3]. Back Surface Field formation on the rear surface of the cell in the form of p^+ layer upon P-type wafer has shown the ability to increase cell's efficency 2.5% from its original value when p^+layer thickness is around 20μm and doping concentration is 10^{19} cm^{-3} [4].

Another type of BSF is using n^+ layer upon N-type wafer. When n^+ layer thickness is around 2μm and its peak doping level is around 4.16×10^{19} cm^{-3} th e maximum efficiency that can be reached is around 19.98% [5].

BSF formation on dendritic web silicon wafer also produced an increase on efficiency. Effciency as high as 17.3% could be achieved when BSF with n+ type layer

* Corresponding author: ferdiansjah@ugm.ac.id

with peak doping level of 3×10^{20} cm^{-3} and thickness of 0.7 µm is applied on the solar cell [6].

Solar cell with PERT and PERL structure also get benefit when BSF is applied. When peak doping level in n+ layer is 5×10^{18} cm^{-3} and its thickess is 1.5 µm, the efficiency could reach 22.7% [7].

In a solar cell when minority carrier generation dominated the process, then the open circuit voltage, V_{OC} could be given as

$$V_{OC} = \frac{kT}{q}\ln\left[\left(\frac{I_{SC}}{I_0}\right) + 1\right] \quad (1)$$

From Eq. 1, it can be seen that V_{OC} is limited by dark saturation current I_0, that is produced within base layer (I_{0b}) and within emitter layer (I_{0e}). For saturation base current I_{ob}, its value will depend on bulk recombination and also BSF saturation current. The value of these variables could be gotten from measurement of bulk life time and Back Surface Recombination Velocity (BSRV). The I_{ob} then could be calculated according the equation

$$I_{ob} = I_{obL}\left(\frac{I_{or}+I_{obL}\tanh\left(\frac{W_B}{L_B}\right)}{I_{obL}+I_{or}\tanh\left(\frac{W_B}{L_B}\right)}\right) \quad (2)$$

where I_{obL} is ideal base saturation current, L_B is minority carrier diffusion length and base, W_B is base width and Ior is saturated current at rear surface [8].

On the other hand the BSRV at BSF layer when structure such as n-p-p$^+$ is used could be estimated by

$$BSRV = \frac{N_A}{N_A^+}\exp\left(\frac{\Delta V_G^+}{kT/q}\right)\cdot\left(\frac{S_0+\frac{D_e^+}{L_e^+}\tanh\left(\frac{W^+}{L_e^+}\right)}{1+\frac{S_0L_e^+}{D_e^+}\tanh\left(\frac{W^+}{L_e^+}\right)}\right) \quad (3)$$

where N_A and N_A^+ is dopant concentration in wafer and BSF layer, respectively, ΔV_G^+ is bandgap narrowing due to heavy dopantt concentration in BSF layer, S_0 is surface recombination velocity without BSF layer, D_e^+/L_e^+ is the electron (minority carrier) diffusivity to diffusion length ratio in BSF layer, and W^+ is BSF thickness [1].

3 Methodology

Solar cell that was under consideration has properties as show in the Table 1. Solar cell performance is simulated under Standard Testing Condition i.e. in ambient temperature of 25°C and AM1.5G spectrum with intensity of 1000 W/m^2.

Simulation was performed to extract variable such as open circuit voltage (V_{oc}), short circuit current density (J_{sc}), efficiency (η), and fill factor (*FF*) as a measure solar cell performance. All of these variables are examines both for P-type wafer and N-type wafer in order to find the effect of different type of wafer against their performance. All data were generated using PC1D v.5.9.

Table 1. Solar Cell properties.

Parameter	**Value**	**Unit**
Surface area	1	cm^2
Junction depth	1.56	µm
Peak doping level emiter (erfc)	***7.5×10^{18}***	***cm^{-3}***
Emitter series resistivity	0.3	Ω
Base series resistivity	0.1	Ω
Cell's thickness	300	µm
Background doping (both for p and n-type wafer)	***1.5×10^{16}***	***cm^{-3}***
Bulk minority lifetime	51	µm
Surface recombination for front area	100	cm.s^{-1}
Surface recombination for rear area	10^5	cm.s^{-1}

4 Result and Discussion

4.1 Effect of peak doping level of BSF (N_{BSF}) on open circuit voltage (V_{OC})

When peak doping level of BSF layer is increased, it will increase V_{oc} value. This happened when doping level on BSF layer is increase up to 2.11×10^{19} cm^{-3} both for p$^+$ and n$^+$ as BSF layers. After the concnetration on BSF layer higher than 2.64×10^{19} cm^{-3}, it show small decrease on V_{oc}. This result could be shown on the Fig. 1.
Another solar cell with wafer thickness of 300 mm that has bulk lifetime of 200 μs and wafer resistivity of 1.3 Ω-cm show the same result. It produces V_{oc} above 660 mV, short circuit current density of 37 mA/ cm^2, and L and L of 8×10^{-14} cm^{-2} and 3×10^{-13} cm^{-2}, respectively [9]. The majority factor is due to bulk minority carrier lifetime. The higher bulk minority carrier lifetime, the higher the V_{oc} will be. Higher minority carrier lifetime could be achieved when BSF layer is applied to reduce Back Surface Recombination Velocity (BSRV). However, at certain point, the V_{oc} will saturate at a high value. In this case, the saturation value for V_{oc} is around 698 mV for P-type wafer and 702 mV for N-type wafer and these condition were reached at BSF doping level of 2.64×10^{19} cm^{-3}.

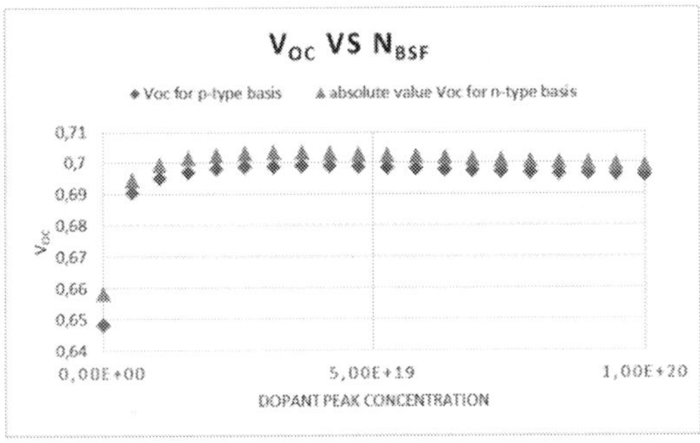

Fig 1. Effect of peak doping level on BSF layer (N_{BSF}) to open circuit voltage (V_{OC})

4.2 Effect of peak doping level of BSF (N_{BSF}) on short circuit current (I_{SC})

It has already been confirmed by many researches that BSF layer will increase short circuit current within solar cells. This effect due to the generation of electric field on the rear surface of the cell. The electric field will help minority carriers that have been generated near the rear surface to escape from recombination process because the electric field will swap the monority carries away from the rear surface of the cell. This mecahnism will enhance carriers collection probability that is produced from long range wavelength because this type of wavelength will tends to be absorbed at the rear side of the cell.

However, BSF usage has limitation when it produces opposite effect. Figure 2 shows that when BSF doping concentration increase there is a tendency that the short circuit current will also decrease.

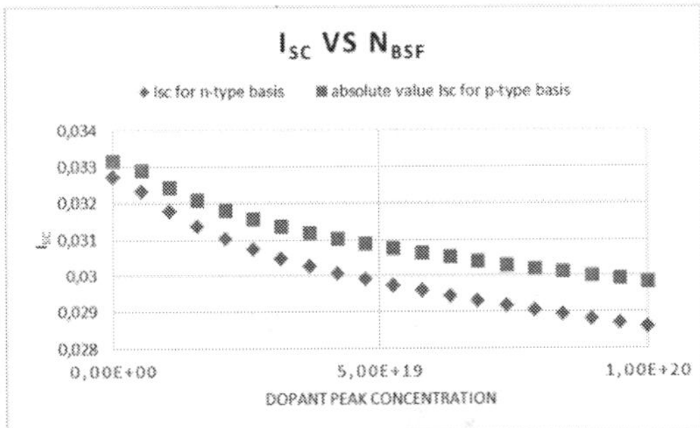

Fig 2. Effect of peak doping level on BSF layer (N_{BSF}) to short circuit current (I_{SC})

Highest short circuit current is achieved when BSF layer has value of 1×10^{17}cm^{-3}. Once this value is increased, the I_{sc} will begin to reduce. Reduction of ISC when dopant concentration on BSF layer increases is attributed to Auger recombination that begin to dominate on higher doping level. From Figure 2 it could be seen that the value of I_{sc} for P-type wafer is higher than N – type wafer. This situation happened because electron as minority carrier in P-type wafer has higher mobility than hole as minority carrier in N-type wafer. The higher carrier mobility will increase the collection probability within the cell, and as the result, it will increase short current.

4.3 Effect of peak doping level of BSF (N_{BSF}) on Efficiency

Efficiency of solar cell measures the effectiveness of the cell to convert photon energy into electricity. The maximum electric power that is generated within the cell will be proportional to its efficiency. This electric power will be greatly affected by its current insofar as the voltage does not change during the operation. This situation is applied when Fig. 1 is considered.

Figure 1 shows that V_{oc} does not change much when BSF doping concentration increased. However, the I_{sc} decreased when BSF doping concentration increased as shown in Fig. 2. This situation will produce efficiency curve against BSF doping concentration that is given in Fig. 3. In Figure 3, it can be seen that efficiency will decrease when BSF doping concentration increase. The pattern on the curve is similar with curve in Fig. 2 so it can be concluded that in this situation short circuit current will dominate the efficiency. Besteffciency for both tipe of solar cell were produced at BSF doping level of 1×10^{17}cm^{-3}. At this level, BSF concentration will high enough to produce electric field while keeping the Auger recombinaation as lower as possible. Increasing the BSF concentration above this level will reduce efficiency because Auger recombination on the back surface will begin to dominate the process.

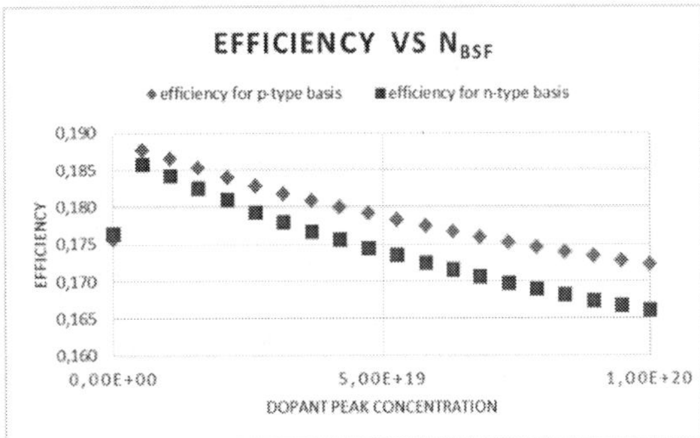

Fig 3. Effect of peak doping level on BSF layer (N_{BSF}) to Efficiency

4.4 Effect of peak doping level of BSF (N_{BSF}) on Fill Factor (*FF*)

Some researches [9,10] show that Fill Factor (*FF*) will be affected by shunt resistance, series resistance, and ideality factor of the diode. All of these factor could be extracted into single factor, that is V_{oc}. Therefore V_{oc} will directly affect FF. Green-Swanson approximation as given in Equation 4 predicts the dependency of *FF* on V_{oc} [9].

$$FF = \frac{v_{OC} - \ln(v_{OC}+0.72)}{v_{OC}+1} \quad (4)$$

where $\upsilon_{\propto}$ is defined as:

$$v_{OC} = \frac{V_{OC}}{(nkT/q)} \quad (5)$$

n being the ideality factor for diode. Figure 4 show how *FF* behaves when BSF doping level changes.

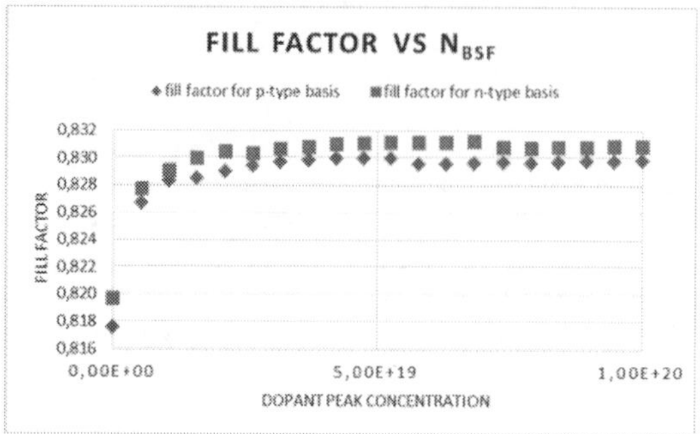

Fig 4. Effect of peak doping level on BSF layer (N_{BSF}) to Fill Factor (*FF*)

In Figure 4 it can be seen that *FF* does not change significantly when $V_{\propto}$ does not change. Figure 1 and Figure 4 together show consistency and in agreement to each other.

5 Conclusions

In general both p-type and n-type wafer will produce higher efficiency when BSF is applied. However, the increase is larger in p-type wafer than in n-type wafer. Better performance for solar cell is achieved when BSF concentration is slightly higher that bulk doping level because at very high BSF concentration the cell's efficiency will be decreased. Best result is achieved when BSF layer doping concentration has value of $1\times10^{17}cm^{-3}$.

References

1. G. Xin, Y. Xuegong, Y. Deren, Efficiency improvement of crystalline silicon solar cells with a backsurface field produced by boron and aluminum co-doping, Scripta Materialia, 66 (2012) pp. 394–397
2. J. Ulrich, S. Dominik, B. Jan, J. Stefan, P. Ralf, A laser based process for the formation of a local back surface field for n-type silicon solar cells, Thin Solid Films, 519 (2011) pp. 3827–3830
3. S. C. Kwang, Parametric Simulation of The Back Surface Field Effect In The Silicon Solar Cell, Solid State Science Journal, 29 (2014) pp. 48-51
4. A. B. Prakoso, Ferdiansjah, Faridah, Optimasi Pembentukan Lapisan Boron P+ Pada Permukaan Belakang Sel Surya Untuk Peningkatan Efisiensi Sel Surya Berbasis Wafer Silikon Monokristal, Teknofisika, Jurnal Ilmiah Bidang Teknik Fisika, 1 (2012) pp. 20-26
5. P. Procel, V. Maccaronio, F. Crupi, G. Cocorullo, M. Zanuccoli, Analysis of The Impact of Rear Side Geometry on Performance of Back-Contact Back-Junction Solar Cells, Fotonica AEIT Italian Conference, (2014) pp.1-4
6. S. Narasimha, G. Crotty, T. Krygowski, A. Rohatgi, D. L. Meier, Back Surface Field And Emitter Passivation Effects Ni The Record High Efficiency N-Type Dendritic High Efficiency N-Type Dendritic Web Silicon Solar Cells, Photovoltaic Specialists Conference, (1997), pp. 235 – 238
7. B. Jan, S. Bernd, M. Ralph, B. Jonas, K. Mathias, M. Andrew, R. Armin, H. Martin, G. Stefan, High Efficiency N-Type PERT and PERL Solar Cells, IEEE Photovoltaics Specialists Conference (PVSC), (2014) pp. 3637 – 3640
8. P. Sungeun, P. Hyomin, K. Yoonmook, L. Hae-Seok, K. Donghwan, Analysis of aluminum back surface field at different wafer specifications in crystalline silicon solar cells, Current Applied Physics 16 (2016) pp. 1062-1068
9. A. G. Martin, Solar Cell Fill Factors: General Graph and Empirical Expression, Solid-State Electronics, 24, 8 (1981) pp. 788-789
10. M. Richard, Swanson, R. Sinton, Advances in Solar Energy, American Solar Energy Society Vol. 1, New York (1983)

https://doi.org/10.1051/e3sconf/20184301007

Analysis of Palm-Oil-Based Biodiesel in Indonesia using Technological Innovation System Approach

Dintani Naimah[1,*] and *Maria* Morgunova[2]

[1]Sustainable Energy Engineering, KTH Kungliga Tekniska Hogskolan, Sweden
[2]Industrial Economics and Management, KTH Kungliga Tekniska Hogskolan, Sweden

Abstract. Biodiesel blending mandate has been established in Indonesia since 2006. After 11 years of implementation, the level of biodiesel penetration on the market is still lower than the target. Technological Innovation System (TIS) approach allows to map factors that influence, either hinder or boost, the diffusion of biodiesel technology. The challenges that should be addressed are tax regulation; price fluctuation due to Crude Palm Oil (CPO) and fossil fuel price; lack of awareness from the society; inconsistency of policy; vulnerable financial funding from palm oil tax levy; lack of sustainability indicators for palm oil; and policy objection regarding palm oil tax levy. We suggest that an agreement, which contains commitment and willingness from all actors involved, should be created to reduce uncertainty towards implementation of biodiesel. Government should not only put incentive to the emerging technology but also reduce competition within the incumbent energy system to increase and stabilize biodiesel demand. Another way to increase biodiesel demand would be to raise awareness about negative consequences of increasing fossil fuel consumption and the benefits of biodiesel for the society.

1 Background

Indonesia is the biggest archipelago country with emerging economy. The country is rich with numerous energy sources, both with renewables and fossils. However, Indonesia still depends on fossil-based energy, as shown in Figure 1.

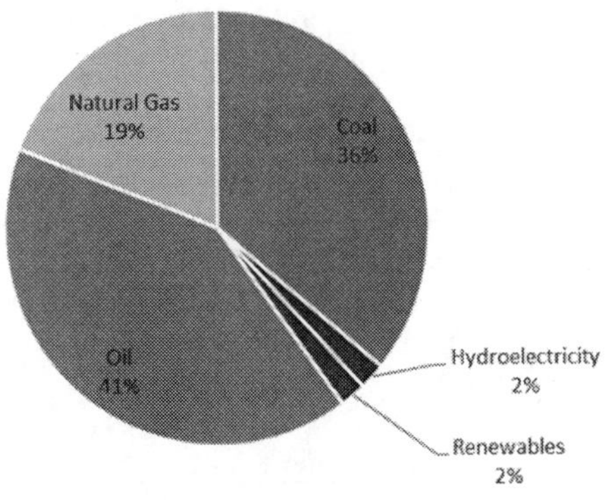

Fig. 1. Indonesia total energy consumption in 2016 [1]

In 2016, oil consumption has reached 41% of total energy consumption in Indonesia. The oil share is slowly but steady increasing [2]. 27% of this share is consumed by transport sector.

Dependence on oil is increasing since the number of road vehicles is growing [3]. During the period 2011-2015 the total number of road vehicles, i.e. passenger cars, bus, trucks, and motorcycles, has increased from 85.6 million to 121.4 million units (9.13% p.a.).

In order to reduce the dependence on oil fuel, the Government of Indonesia (GoI) enforced a blending mandate for biodiesel in 2006. A target was to increase share of biofuel to 5% of a total primary energy supply (TPES) by 2025. The blending mandate for biodiesel kept evolving over the years, where according to the latest version dated December 2015 (enforced through the Ministry of Energy and Mineral Resources, MEMR) the fossil-based diesel in transport sector should contain 20% of biodiesel by 2016. The blending composition between diesel and biodiesel should be increased to 30% by 2020.

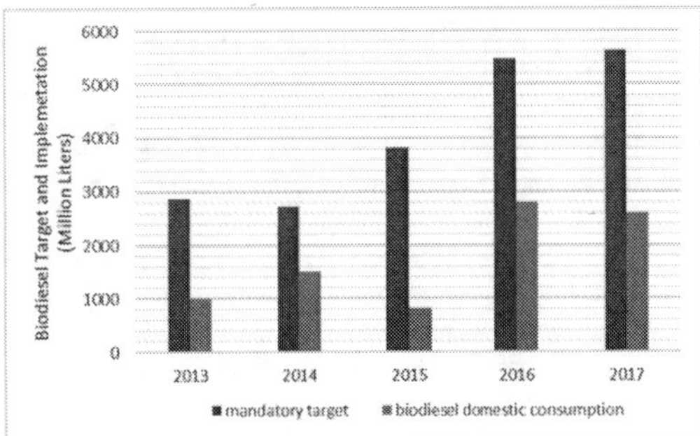

Fig. 2. Biodiesel target and actual implementation [4]

* Corresponding author: naimah@kth.se

Even though the level of blended biodiesel has significantly increased in the past few years, the actual blending and the produced volumes of biodiesel is still lower than it is prescribed by the mandate [4]. This study aims to determine the factors that hamper and foster diffusion of biodiesel in Indonesian market.

Factors that hinder biodiesel development could be grouped into technical, market, social, and policy factors [5]. Some of those challenges should have already been overcome, due to implementation of advanced technology in biorefineries, standard quality regulation for biodiesel and many other now existing supporting regulations. The attempts to improve the biodiesel implementation also included in-depth analysis regarding network of actors in biodiesel business. For example, Purwanto et al [6] proposed a long-term strategy for sustainable biodiesel development in Indonesia based on the SWOT-analysis and Multi Actor Approach. According to the study, GoI has to secure palm oil for biodiesel production, enhance sustainability of palm oil plantation, and improve collaboration with other stakeholders. Purwanto et al [4] developed the recommendation by using Duri, Riau Province, as case study.

Nevertheless, the key challenges are still present. First, uncertainty of biodiesel implementation, including inconsistency of pricing mechanism and its policy. Second, lack of awareness in society regarding the biodiesel and its advantages. Finally, biodiesel market in Indonesia is still on its early stage.

With aim to study the bigger system perspective and to discover the future development possibilities we suggest to examine the deployment of biodiesel in Indonesia using Technological Innovation System (TIS) approach. Thus, the biodiesel industry can be studied as a system with many actors connected through networks. Each actor can be evaluated whether it fulfils its functions on a way to successful biodiesel penetration. This paper results in policy recommendations for successful deployment of biodiesel in Indonesia.

2 Method

TIS approach allows to analyse an emerging technology as a system, in which each actor does their key functions and is connected to other stakeholders. TIS is an effective approach to map dynamics of technological innovation process.

According to TIS, the innovation system is divided into three structural components, i.e. actors, networks and institutions. The actors are involved within the whole value chain, which in this case is biodiesel. Actors are linked through the network, which allows them to exchange knowledge in order to deploy the technology. The final component, institutions, is legal framework where the emerging technology gains ground. Actors within TIS are not only competing on the market, but also influence the institutional alignment. Changing the institutional context could either foster or limit the diffusion of technology. All TIS components should be evaluated to determine its strengths and weaknesses, so that guiding policy to succeed the diffusion of technology could be developed.

Bergek et al. (2008) and Hekkert et al. (2007) proposed evaluation of functions of innovation systems to understand the actual process that is happening within the emerging technology system [7-8]. The seven functionalities are summarized in the Table 1. A better fulfilment of system functions means higher chances for development, diffusion and implementation of the emerging technology.

Many studies using system functions analysis to identify factors that accelerates and limits diffusion of an emerging technology. The analysis is done through evaluating indicators for each function. It allows to identify, which functions in technology diffusion is lagging more than the other. The functions analysis can be used in order to develop effective guiding policy with specific targets. For example, Negro (2007) investigated biomass energy in The Netherlands by using TIS approach [9], while Vasseur et al. (2013) using the same approach to compared penetration of photovoltaic (PV) technology in Japan and The Netherlands [10]

Table 1. Functions and Indicators in TIS [10]

Function	Indicators
Entrepreneurial Activities	1. Organizations or companies entering the market 2. Export activities
Knowledge development	1. Research and technological projects 2. Demonstration and pilot projects 3. Learning by doing and learning by using
Knowledge exchange	National knowledge exchange between organizations
Guidance of the search	Target set by the government or industry
Market formation	1. Financial market incentives 2. Regulation/tax regimes 3. Market size
Resources Mobilization	1. Financial resources 2. Natural resources
Creation of legitimacy	1. Promotion of the technology by organizations 2. Lobby activities against or for the technology

This study is focusing on the system functions of biodiesel as transport fuel in Indonesia. Export activities in entrepreneurial function is the only international activity that is considered, since it might influence supply of biodiesel on the domestic market. The other international activities, such as international knowledge exchange and foreign direct investment are out of the scope of this study. Data for this study is collected via

https://doi.org/10.1051/e3sconf/20184301007

desk research (journals articles, newspapers, reports, and websites).

3 Result

The result section subsequently describes the results of the study based on TIS approach. First, TIS structure is described to understand actors, networks and institutional framework of the system. Secondly, the next part of the result section would cover the fulfilment of the system functions analysis. Finally, policy implications for biodiesel penetration is presented in the last part of this section.

3.1 TIS Structure

Biodiesel TIS in Indonesia is started with Regulation 1/2006, which enforced its distribution and implementation. To foster the goal, National Biofuels Development Team (Tim Nasional BBN, NBDT) is established through Presidential Decree 20/2006. The first biodiesel mandate was established through MEMR Regulation 32 2008, which enforce biodiesel (B100) implementation by 5% from total demand of fuel transportation in 2015, 10% in 2020, and 20% in 2025. The regulation is revised several times and the latest MEMR Regulation 12 2015 enforces blending diesel with biodiesel by 20% (B20) in 2016 and 30% (B30) in 2025. In addition, GoI sets target for biodiesel production through Regulation 22/2017, which is 11.6 million kilo-litre of biodiesel for transport, industry and electricity sectors.

Howlett et al [11] divided cathegorized actors in Indonesia into eight groups, which is government; international institution; producer or private company; academia; certification agency; state owned enterprise, and non-government organization (NGO). More detail regarding actor and network within TIS is presented in the next section.

3.2 System Functions Analysis

3.2.1 Entrepreneurial Activities

Entrepreneurial activities turn knowledge of emerging technology into business practice and market-oriented activities. By entrepreneurial activities, the practicality and value proposition of the emerging technology could be proven. However, the emerging technology often follows by unpredictable consequences when it is applied in the society. These consequences create changes to the emerging technology and its institutional context so that it could adapt to the environment and vice versa. This adaptation also required integration from other system function, such as guidance of the search. Thus, entrepreneurial activities are seen as the core activities in TIS framework.

Entrepreneurial activities of biodiesel were mainly driven by Regulation No 1/2006. To help biodiesel industry to adjust , Government of Indonesia (GoI) built the first biodiesel plant with capacity 8,000 tonnes per day in 2005 [12]. Then, many biodiesel companies were established from 2005 until 2008. During 2011, many biodiesel companies were phased out due to high price of CPO and biodiesel competition with fossil-based diesel; the number were decreased from 16 to only 5 [13]. Nevertheless, biorefinery industries were strengthen back in 2009, and further in 2016, when the blending mandate for biodiesel were increased from 15% to 20%. By 2016, total production capacity of all the biorefinery plants has reached 10,898 million liter of biodiesel. However, the total production only accounts for 33.5% of the total capacity.

By 2014, realization of biodiesel blending fuel only reached 73% from the 1,5 million kilolitres target. Penalties for non-compliance institution is ranging from official warning letter until revocation of business permit. International Council on Clean Transportation (ICCT) reported that none of penalties was given to non-compliance company. Thus, this regulation is still non-binding for the actors involved within the system [14].

Export demand for biodiesel in 2015 decreased to 75%, from 1,350 million litres to 343 litres, due to low price of fossil fuel and anti-dumping policy by European Union (EU). Export of biodiesel was slightly increased to 478 million litres by 2016. Indonesian biodiesel export in 2017 and 2018 is predicted to be stable, especially after deferment of EU anti-dumping policy [15].

Even though entrepreneurial activities are the core of TIS, it still highly depends on constitutional framework and incentives from the government. The government should keep the incentives high enough, despite the fluctuation of CPO and diesel prices, so that the entrepreneurial activities keep running.

3.2.2 Knowledge development

Knowledge development involves learning activities, not limited to the technology, but also the actors, its network, and the market. Main learning activities could be divided into two, i.e. learning by searching and learning by doing. Learning by searching is mostly done by academia and research institution, while learning by doing is conducted by practitioners like entrepreneurs. Knowledge development creates many options for development of emerging technology, which caused uncertainty. However, other system functions, such as guidance of the search and entrepreneurial activities, contribute to the knowledge alignment to fit the environment.

Several universities in Indonesia are involved in knowledge development of biodiesel through conducting several researches. For instance, research about second generation of biodiesel, diversification of feedstock, and laboratories test for biodiesel [11-12]. Universities, such as Universitas Gadjah Mada (UGM) and Institut Teknologi Bandung (ITB), also collaborated with

 https://doi.org/10.1051/e3sconf/20184301007

automotive industry to use biodiesel in passenger cars [13].

All indicators of knowledge development activities are done by government institution. Research, technological and pilot project on biodiesel are mostly conducted by Agency for the Assessment and Application of Technology (Badan Pengkajian dan Penerapan Teknologi, AAAT) and Indonesian Institute of Science (Lembaga Ilmu Pengetahuan Indonesia, IIoS). While demonstration, learning by doing and learning by using are mostly done by cooperation between Ministry of Energy and Mineral Resources (MEMR); Research and Development Centre for Oil and Gas Technology (Lemigas, RDCOGT); and Ministry of Research, Technology, and Higher Education (MRTEH).

GoI also cooperated with private companies, especially for test drive of vehicle which using biodiesel. For example, performance assessment for vehicle which using biodiesel were done with collaboration with The Association of Indonesia Automotive Industries (Gabungan Industri Kendaraan Bermotor Indonesia, TAIA). The result of demonstration activities is used to improve quality of biodiesel and performance of the vehicle.

Knowledge development functions also aims to create various options for the emerging technology and its applications. For biodiesel in Indonesia, MEMR, IIoS, AAAT and some universities also try to produce biodiesel from raw material other than palm oil, such as microalgae, *Schleichera oleosa*, *Ricinus communis*, and *Reutealis trisperma* [18–21]. Researches for diversity of raw material is limited and still on its early stage. The only pilot project that developed is biodiesel from *Ricinus communis*, while other material for biodiesel is still in lab-scale research.

Currently, knowledge development of biodiesel in Indonesia focuses on compatibility test and performance assessment of automotive vehicle which using biodiesel. While diversification of raw material beside palm oil is still on its early stage. Collaboration between government institutions, academia, and private companies in these activities have been run. This show that knowledge development function for biodiesel is already fulfilled.

3.2.3 Knowledge exchange

Characteristics of TIS organisational structure is the network, in which all actors can interact and have knowledge exchange. Knowledge exchange is expected to engage many actors with different background, such as government, private companies, and researcher. Through frequent knowledge exchange, mutual understanding of the emerging technology could be developed. Furthermore, policy maker can plan the future regulations according to the latest technological insight.

First seminar on bioenergy in general is held on 2008 with the main agenda to discuss the blending mandate [22]. The speakers were RDCOGT, Pertamina EP, Association of Indonesia Biofuel Producers (Asosiasi Produsen Biofuel Indonesia, AIBP), Indonesia Chambers of Commerce and the committee from Energy-Independent Village. The second bioenergy seminar was held in 2011, which connected more actors than the previous one [23]. After that in 2015 the biodiesel seminar is more focused on the technical assessment and implementation planning for B20 fuel [16,17].

Even though seminar and exhibition between policy makers and researchers have been done, knowledge exchange between government and society at large, especially consumers, seems to be lacking [24]. Therefore, seminars and knowledge exchange activities to vehicle users and community at large regarding benefit and characteristics of biodiesel should be conducted at a bigger scale.

3.2.4 Guidance of the search

Guidance of the search refers to any activities that align the development of emerging technology from all actors, for example, establishment of policy directives, expectations, and promises. Without clear guidance of search, the development of emerging technology will lose focus and other functions, such as knowledge development; knowledge diffusion, and entrepreneurial activities.

The first enforcement for implementation of biodiesel was established through Government Regulation No 1 2006. Then this regulation was followed by other supporting policies, such as Presidential Decree 20/2006 about establishment of biodiesel supervision and Energy Law 2007.

The principal guidance of the search for biodiesel TIS is MEMR Regulation 12/2015, which enforced B20 and B30 program. The target is to produce 11.6 million kilo-litre of biodiesel for transport, industry and electricity generation in 2025. Currently, GoI focus to produce biodiesel from palm oil, although there are also potential to produce it from Jatropha, sugarcane, and cassava. Palm oil was chosen due to security of supply [14].

However, Ministry of Industry published Regulation 33/2013, which allows the distribution of low cost green car (LCGC) in Indonesia. Enforcement of this regulation created competition between biodiesel and fossil fuel, especially for passenger cars. Additionally, guidance for quality standard of B20 fuel is still missing in the regulation [9, 16, 17].

3.2.5 Market formation

Market formation is required to support the emerging technology compete with the incumbent technology. Activities in market formation can help creating demand for the emerging technology. Market formation is important, especially for sustainable energy technology, because fossil-based energy does not internalize external cost, like its environment effect. Some examples of market formation are marketing campaign with temporary discount, subsidy, institutional framework to account for external cost, etc.

Since 2015, the current blending mandate is funded from tax levy on exported crude palm oil (CPO), which is enforced through Regulation 24/2015 and 61/2015. CPO exporters requires pay for the levy if the price of CPO is under USD 750. If the price of CPO goes beyond USD 750, the exporters are expelled to pay the levy, but instead they required to pay for regular export tax. This CPO export fund is utilized to offset the price gap between biodiesel and fossil-based diesel and support research and development of biodiesel.

In May 2017, MEMR reduced amount of paid per unit for biodiesel blenders. The regulation imposed that the amount paid to the blenders were summation of average CPO Price and USD 125/MT of biodiesel, which became USD 100/MT in the current regulation. Reduction of biodiesel revenue might weaken the domestic market and made international market more attractive.

Palm oil tax levy is the only market formation regulation that enforced in Indonesia. No other incentive and tax beside palm oil export levy have been issued to the current date.

3.2.6 Resources Mobilization

Resources mobilization refers to allocation of financial and natural resources to support the emerging technology. Typical activities in resources mobilization are investments and subsidies.

By 2016, total revenues from palm oil export tax reached 9.5 trillion Rupiah (USD 698 million) which is claimed to be enough to fund all blending mandate program, including to offset the price gap between diesel and biodiesel, and to fund research and development of biodiesel. Global Agricultural Information Network (GAIN) reported that the actual blending levels of biodiesel in the market were correlated with the amount of funding from palm oil levy. Thus, fluctuation of palm oil export will influence amount of biodiesel blending.

Biodiesel mandate increased demand of palm oil, which leads to more area required for palm oil plantation. However, the biofuel production currently causes 6.5% of annual deforestation in Indonesia [26]. With biodiesel mandate, Indonesia requires 5.15-6 million ha additional land for palm oil plantations [20, 21]. The required area for palm oil expansion for biodiesel is 11 – 17 Mha without jeopardizing policy on agriculture, forestry, and climate mitigation resources [28], while the available area for biodiesel is 20.9 Mha [29]. The required area for biodiesel could be reduced if there is an increased palm oil yield per area of plantation (tonne of palm oil per plantation area). Palm oil yields could be increased by 20% with intensification of plantation, such as crop recovery, canopy management and soil, moisture, and nutrient management [30].

Biodiesel resources mobilization consist of the allocation of financial and natural resources. Palm oil tax levy is the main funding source for blending mandate program. Although the available land in Indonesia is sufficient to meet biodiesel target, sustainability indicators for biodiesel production is still missing.

3.2.7 Creation of legitimacy

Emerging of new technology into the system is followed with resistance by some actor in the incumbent energy system [31]. The actors which against the change usually do political lobbies or advice activities to hinder penetration of the emerging technology. In order to diffuse the technology, other supporting actor should hold counteract lobbies and promotion activities. These creations of legitimacy will push the decision maker to fasten or limit the technology.

Despite the solid financial funding, the gap between biodiesel and fossil-based diesel price has greatly increased, as shown in Figure 4. Thus, biodiesel blenders asked the Government to revise the subsidy in response to the change.

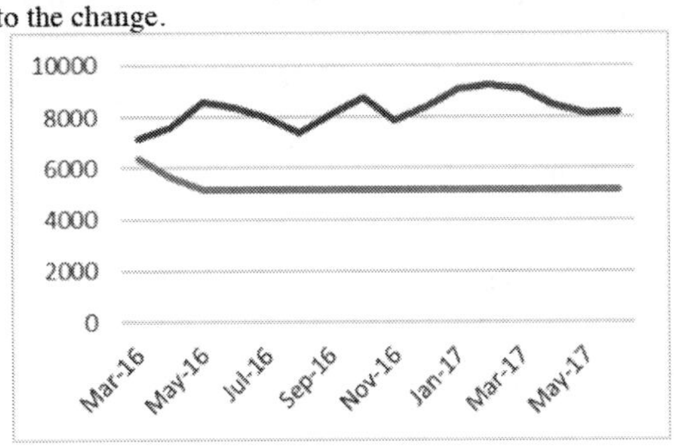

Figure 3. Biodiesel Market Price Compared to Fossil-based Diesel (Unit: IDR/liter) [4]

Enforcement of CPO tax levy reduced income of palm oil farmers. Indonesian Palm Oil Association (Gabungan Pengusaha Kelapa Sawit Indonesia, IPOA) continually asks the government to reduce the palm oil tax levy and replace it with other tax regime, such as carbon credit for fossil fuel user. In addition, palm oil farmers, under Association of Plasma Palm Oil Farmers (Asosiasi Petani Plasma Kelapa Sawit, APPOF), sent objection and judicial review regarding CPO levy fund to Indonesia Supreme Court [28-29]. APPOF demands the abolishment of the tax levy due to allegation that there is a corruption practice in the levy fund [34]. These cases have been running since 2016 and are still under investigation. Despite all the projections, GoI is still determined to meet the biodiesel mandate by 2025.

4 Discussion

Institutional framework for implementation of biodiesel seems to be clear, where the specific target of blending mandate exists. However, several upcoming policies by the GoI leads to uncertainty towards diffusion of biodiesel technology. For example, reduction of unit paid for biodiesel blenders and support of LCGC. This uncertainty has also created more risk for the entrepreneurs.

Taking Sweden as a benchmark, uncertainty in implementation of biofuels could be removed by having a clear guidance of the search on its early stages, which is called the Three-Party Agreement 1991 [31]. This agreement is followed by multi-sectoral strategy to support biofuel deployment.

Besides supporting the emerging technology, government also requires to put tax for the incumbent energy source to reduce competition. This tax could increase biodiesel demand and reduce the risk of biodiesel producers. Tax on fossil fuel would also help the government to have financial sources for blending mandate program beside palm oil tax levy. Thus, putting tax on fossil fuel would influence system functions in Market Formation; Resources Mobilization and Entrepreneurial Activity.

Other challenge that hinder diffusion of technology is the lack of awareness by the end-users. To tackle this challenge promotion and broader knowledge exchange about biodiesel product should be initiated by the government. Knowledge Exchange regarding negative consequences of fossil-based fuel would also increase people awareness. With higher awareness from the society, the urge to shift to renewable energy, such as biodiesel, is expected to increase [31]. This activity would advance system functions in Knowledge Development and Exchange, Creation of Legitimation, and Market Formation.

Enforcement of policy and regulation should be accompanied with commitment and willingness from all the actors. Government should also be consistent, especially in giving penalties to the non-compliance actors. Funding management, tenders, projects, and other activities within the system should be transparent to gather trust from the society and in order to reduce activities that against the diffusion of the technology.

5 Conclusion and Policy Implications

Biodiesel blending mandate has been established in Indonesia since 2006. But after 11 years of implementation, penetration of biodiesel in the market is still under the target. TIS approach allows us to map all the lagging aspects during the diffusion of biodiesel technology through system function analysis. Some challenges that should be addressed are lax regulation; price fluctuation due to CPO and fossil fuel price; lack of awareness from the society; inconsistency of policy lead to competition with fossil fuel; vulnerable financial funding from palm oil tax levy; lack of sustainability indicators for palm oil; and policy objection regarding palm oil tax levy.

An agreement between all the actors is required to clarify the situation about the future of the technology. The agreement should be followed by the strategy or guiding policy to support the technology. In biodiesel TIS, GoI has already put incentives for biodiesel blenders to offset the price between biodiesel and fossil-based diesel. Nevertheless, negative consequences of fossil fuel in Indonesia has not been internalized in the price; instead, the fossil fuel is still subsidized. Additionally, tax on fossil fuel could be the financial source for blending mandate program other than the palm oil tax levy. A broader knowledge exchange and promotion of biodiesel should also reach everyone in the society, and not be limited to researchers and private company owners. If the society has higher awareness about biodiesel, it is expected that the demand will increase.

With diffusion of biodiesel, GoI would be able to reduce the dependence on fossil fuel, increase the share of renewable energy and reduce the greenhouse gas emission in the country. However, sustainability indicators for palm oil production should be introduced to minimize environment effect of palm oil plantation.

References

1. British Petroleum, "BP Statistical Review of World Energy," (2017).
2. International Energy Agency, "IEA Sankey Diagram," 2014. [Online]. Available: https://www.iea.org/Sankey/#?c=Indonesia&s=Final consumption. [Accessed: 24-Aug-2017].
3. Statistics of Indonesia, "Land Transportation Statistics 2015," Jakarta, 8302004, (2016).
4. T. Wright, A. Rahmanulloh, and A. Abdi, "Indonesia Biofuels Annual Report 2017," Global Agriculture Information Network (GAIN), Jakarta, ID1714, (2017).
5. A. Daryanto, "Urgensi pengembangan sumber energi terbarukan," Bogor Agriculture Institute (IPB), Bogor, (2010).
6. B. A. Purwanto, E. Hambali, Y. Arkeman, and H. Wijaya, "Formulating a Long Term Strategy for Sustainable Palm Oil Biodiesel Development in Indonesia," *J. Sustain. Dev.*, vol. 9, no. 4, pp. 124–132, (2016).
7. A. Bergek, S. Jacobsson, B. Carlsson, S. Lindmark, and A. Rickne, "Analyzing the functional dynamics of technological innovation systems: A scheme of analysis," *Res. Policy*, vol. 37, no. 3, pp. 407–429, (2008).
8. M. P. Hekkert and S. O. Negro, "Functions of innovation systems as a framework to understand sustainable technological change: Empirical evidence for earlier claims," *Technol. Forecast. Soc. Change*, vol. 76, no. 4, pp. 584–594, (2009).
9. S. O. Negro, M. P. Hekkert, and R. E. Smits, "Explaining the failure of the Dutch innovation system for biomass digestion—A functional analysis," *Energy Policy*, vol. 35, no. 2, pp. 925–938, (2007).
10. V. Vasseur, L. M. Kamp, and S. O. Negro, "A comparative analysis of Photovoltaic Technological Innovation Systems including international dimensions: the cases of Japan and The Netherlands," *J. Clean. Prod.*, vol. 48, pp. 200–210, (2013).
11. M. Howlett, I. Mukherjee, and J. Koppenjan, "Policy learning and policy networks in theory and practice: the role of policy brokers in the Indonesian biodiesel policy network," *Policy Soc.*, vol. 36, no. 2, pp. 233–250, (2017).

12. Merdeka, "Menristek Resmikan Pabrik Biodisel Pertama di Indonesia," *merdeka.com*, Jan-2005. [Online]. Available: https://www.merdeka.com/uang/menristek-resmikan-pabrik-biodisel-pertama-di-indonesia-x0a1bv6.html. [Accessed: 02-Sep-2017].
13. N. Panjaitan, "Development of Biofuel Technology in Indonesia," TU Delft, Delft, 2013.
14. A. Kharina, C. Malins, and S. Searle, "BIOFUELS POLICY IN INDONESIA: OVERVIEW AND STATUS REPORT," (2016).
15. Reuters, "EU postpones decision on biodiesel anti-dumping duties," Jul-2017. [Online]. Available: https://www.reuters.com/article/us-eu-biodiesel-idUSKBN1AC2A3. [Accessed: 28-Aug-2017].
16. SB, "Kemenristekdikti Danai 7 Inovasi Perguruan Tinggi," *Technology Indonesia*, 2016. [Online]. Available: http://www.technology-indonesia.com/index.php/component/content/article/129-umum/974-kemenristekdikti-danai-7-inovasi-perguruan-tinggi. [Accessed: 03-Sep-2017].
17. A. Susanti, "Enam Kampus Siap Kembangkan Inovasi ke Industri," *Okezone*, Mar-2016. [Online]. Available: https://news.okezone.com/read/2016/03/07/65/1329569/enam-kampus-siap-kembangkan-inovasi-ke-industri. [Accessed: 03-Sep-2017].
18. Y. K. Irawan, "Kalbar Kembangkan Kemiri Sunan untuk Bahan Baku Biodiesel," *KOMPAS.com*, Jun-2016. [Online]. Available: http://bisniskeuangan.kompas.com/read/2016/06/29/160000626/kalbar.kembangkan.kemiri.sunan.untuk.bahan.baku.biodiesel. [Accessed: 10-Sep-2017].
19. Ministry of Energy and Mineral Resource (MEMR), "Pengembangan Kemiri Sunan di Daerah Pertambangan," *Pengembangan Kemiri Sunan di Daerah Pertambangan*, Dec-2013. [Online]. Available: http://litbang.esdm.go.id/berita/penelitian-dan-pengembangan-mikroalga-sebagai-bahan-baku-biodiesel. [Accessed: 10-Sep-2017].
20. Indonesian Institute of Science, "LIPI dan IPB Lirik Biji Kesambi NTT," Mar-2011. [Online]. Available: http://lipi.go.id/berita/single/LIPI-dan-IPB-Lirik-Biji-Kesambi-NTT/6469. [Accessed: 10-Sep-2017].
21. Indonesian Institute of Science, "Biodiesel Alga Hijau Biru Masuk PPRI LIPI," Oct-2007. [Online]. Available: http://lipi.go.id/berita/single/Biodiesel-Alga-Hijau-Biru-Masuk-PPRI-LIPI/1476. [Accessed: 10-Sep-2017].
22. H. S. Dillon, T. Laan, H. S. Dillon, International Institute for Sustainable Development, Global Subsidies Initiative, and International Institute for Sustainable Development, *Biofuels, at what cost?: government support for ethanol and biodiesel in Indonesia*. Winnipeg, Man.: International Institute for Sustainable Development, (2008).
23. Ministry of Energy and Mineral Resource (MEMR), "Indo-Bioenergy 2011," Jun-2011. [Online]. Available: http://www3.esdm.go.id/siaran-pers/55-siaran-pers/4528-indo-bioenergy-2011.html. [Accessed: 03-Sep-2017].
24. Bioenergy Division, "Kajian Teknis dan Uji Pemanfaatan Biodiesel 20% Pada Kendaraan Bermotor dan Alat Besar," presented at the Seminar on Technical Assessment and B20 Implementation, Jakarta, (2015).
25. B. Restituta, I. Paryanto, and R. Riza, "Spesifikasi Mutu B-20 di Indonesia dan Perbandingannya dengan Spesifikasi Biodiesel, Minyak Solar dan Standard International," presented at the Seminar on Technical Assessment and B20 Implementation, Jakarta, (2015).
26. K. Obidzinski, R. Andriani, H. Komarudin, and A. Andrianto, "Environmental and Social Impacts of Oil Palm Plantations and their Implications for Biofuel Production in Indonesia," *Ecol. Soc.*, vol. 17, no. 1, (2012).
27. A. Rahmadi, L. Aye, and G. Moore, "The feasibility and implications for conventional liquid fossil fuel of the Indonesian biofuel target in 2025," *Energy Policy*, vol. 61, pp. 12–21, (2013).
28. F. Harahap, C. Palmen, S. Silveira, and D. Khatiwada, "Conditions for a sustainable development of palm-oil-based biodiesel in Indonesia," (2016).
29. F. Harahap, S. Silveira, and D. Khatiwada, "Land allocation to meet sectoral goals in Indonesia—An analysis of policy coherence," *Land Use Policy*, vol. 61, pp. 451–465, (2017).
30. C. R. Donough, C. Witt, T. H. Fairhurst, and others, "Yield intensification in oil palm plantations through best management practice," *Better Crops*, vol. 93, no. 1, pp. 12–14, (2009).
31. R. A. A. Suurs, "Motors of Sustainable Innovation," Utrecht University, Utrecht, (2009).
32. M. S. Widodo and B. Febriana, "Murah Harga Pungutan Digugat," *GATRA*, Jakarta, (2016).
33. D. Ningsih, "Petani Plasma Protes Pungutan Sawit," *Berita Satu*, Jakarta, (2016).
34. Global News, "KAKI Desak KPK Audit Investigasi Penyelewengan Dana Pungutan CPO," *Global News*, (2017). .

https://doi.org/10.1051/e3sconf/20184301008

Characterization of Natural Zeolite and Chicken Manure Derived Biochar for Carbon Dioxide Adsorption in Biogas

Margaretha Arnita Wuri[1], *Ambar* Pertiwiningrum[2,*], *Rachmawan* Budiarto[3], and *Cahyono* Agus Dwi Koranto [4]

[1]Master Program of Technology for Sustainable Development, Universitas Gadjah Mada, Indonesia
[2]Faculty of Animal Science, Universitas Gadjah Mada, Indonesia
[3]Faculty of Engineering, Universitas Gadjah Mada, Indonesia
[4]Faculty of Forestry, Universitas Gadjah Mada, Indonesia

Abstract. Although biogas is clean and environmentally friendly, its existence is often unsustainable, especially in rural area. Consumption of fossil fuel energy still dominates because cheaper, higher in energy efficiency and ready to use than biogas. One of the efforts to make biogas can compete with fossil fuel energy is by increasing calorific value. Reduction of carbon dioxide by adsorption enhances the calorific value that evidenced by methane enrichment. In this study, we investigated the effect three types of combined adsorbent: pure natural zeolite (Z-Z), zeolite-biochar from chicken manure (Z-B1) and zeolite-modified biochar (Z-B2) on methane content. The result showed that adsorption of biogas using Z-Z, Z-B1 and Z-B2 increased methane level with enhancement of 1,36; 28,92; and 11,27% respectively compared before adsorption. To confirm whether these results were influenced by adsorbent's character or not as consequence adsorbent characterization have been done in this study.

1 Introduction

Today global warming becomes the global disaster that threats all countries. The cause of global warming is the continuous increase of green house gases (GHGs), carbon dioxide, methane and dinitrogen oxide in the atmosphere. The increase of GHGs has been observed by *National Oceanic and Atmospheric Administration* (NOAA) since 1958 and the peak at 2016 [1]. Moreover, methane and dinitrogen oxide showed the same phenomenon [2]. As a matter of fact, methane and dinitrogen oxide have 21 times and 300 times greater potency than carbon dioxide to cause global warming. Some reports said that total methane and dinitrogen oxide in atmospheric are 10-12% of global GHG emissions. These emissions are dominated by agriculture and livestock sectors [3-5].

The fact becomes driving force renewable energy, for example, biogas technology. Biogas is able to isolate GHG emissions from anaerobic decomposition of agriculture and livestock waste and convert it directly into energy [6]. Although biogas has benefits in mitigating GHGs and overcoming the waste issue, implementation of biogas technology is often unsustainable. Lower calorific value of biogas becomes one of the barriers in developing of biogas technology. The decrease calorific value is caused by the present non-CH4 gases especially CO2 [7-8].

An innovation to improve calorific value is through biogas purification. Biogas purification is methane enrichment method by separating methane from decreasing calorific value gases [9]. The other methods of biogas purification such as water scrubbing, cryogenic, membrane separation, chemical absorption and adsorption have been developed and evaluated [7-8]. But in a rural area, a cheap, simple and viable method becomes more attractive so that implementation can be sustainable [9]. Adsorption becomes the most choosable option to be selected because adoptable and inexpensive [10].

In this study, we investigated the effect of combinations adsorbent, natural zeolite and chicken manure-derived biochar against methane enrichment in biogas. Zeolite has good performance in carbon dioxide adsorption [11] but zeolite has a drawback that no every rural area have natural zeolite resources. As consequence, adsorbent cost becomes expensive because of packaging and distribution process. Biochar has been purposed as a substitute adsorbent for partial natural zeolite due to low cost and environmental friendly [12-13]. Surface modification on biochar was also carried out in this study to improve CO_2 adsorption capability [14-15].

* Corresponding author: artiwi@mail.ugm.ac.id

https://doi.org/10.1051/e3sconf/20184301008

2 Methodology

2.1. Materials

Natural zeolite with an average particle size between 1-3 mm (Z) and ammonia solution from Bratachem. Natural zeolite pretreated by calcinating at 255°C for 3 hours. Chicken manure for biochar production was supplied from PT. Charoend Phokpand. Ammonia solution 25% was dissolved into 10% v/v for modification of biochar. Biogas was supplied by Gama Biogas, Agrotechnology Innovation Center, UGM.

2.2 Synthesis of biochar and its modification

After being dried by aeration for the full day, chicken manure was pyrolyzed at 500°C under N_2 atmosphere for 4 hours (B1). Modification of biochar was done by immersed biochar in ammonia solution 10% (v/v) for 24 hours. Then the mixture was separated and oven dried at 105°C for 4 hours (B2).

2.3 Characterization of Adsorbents

Surface area, total pores volume and pore size were determined by Surface Area Analyzer (SAA). Pores surface area was determined by nitrogen (N_2) adsorption isotherms at 77K. by calculating using Braunauer-Emmett-Teller (BET) method. Total pores volume were calculated from the amount N_2 adsorption at the relative pressure of 0,99. Biochars that produced from chicken manure and its modification were characterized using infrared (IR) spectroscopy with the potassium bromide (KBr) technique. Sample preparation as KBr pellets by mixing samples with KBr at 1,3-1,5 wt% using ball mill. The FTIR spectra scanned range of 4000-400 cm^{-1}.

2.4 Biogas Purification by Adsorption

The experimental for adsorption test consists of two interconnected fix bed columns that filled 80 grams adsorbents for each column. The mass ratio of the adsorbent of 1:1 (40 grams:40 grams) in each column. Details of the formulation of adsorbents given in **Table 1**.

Table 1. Combinations of adsorbent

Code	1st column (40:40 grams)	2nd column (40:40 grams)
Z-Z	Natural zeolite	Natural zeolite
Z:B1	Natural zeolite : biochar	Natural zeolite : biochar
Z:B2	Natural zeolite : modified biochar	Natural zeolite : modified biochar

For adsorption process, the first column was connected to the compressor which related to biogas holder from Gama Biogas and the other column was connected to the tapping biogas holder. The adsorption process was done at a gas pressure ranging 2 to 7 kg/cm^2. To know the effect of adsorbent against methane composition, biogas before and after adsorption were analyzed with gas chromatography.

3 Results and Discussion

3.1. Adsorbent Characteristics

The surface area of natural zeolite was calculated from N_2 adsorption isotherms using Braunauer-Emmett-Teller (BET) method. Natural zeolite had the specific surface area of 27,9 m^2/g. Pores consisted mainly mesopores and dominated size pores of 2,67 nm. **Fig. 1**. showed N_2 sorption isotherms curve of natural zeolite. The curve exhibited type III according to International of Pure Applied and Chemistry (IUPAC) classification. The weak interaction between adsorbents and adsorbate made curve rose slightly at low relative pressure. When molecules have begun adsorped, the interaction between adsorbate get stronger and accelerated the rising curve at high relative pressure. It concluded that at high pressure is related to the higher loading of adsorbate on the natural zeolite. **Fig. 1.** showed that volume of N_2 was adsorbed and desorbed on pores of natural zeolite almost same. It means natural zeolite can adsorb reversibly and has potential to be used for several adsorption processes with regeneration [16].

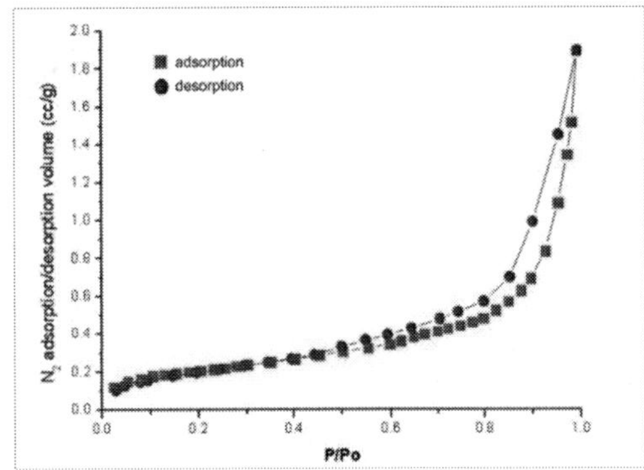

Fig. 1. N_2 adsorption-desorption isotherm on natural zeolite

To investigate effect surface modification on the chemical characteristic of adsorbents, biochar and modified biochar (B1 and B2) were characterized using infrared (IR) spectroscopy. Spectrum at **Fig. 2.** showed no significant difference between B1 and B2. Even it presented some similarities of typical bands. The bands between 3440-3457 cm^{-1} confirmed vibration of O-H stretching of phenol, alcohol, and carboxylic acid on the surface of biochar [17-18]. The vibration of C=C aromatic stretching was identified at 1400-1500 cm^{-1}. The peak represented C-O from alcohol and C-C aromatic bending confirmed at 1000-1200 and 873-875 cm^{-1} respectively [19].

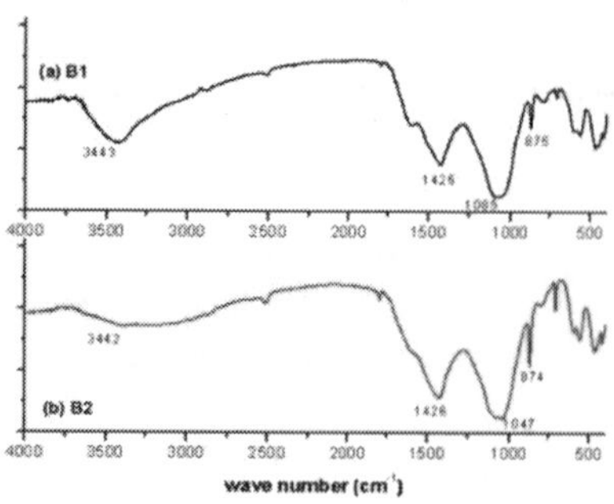

Fig. 2. IR Spectra (a) B1, biochar and (b) B2, modified biochar

3.2. Biogas purification

So far investigation effect combinations of adsorbent against methane composition in biogas, adsorption process have been done. **Table 2**. represents methane composition before and after adsorption.

Table 2. Methane composition in biogas before and after adsorption

Code	Methane composition (%)		Enhancement (%)
	Before	After	
Z-Z	42,5	43,8	1,36
Z-B1	25,35	32,68	28,92
Z-B2	32,56	36,23	11,27

In this study, biogas was considered composed of methane and the remaining carbon dioxide since the other gases composition were very small [20]. The decrease of carbon dioxide by adsorption was known from an enhancement in methane content. Methane composition in raw biogas fluctuated between 25 to 42,5%. After adsorption with contact time for 10 minutes, methane content in purified biogas was analyzed. The best result in carbon dioxide adsorption was performed by combination adsorbent Z-B1 with a methane enrichment level of 28,92%, followed by Z-B2 and Z-Z at 11,27% and 1,36% respectively.

The result from **Table 2**. was explained in this paper. Carbon dioxide adsorption on natural zeolite performed lowest methane enrichment. However, according to some reports zeolite showed better performance [11,19] than biochar. Adsorption is affected by the specific surface area where the higher specific surface area, more the number of adsorbate. But in this study natural zeolite has a low specific surface area of 27,9 m^2/g, smaller than surface area that reported by Hauchhum and Mahanta, and Alonso-Vicaro [11,22]. The absence of further activation was indicated micropore cavity clogged by impurities that should be dissolved during thermal activation. These clogged pores could not facilitate carbon dioxide adsorption that related to the low methane enhancement. It is also possibly caused by the presence of considerable moisture content in biogas inhibited the carbon dioxide capture because it competes with carbon dioxide in adsorption on zeolite [23].

In this study, substitution of some zeolite with biochar (50% of total mass) can increase methane content in biogas. Biochar is composed of hollow aromatic carbon strictures [24] which can capture carbon dioxide molecules. To confirm the effect of the effectiveness of specific surface area on methane enrichment compared to zeolite, SAA analysis is also required. In addition, biochar is not sensitive to water vapor [18] compared with natural zeolite so in this study was indicated that the absence of capturing competition between water vapor and carbon dioxide on biochar. Xu, et al. [25] also reported that biochar derived from livestock manure are rich in minerals which can facilitate carbon dioxide capture. the content of alkali metals such as K, Na, Ca and Mg increase the alkalinity of biochar. Based on Lewis acid-base theory, carbon dioxide is acidic molecule so that when biogas is passed through biochar then carbon dioxide will be adsorbed on biochar's pores and formed weak interaction with biochar [26]. This is the reason why partial replacement of natural zeolite with biochar actually raised methane level to the greatest of 28,92% during 10 minutes of contact time.

Surface modification with doping nitrogen can increase the alkalinity of biochar so that improve the capability of carbon dioxide capture on biochar [13-14] but the contrasting result from previous authors was also reported in this paper. In this study, methane enrichment by adsorption with Z-B2 was smaller (11,27%) than adsorption using combination adsorbent Z-B1 with unmodified biochar (28,92%). The lower methane enrichment of this sample than Z-B1 indicated that incorporation of nitrogen groups on biochar's surface didn't cause an increment capacity of carbon dioxide adsorption. It was caused some pore partially blockage by ammonia. The fact was evidenced from FTIR spectra of B2 that there is no new functional group formed (**Fig. 2**). Shafeeyan, et al. [27] also demonstrated that introduction of base functional at room temperature can lead to decrease micropore volume as consequence lowered carbon dioxide adsorption and then lowered methane enrichment. Formation of nitrogen-containing functional groups on biochar's surface required heat treatment. Heat treatment in doping nitrogen plays important role in the formation of free radicals from ammonia so that interaction between free radicals and carboxylic sites on biochar occurred and formed base functional groups [17].

4 Conclusion

Carbon dioxide adsorption in biogas that was represented by methane level was conducted using combinations of adsorbent. Although many authors claimed that zeolite has good performance in carbon dioxide but in this study, natural zeolite demonstrated the worst performance. Surface modification on biochar was not synthesized successfully. It can be confirmed from the result of adsorption using Z-B2 that perform lower methane enhancement than Z-B1. Heat treatment is required in surface modification on the biochar. The highest percentage of methane enhancement were 28,92% that performed by Z-B1. Based on this findings, chicken manure-derived biochar can be used an alternative adsorbent for carbon dioxide removal in biogas. Before implementation in a rural area, optimization of adsorption with chicken manure-derived biochar should be done.

This study was supported financially by grant funding from Graduate School, Universitas Gadjah Mada, Indonesia.

References

1. D. Dlugokencky, P. Tans, *Trends in Atmospheric Carbon Dioxide: Recent Global CO_2*, <https://esrl.noaa.gov/gmd/ccgg/trends/global.html> (2004)
2. J. H. Buttler, S. A. Montzka, *The NOAA Annual Greenhouse Gas Index (AGGI)*, <https://esrl.noaa.gov/gmd/aggi/aggi.html> (2016)
3. P. Smith, D. Martino, Z. Cal, D. Gwary, H. Janzen, P. Kumar, B. McCarl, S. Ogle, F. O'Mara, C. Rice, B. Scholes, O. Sirotenko, *Agriculture in Climate Change 2007: Mitigation* (Cambridge University Press, Cambridge, 2007)
4. F. N. Tubiello, R. D. C. Golec, A. Ferrara, S. Rossi, R. Biancalani, S. Federici, H. Jacobs, A. Flammini, *Agriculture, Forestry and Other Land Use Emissions by Sources and Removals by Sink: 1990-2011* (FAO, Rome, 2014)
5. IPCC (Intergovernmental Panel on Climate Change), *Climate Change 2014: Mitigation of Climate Change Contribution of Working Group III to the Fifth Assessment Report* (Cambridge University Press, Cambridge, 2014)
6. A.D. Cuellar, M. E. Webber, Env. Res. Lett **14** (2008) 1-8
7. Q. Sun, H. Li, J. Yan, L. Liu, Z. Yu, X. Yu, Renew. and Sustainable Energy Reviews **51** (2015) 521-532
8. W. M. Buszianowski, Renew. and Sustainable Energy Reviews **54** (2016) 1148-1171
9. M. R. A. Mamun, M. R. Karim, M. M. Rahman, A. M. Asiri, S. Torri, J. of The Taiwan Institute of Chem. Eng. **58** (2016) 476-481
10. B. Boulinguiez, P. L. Cloirec, Energy Fuels **24** (2010) 4756-4765
11. L. Hauchhum, P. Mahanta, Inter. J. of Energy, Env. and Eng. **5**, 4(2014) 349-356
12. F. Lian, G. Cui, Z. Liu, L. Duo, G. Zhang, B. Xing, J. of Env. Management **176** (2016) 61-68
13. H. A. Alhashimi, C. B. Aktas, Res., Conversation and Recycling **118** (2017) 13-26
14. X. Zhang, S. Zhang, H. Yang, Y. Feng, Y. Chen, X. Wang, H. Chen, Chem. Engineering J. **257** (2014) 20-27
15. M. V. Nguyen, B. K. Lee, Process, Safety and Env. Protec. **104** (2016) 490-498
16. E. Wibowo, M. Rokhmat, Sutisna, Khairurrijal, M. Abdullah, Desalination **409** (2017)
17. A. Heidari, H. Younesi, A. Rashidi, A. A. Ghoreyshi, Chem. Eng. J. **254** (2014)
18. Y. Huang, P. Chiueh, C. Shih, S. Lo, L. Sun, Y. Zhong, C. Qiu, Energy **84** (2015)
19. M. Stefaniuk, P. Oleszcuk, J. of Analytical and Applied Pyrolysis **115** (2015)
20. P. J. Jorgensen, *Biogas-Green Energy* (Digisource Danmark A/S, Aarhus, 2015)
21. D. P. Bezerra, R. S. Oliveira, R. S. Vieira, C. L. Cavalcante Jr., D. C. S. Azevedo, Adsorption **17** (2011) 235-246
22. A. Alonso-Vicaro, J. R. Ocha-Gomez, S. G. Gil-Rio, O. Gomez-Jimenez-Aberasturi, C. A. Ramirez-Lopez, J. Torrecilla-Soria, A. Dominguez, Microporous and Mesoporous Mater. **134** (2010) 100-107
23. D. Bonenfant, M. Kharoune, P. Niquette, M. Mimeault, R. Hausler, Sci. Technol. Adv. Mater **9** (2008) 1-7
24. A. Jain, R. Balasubramanian, M. P. Srinivasan, Chem. Eng. J. **283** (2016) 789-805
25. X. Xu, Y. Kan, L. Zhao, X. Cao, Env. Pollution **213** (2016) 533-540
26. Y. Yao, B. Gao, M. Inyang, A. R. Zimmerman, X. Cao, P. Pullammanappallil, L. Yang, J. Hazardous Mater. **190** (2011) 501-507
27. M. S. Shafeeyan, W. M. A. W. Daud, A. Houshmand, A. Arami-Niya, Appl. Surf. Sci. **257** (2011) 3936-3942

 https://doi.org/10.1051/e3sconf/20184301009

Design and Implementation Buck Converter for 540WP Solar Charger Using Fuzzy Logic Control

Sutedjo[1,*], *Ony* Asrarul Qudsi[1], *Andi* Ardianto[1], *Diah* Septi Yanaratri[1], *Suhariningsih*[1], *Moch.* Darwis[1]

[1]Electrical Engineering Department Politeknik Elektronika Negeri Surabaya, Indonesia

Abstract. This paper presents the details of design and implementation of DC-DC Buck converter as solar charger. This converter is designed for charging a battery with a capacity of 100 Ah (Ampere Hours) which has a charging voltage of 27.4 volts. The constant voltage method is selected on battery charging with the specified set point. To ensure the charging voltage is always on the set point, the duty cycle control of buck converter is set using Fuzzy Logic Control (FLC). The design implementation has been tested on PV (photovoltaic) with 540WP capacity. Based on the test results, this method is quite well implemented on the problem charger.

1 Introduction

The increasing energy demand and the declining availability of fossil energy sources led to the rapid development of alternative energy. One of the most widely used alternative energy sources is solar energy [1]-[3]. The transformation of energy sources from fossils into solar energy sources has had a considerable impact on the development of power electronics products. Batteries are energy storage that is included in power electronics products. Batteries are widely used in various applications, such as telecom power supplies, electric vehicles, UPS (uninterruptible power supplies), photovoltaic systems, portable electronic products, and others [12]-[13]. The charging process of a battery plays an important role because it can affect the durability and capacity of the battery that indirectly also affect the battery life. Many engineers and researchers look for simple, effective, and economical methods to solve this problem. Different types of battery charging approach show different efficiency values.

Conventional charging will result in losses due to power dissipation in the process of charging the battery. The conventional charging process is done by converting power from AC source to DC source [4]. The use of this method requires the rectifier of the thyristor converter which will convert the AC source into DC, with the addition of series resistance to control the flow of power in the charging process. This battery charging method will of course produce a very high ripple current. This method has poor efficiency when applied on a large scale. Such conditions cause concerns about the quality of the battery charger to increase, so it takes a battery charging circuit that can reduce the ripple and prolong battery life.

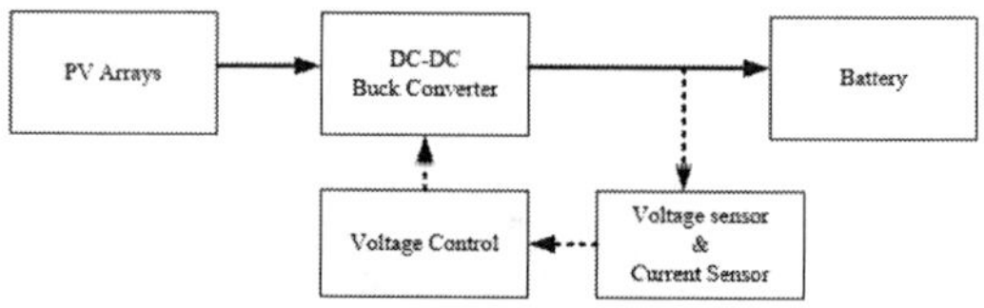

Fig. 1. The topology of battery charging

This research proposes a battery charging process by utilizing solar energy sources. The topology of proposes a battery charging show in Figure 1. Generally, the energy collected in PV arrays is used as a charger source. However, the PV output voltage arrays vary according to the existing sunlight. Unstable PV arrays output voltage can shift the operating point of the charger. Therefore, this study designed a new approach with Buck Converter as DC-DC converter which has been widely applied to DC motor speed regulator and energy backup system will be implemented in battery charging system [5]-[11]. In this research, the implementation of DC-DC converter is used as battery charging system for battery charging using constant voltage method where the output voltage is set at the desired set point so that the output voltage becomes fixed and the current generated follows the input power.This means that the output voltage of the buck converter that always at setpoint will make the input voltage for battery charging stable. Such conditions will indirectly affect the battery power and battery life. Figure 1 illustrates the topology of a battery charging system with a solar energy source. Buck Converter is used for battery charging with capacity of 100 Ah and charging voltage of 27.4 V with an input power supply of 540WP PV. To keep the charging voltage always on the setpoint, the duty cycle on the buck converter will be set using Fuzzy Logic Control (FLC). Of the various

sutedjo@pens.ac.id, ony@pens.ac.id, andiardianto.id@gmail.com,
diahsepti@pens.ac.id, nuning@pens.ac.id, mdarwis@staff.pens.ac.id

intelligent controllers, the FLC is the simplest and better control than conventional controllers in terms of insensitivity to parameter variations, time response, and resilience [16]-[18]. Furthermore FLC is an adaptive control technique that allows to change the duty cycle buck converter parameters under unstable PV voltage output conditions. The design then implemented so that prototype solar charger can be used for charging the battery with the stable input voltage.

2 Buck Converter Design for Solar Charger

Buck converter has the characteristic of lowering DC voltage by adjusting duty cycle (D) by switching process using a semiconductor. The buck converter circuit is shown in Figure 2.

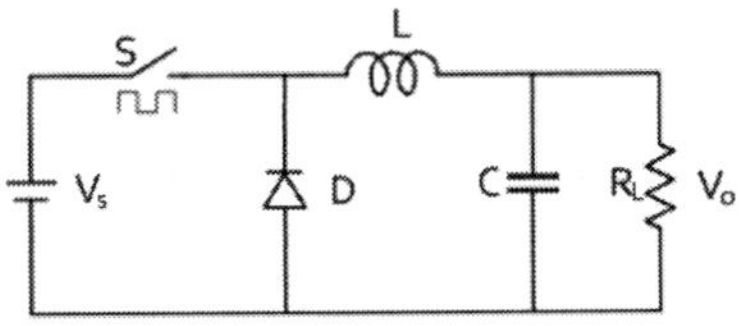

Fig. 2. Buck converter circuit.

When the switch is open, the output gets energy from the capacitor and from the inductor. The circuit when the switch is open is shown in Figure 3 (a).

$$V_L = \text{-}V_o \tag{1}$$

$$V_L = L\frac{di_L}{dt} \tag{2}$$

$$\frac{di_L}{dt} = \frac{\text{-}V_O}{L} \tag{3}$$

Changes in the inductor current are:

$$\frac{di_L}{dt} = \frac{\Delta i_L}{\Delta t} \tag{4}$$

$$\frac{di_L}{dt} = \frac{\Delta i_L}{(1-D)T} \tag{5}$$

The solution for ΔiL for open switches is:

$$\Delta i_L(open) = \frac{(\text{-}V_O)(1-D)T}{L} \tag{6}$$

When the switch is closed the diode is in reverse biased position. The output is isolated and the input supplies energy to the inductor, the circuit shown in Figure 3 (b).

The inductor voltage is:

$$V_L = V_S\text{-}V_O \tag{7}$$

$$\frac{di_L}{dt} = \frac{V_S\text{-}V_O}{L} \tag{8}$$

The change in the inductor current is:

$$\frac{di_L}{dt} = \frac{\Delta i_L}{D\,T} \tag{9}$$

The solution for ΔiL for closed switches is:

$$\Delta i_L(closed) = \frac{(V_S\text{-}V_O)\,D\,T}{L} \tag{10}$$

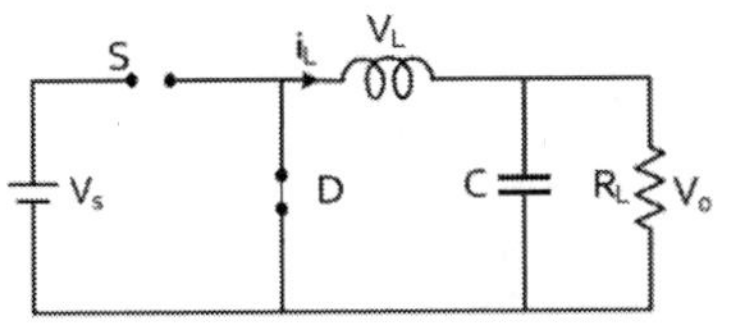

(a)

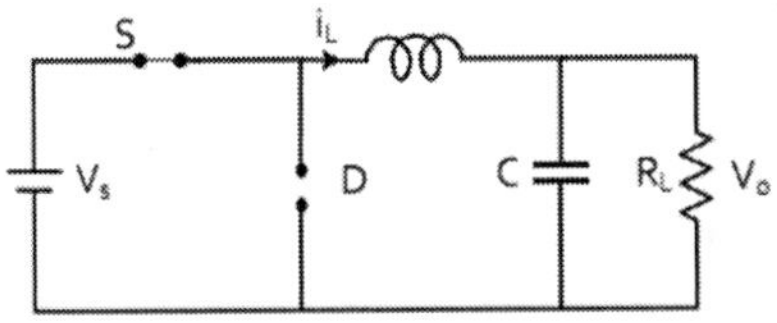

(b)

Fig. 3. (a) Buck converter circuit when the switch is open, (b) Buck converter circuit when the switch is closed

The method of battery charging proposed in this research is constant voltage method. This method is chosen to keep the dry battery from over-charged. Therefore, definite voltage control and proper charging time in the battery charging method is essential to maintain expected battery life time. The charging voltage for a 100Ah battery is 13.8V-14.7V. Determination of the charging voltage is obtained from the battery capacity by State of Charge (SoC). The charging voltage of 13.7 Volts has reached 90% of the battery voltage in full condition. The design of buck converter output power is based on charging current and voltage charging of the battery. Parameter of buck converter show in table 1.

Table 1. Basic Parameter of Buck Converter

Parameter	Unit	Value
Input Voltage(V_s)	Volt	55.4
Output Voltage (V_o)	Volt	27.4
Output Current (I_o)	Ampere	13.795
Voltage Ripple	%	0.1
Current Ripple	%	20
Maximal Input Power	WP	361
Switching Frequency	kHz	40

https://doi.org/10.1051/e3sconf/20184301009

2.1 Inductor Design

The inductor can be modeled like the equivalent circuit in Figure4. *R* is the DC resistance of the coil. Then the inductor can produce inductance and resistance *R* on the coil. The inductor is not saturated when using the worst case of I_{max}'s peak current. As a note the relationship between *R*.

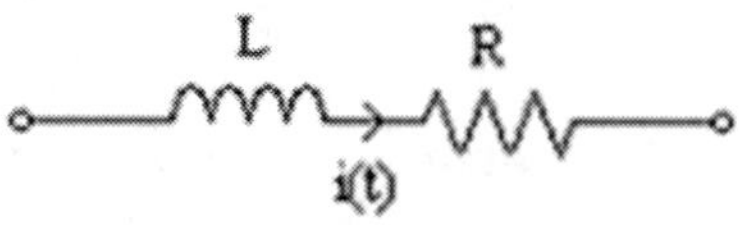

Fig. 4. Inductor model

The resistance of the inductor coil affects the efficiency and output voltage of the converter. Thus, the converter design requires the construction of an inductor with a fairly small coil resistance.
Iinductor values isdetermined using equation (11):

$$L = \left(\frac{1}{f}\right)x(Vin\_\max - Vo)x\left(\frac{Vo+Vf}{Vin\_\max + Vf}\right)x\left(\frac{1}{\Delta_{IL}}\right) \quad (11)$$

Where:
L : Inductance value (Henry)
Vin : Input Voltage (Volt)
Vo : Output Voltage (Volt)
Vf : Diode *falling*Voltage(Volt)
ΔiL :*Ripple* current (Ampere)

The number of inductor windings is obtained using equation (12):

$$N = \frac{L \times \mathrm{Imax}}{B\max \times Ac} \quad (12)$$

Where :
N : Number of turn
Bmax : Magnetic field (Tesla)
Ac : core cross-section area (mm^2)

The average current and the maximum current of the inductor can be calculated using the equations (13) and (14):

$$I_L(avg) = I_o = \frac{Vo}{R} \quad (13)$$

$$\mathrm{Im}ax = I_L + \frac{\Delta_{IL}}{2} \quad (14)$$

Where:
I_L (avg) : Inductor average current (Ampere)
Io : Output current (Ampere)
Vo : Output Voltage (Volt)
R :Resitance (Ohm)
Imax : Inductor maximum current (Ampere)
ΔI_L : Inductor ripple current (Ampere)
I_L : Inductor current (Ampere)

To calculate the length of the wire of each split, equation (16) can be used:

$$AWG = \frac{\sqrt{I_{in}^{2} + \left(\frac{\Delta I_L/2}{\sqrt{3}}\right)^2}}{n} \quad (15)$$

$$l = ((n \times k) + 40\%(n \times k)) \times n \quad (16)$$

Where:
l :length of the wire (m/split)
n : number of split
k : Bobbin's Circumference (mm)
AWG : standard wire gauge (Ampere)

Cross Sectional Area of wire (qw) and wire diameter can be calculated using the following equations (17) and (18):

$$qw = \frac{I_L}{J} \quad (17)$$

$$dw = \sqrt{\frac{4}{\pi} \times qw} \quad (18)$$

Where:
J : Magnetic Energy (A/mm^2)
Qw : Cross Sectional Area of wire (mm^2)
Dw : Wire diameter (mm)

From the design results obtained parameters of the inductor according to Table 2.

Table 2. Parameter of Inductor

Parameter	Unit	Value
IL (avg)	Ampere	13.79
Vo	Volt	27.4
R	Ohm	1.987
L	μH	128.57
Imax	Ampere	15.169
N	Turn	11
AWG	Ampere	1.392
l	m/split	1.155
qw	mm^2	0.309
dw	mm	0.65

2.2 Capasitor Design

In this research, the design of capacitor parameters includes the output voltage ripple (ΔVo) and capacitor capacitance (Co). Equation (20) shows the relation of the output voltage (ΔVo) and the capacitance value of the capacitor (Co).

$$\Delta V_o = \frac{\Delta Q}{C_o} = \frac{\Delta_{IL} T}{8C_o} \quad (20)$$

Where:

ΔV_o : output voltage ripple (Volt)

T :switching period (Second)

C_o : output capacitor (Farad)

From the design results obtained parameters of the capacitor according to Table 3.

Table 3. Parameter of capacitor

Parameter	Unit	Value
ΔV_o	Volt	27.4
C_o	µF	47

3 Fuzzy Logic Control for Buck Converter

In this research, input from FLC used is input voltage which consist of input error and Δerror. Figure 5 shows the membership function of Δerror. The rules base is following the general behavior of the system written in the linguistiq label membership function pattern. The system output of the planned rule consists of 25 rules shown in the table 4.

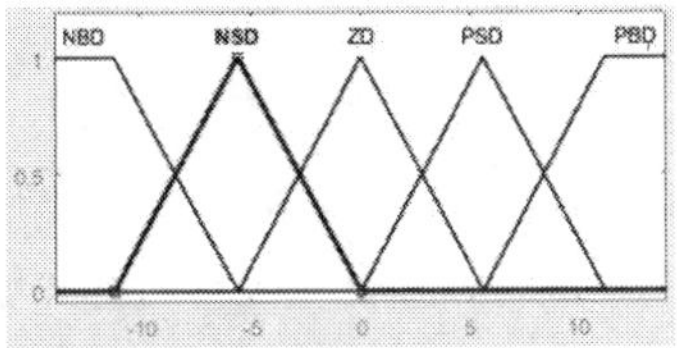

Fig. 5. Membership function Δerror

Table 4. Rule Base FLC For Buck Converter

RULE BASE FUZZY LOGIC					
Delta Error	NBE	NSE	ZE	PSE	PBE
NBD	NXO	NXO	NBO	NSO	ZO
NSD	NXO	NBO	NSO	ZO	PSO
ZD	NBO	NSO	ZO	PSO	PBO
PSD	NSO	ZO	PSO	PBO	PXO
PBD	ZO	PSO	PBO	PXO	PXO

If the error condition is NBE and Δerror NBD, then the output voltage is now greater than the requested output voltage. Therefore the voltage must be lowered larger. So for the first rule is: if error is NBE and Derr is NBD then PWM is NXO.

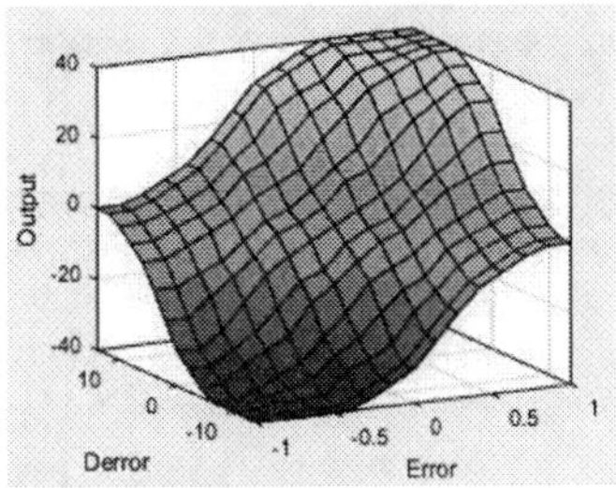

Fig. 6. Surface of rule base

From the results of the rule base obtained then we can see linearization of rules that we make that seen in Figure 6.

4 Experimental Result

Testing is intended to get an evaluation of the circuit, in order to obtain better performance. Better performance is achieved by making improvements to the composition of the circuit that experienced a known error during the test. The test method is done into two stages, they are partial testing and integration testing. Partial testing is intended to test the suitability of each part of the design. Partial testing includes solar panel testing and buck converter testing. Further testing is done by integrating PV, buck converter, and battery.

4.1 PV Testing

Testing of PV in this study using 4 pieces of 135 WP solar panels that are connected parallel between the two solar panels. The purpose of this parallel relationship is to obtain the maximum current value between the two PV modules. From the PV testing is found that the value of the voltage on average bright conditions is 55.4-35.9 volts.

4.2 Buck Converter Testing

The buck converter circuit testing aims to determine the performance of the buck circuit that has been designed and made. Figure 7 is a trigger waveform after being inserted on the Gate-Source MOSFET pin with a 30% duty cycle. There is a voltage ripple that can reduce the efficiency of a device. In Figure 7 it can also be seen that the buck converter circuit has worked. Where the characteristics of the buck converter work are the occurrence of the oscillation waveform before the drain-Source wave touches the zero value.

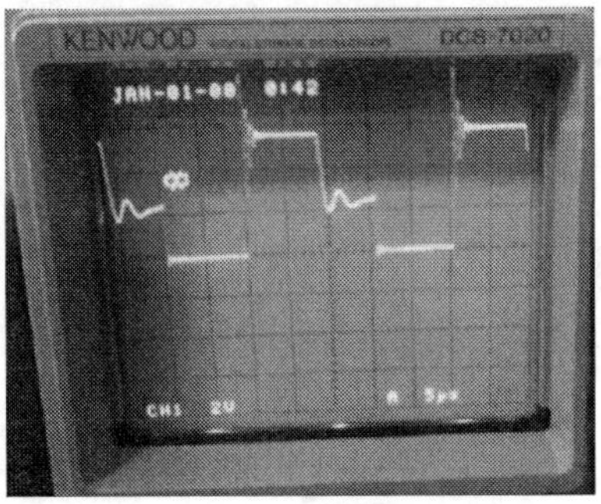

Fig. 7. Drain-Source MOSFET Buck Converter (D=30%)
Volt/div : 2 V, Time/div:5us

After the trigger waveform can be generated, the test buck converter circuit will use an input voltage supply of 55.4V with a resistive load of 100W and use the digital PWM of the microcontroller as the switching trigger of the buck converter circuit.

Table 5. buck converter circuit testing result

D (%)	Vin (V)	Iin (A)	Pin (W)	Vo (V)	Io (A)	Po (W)	ή (%)
30	55,4	0,18	10,24	15,84	0,57	9,02	88,09
40	55,4	0,35	19,72	27,55	0,69	19,00	96,38
41,4	55,4	0,37	20,77	28,76	0,70	20,13	96,90
50	55,4	0,48	26,59	34,46	0,74	25,67	96,54
60	55,4	0,55	30,52	38,06	0,77	29,49	96,63
70	55,4	0,65	36,28	43,70	0,81	35,39	97,54
80	55,4	0,88	49,25	55,00	0,89	48,95	99,39

Based on Table 5, the efficiency generated in the buck converter circuit is greater when the dutycycle used is greater. This is because the value of DC resistance of the inductor has a large enough value so that in the small dutycycle has big losses, thus affecting the efficiency of the buck converter. However, based on the partial experimental data that has been done, the buck converter circuit works well.

4.3 Integrated Test Buck Converter for Solar Charger

Integrated testing is done by connecting the PV as the input source of the Buck converter, then the output voltage of the buck converter will serve as the voltage for charging the battery. The initial battery voltage is 24.28 Volts and the rated charging voltage is set at 27.4 Volts, with duty cycle control to adjust the output voltage of the buck converter. Table 6 and 7 shows the results of battery charging using 4 pieces of PV with each 135WP PV capacity.

From the test data shown by table 6, the test uses manual duty cycle arrangement. Adjustment of duty cycle with the desired output voltage resulted in less good value of efficiency.

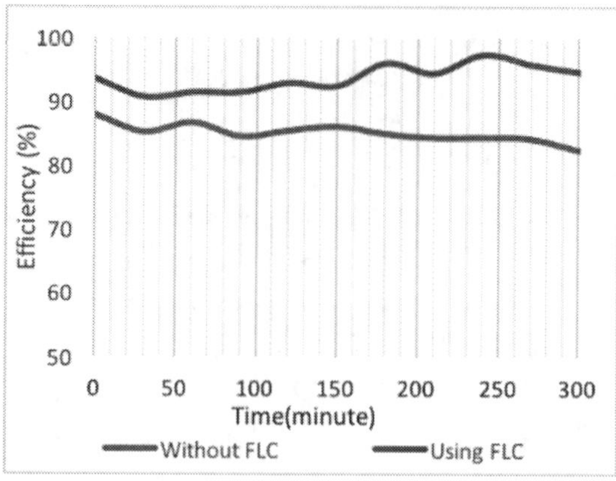

Fig. 8. Efficiency of the buck converter without FLC and using FLC

Table 6. open loop testing result Of Buck Converter (without FLC)

Vin (V)	Vo (V)	Iin (A)	Io (A)	Pin (W)	Pout (W)	ή (%)	lux	D
47.5	27.40	4.0	6.10	190.0	167.1	87.97	892	68
48.5	27.35	3.8	5.75	184.3	157.2	85.33	897	65
48.4	27.38	3.6	5.46	174.2	149.4	86.79	902	64
49.1	27.32	3.4	5.25	169.4	143.4	84.67	892	64
49.7	27.38	3.1	4.85	154.0	132.7	86.19	872	63
49.8	27.42	2.9	4.48	144.4	122.8	85.06	841	63
49.9	27.41	2.7	4.15	134.7	113.7	84.43	824	62
50.9	27.42	2.5	3.92	127.2	107.4	84.46	795	61
50.5	27.38	2.4	3.73	121.2	102.1	84.26	813	60
51.3	27.38	2.1	3.31	107.7	90.63	84.12	875	59
51.1	27.40	2.0	3.04	102.2	83.30	81.50	798	58
50.7	27.42	1.7	2.57	86.19	70.47	81.76	756	59
51.1	27.42	1.6	2.46	81.76	67.45	82.50	747	57
51.0	27.38	1.6	2.40	81.60	65.71	80.53	751	57
50.1	27.40	1.5	2.34	75.15	64.12	85.32	536	59
50.0	27.41	1.5	2.29	75.00	62.77	83.69	612	56
51.0	27.39	1.4	2.18	73.95	59.71	80.74	721	56

Controlled close loop system testing is done to charge the battery. Table 7 shows the results of controlled charging. From these results show fuzzy control is running to make the output voltage constant. From the results of the tests performed, the comparison of efficiency values on a controlled system (using FLC) and without controls looks different. The highest efficiency value when using a controlled system. Figure 8 below is a graph of efficiency value comparison of testing using FLC and without FLC.

Table 7. testing result Of Buck Converter Using FLC

Vin (V)	Vo (V)	Iin (A)	Io (A)	Pin (W)	Pout (W)	η (%)	lux	D
42.3	27.39	6.75	8.42	285.53	230.62	80.77	736	84
43.7	27.24	5.64	7.65	246.47	208.39	84.55	840	76
46.1	27.23	4.73	6.73	218.05	183.26	84.04	790	70
48.4	27.32	4.13	6.03	199.89	164.74	82.41	837	68
50.0	27.26	3.65	5.60	182.50	152.66	83.65	875	65
51.2	27.21	3.10	4.80	158.72	130.61	82.30	846	61
51.8	27.39	2.50	4.10	129.50	112.30	86.72	823	59
52.1	27.19	2.13	3.40	110.97	92.45	83.31	942	58
52.4	27.24	1.98	3.12	103.75	84.99	81.91	921	56
52.3	27.32	1.53	2.64	80.01	72.12	90.14	863	56
52.8	27.23	1.34	2.14	70.75	58.27	82.40	738	55

From the tests performed, the efficiency value increases when using a controlled system. The fuzzy controls used are effective enough to increase the efficiency of the system.

4 Conclusion

Buck converter is very suitable to be implemented as a solar charger for batteries. The results of this research prove that buck converter can provide a stable output voltage. The output voltage generated by the buck converter can be used as a battery charging voltage by constant voltage method. Thus, such conditions can affect the durability and capacity of the battery that indirectly also affect the battery life. The addition of FLC causes the efficiency of the buck converter to increase from 84% to 91%. This shows FLC is quite effective to improve the efficiency of solar charger.

References

1. P. K. C. Wong, A. Kalam and R. Barr, "Modelling and analysis of practical options to improve the hosting capacity of low voltage networks for embedded photo-voltaic generation," in IET Renewable Power Generation, vol. 11, no. 5, pp. 625-632, 4 12 2017. J. Clerk Maxwell, A Treatise on Electricity and Magnetism, 3rd ed., vol. 2. Oxford: Clarendon, 1892, pp.68-73.
2. Wahjono, Endro, Dimas Okky Anggriawan, Epyk Sunarno, Syechu Dwitya Nugraha, and Anang Tjahjono. "Maximum Power Point Tracking of Photovoltaic Module for Battery Charging Based on Modified Particle Swarm Optimization." International Review on Modelling and Simulations (IREMOS) 10, no. 1 (2017): 77-84.
3. A. Tjahjono, O. A. Qudsi, N. A. Windarko, D. O. Anggriawan, A. Priyadi and M. H. Purnomo, "Photovoltaic module and maximum power point tracking modelling using Adaptive Neuro-Fuzzy Inference System," 2014 Makassar International Conference on Electrical Engineering and Informatics (MICEEI), Makassar, 2014, pp. 14-19.
4. G. Buja, R. K. Jha, M. Bertoluzzo and M. K. Naik, "Analysis and comparison of two wireless battery charger arrangements for electric vehicles," in Chinese Journal of Electrical Engineering, vol. 1, no. 1, pp. 50-57, Dec. 2015.
5. H. Sira-Ramirez and M. A. Oliver-Salazar, "On the Robust Control of Buck-Converter DC-Motor Combinations," in IEEE Transactions on Power Electronics, vol. 28, no. 8, pp. 3912-3922, Aug. 2013.
6. E. Guerrero, J. Linares, E. Guzman, H. Sira, G. Guerrero and A. Martinez, "DC Motor Speed Control through Parallel DC/DC Buck Converters," in IEEE Latin America Transactions, vol. 15, no. 5, pp. 819-826, May 2017.
7. H. K. Samitha Ransara, U. K. Madawala and T. H. Liu, "Buck converter based model for a brushless DC motor drive without a DC link capacitor," in IET Power Electronics, vol. 8, no. 4, pp. 628-635, 4 2015.
8. R. Silva-Ortigoza, V. M. Hernández-Guzmán, M. Antonio-Cruz and D. Muñoz-Carrillo, "DC/DC Buck Power Converter as a Smooth Starter for a DC Motor Based on a Hierarchical Control," in IEEE Transactions on Power Electronics, vol. 30, no. 2, pp. 1076-1084, Feb. 2015.
9. R. Silva Ortigoza, J. N. Alba Juarez, J. R. Garcia Sanchez, M. Antonio Cruz, V. M. Hernandez Guzman and H. Taud, "Modeling and Experimental Validation of a Bidirectional DC/DC Buck Power Electronic Converter- DC Motor System," in IEEE Latin America Transactions, vol. 15, no. 6, pp. 1043-1051, June 2017.
10. R. Silva Ortigoza, J. N. Alba Juarez, J. R. Garcia Sanchez, V. M. Hernandez Guzman, C. Y. Sosa Cervantes and H. Taud, "A Sensorless Passivity-Based Control for the DC/DC Buck Converter- Inverter- DC Motor System," in IEEE Latin America Transactions, vol. 14, no. 10, pp. 4227-4234, Oct. 2016.
11. J. Linares-Flores, J. L. Barahona-Avalos, H. Sira-Ramirez and M. A. Contreras-Ordaz, "Robust Passivity-Based Control of a Buck–Boost-Converter/DC-Motor System: An Active Disturbance Rejection Approach," in IEEE Transactions on Industry Applications, vol. 48, no. 6, pp. 2362-2371, Nov.-Dec. 2012.
12. N. Anandh, P. A. D'sa, M. V. Gautam and V. S. Sandeep, "Power quality estimation, analysis and improvement for uninterrupted power supply," 2016 International Conference on Control, Instrumentation, Communication and Computational Technologies (ICCICCT), Kumaracoil, India, 2016, pp. 26-31.
13. O. Gomozov, J. P. F. Trovão, X. Kestelyn and M. R. Dubois, "Adaptive Energy Management System Based on a Real-Time Model Predictive Control

With Nonuniform Sampling Time for Multiple Energy Storage Electric Vehicle," in IEEE Transactions on Vehicular Technology, vol. 66, no. 7, pp. 5520-5530, July 2017.

14. Windarko, N. A., Qudsi, O. A., Tjahjono, A., Dimas, O. A., & Purnomo, M. H. (2014, November). Optimized PI constant for current controller of grid connected inverter with LCL filter using genetic algorithm. In Electrical Engineering and Informatics (MICEEI), 2014 Makassar International Conference on (pp. 9-13). IEEE.
15. J. L. Russi, M. L. S. Martins, H. A. Grundling, H. Pinheiro, J. R. Pinheiro and H. L. Hey, "A unified design criterion for ZVT DC-DC PWM converters with constant auxiliary Voltage source," in IEEE Transactions on Industrial Electronics, vol. 52, no. 5, pp. 1261-1270, Oct. 2005.
16. M. E. Sahin and H. I. Okumus, "A fuzzy-logic controlled PV powered buck-boost DC-DC converter for Battery-Load system," 2012 International Symposium on Innovations in Intelligent Systems and Applications, Trabzon, 2012
17. G. Gupta and P. Gaur, "Fuzzy logic controlled-power factor corrected bridgeless buck boost converter-fed brushless DC motor drive," 2015 International Conference on Computer, Communication and Control (IC4), Indore, 2015
18. N. Ikken, A. Bouknadel, H. El Omari and H. El Omari, "Design and implementation of intelligent PI-fuzzy logic control for grid connected inverters," 2016 International Renewable and Sustainable Energy Conference (IRSEC), Marrakech, 2016

Design And Implementation Soft-switching MPPT SEPIC Converter Using P&O Algorithm

Luki Septya M[1,*], *Indhana* Sudiharto[1], *Syechu* Dwitya N[1], *Ony* Asrarul Qudsi[1], *Epyk* Sunarno[1]

[1] Electrical Engineering Department Politeknik Elektronika Negeri Surabaya, Indonesia

Abstract. This paper presents details of design and implementation of Maximum Power Point Tracking (MPPT) SEPIC Converter to maximize Photovoltaic (PV) power output. Perturb and observe (P & O) algorithm has been used as a tracking technique to determine MPPT. This algorithm are simple and efficient to determine the maximum power point (MPP). The soft-switching method will be added to the implementation of MPPT SEPIC Converter to reduce switching losses. The results show that the proposed MPPT method implementation can increase the output power of PV.

1 Introduction

In recent years, solar energy has gained great attention and popularity. This condition caused solar energy is considered the best solution for long-distance power applications that demand low voltage DC supplies that utilize alternative energy sources such as chargers for small electronic gadgets, laptops, uninterruptible power supply (UPS) and electric vehicles.

The solar energy sources conversion into electrical energy usually uses PV modules. Until now, PV is considered the best solution for remote area applications. PV has two major weakness, the power generated from PV is highly dependent on weather conditions and the conversion efficiency of PV systems is very low. However, the output characteristics of a non-linear PV module represent a unique operating point where there is maximum power generated in each irradiation. One common method to improve the efficiency of PV systems is develop MPPT controllers that continuously track MPP PV modules on all irradiation and temperature.

Many MPPT methods can be used to obtain optimal PV output power [1] - [6]. This adjusts to the complexity and needs of the system. Direct MPPT search methods such as Perturb and Observe (P&O), Incremental Conductance (INC) and Hill Climbing algorithms can be applied using spontaneous PV voltage (VPV) and PV current (IPV) values to generate control signals.

This paper focuses on the P&O algorithm because of its advantages, for example it can be implemented in analog and digital form, does not require periodic, efficient and easy to implement [2] [3] [8]. The P & O algorithm depend on the slope of the PV characteristic. In this paper, the SEPIC converter is used to validate the proposed P&O MPPT algorithm. Much research has been done in this region that uses buck, boost, and buck boost converters.

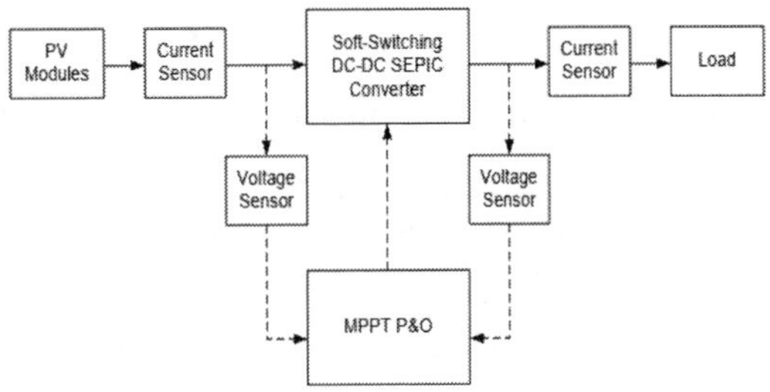

Fig. 1. Topology of soft-switching MPPT SEPIC converter

Although the boost converter is more efficient than the SEPIC converter, but the SEPIC buck-boost property makes it more adaptable [9] - [11]. In this paper we also propose soft switching on SEPIC converters that have high efficiency to improve the overall efficiency of PV systems. A Soft Switching SEPIC converter is a converter that has a non-inverted buck-boost output and has a very high efficiency.

Figure 1 shows the topology of soft switching MPPT SEPIC converter. The voltage and current of the PV output will be sensed by the voltage sensor and current sensor. These currents and voltages are used to adjust the duty cycle switching converter and are used for the MPPT P & O algorithm. Furthermore, soft-switching DC-DC SEPIC converter as a circuit that serves to increase and decrease the voltage without changing the polarity of the resulting voltage. The output voltage sensor is used for monitoring the output voltage of the converter [12] - [14]. Then the proposed method will be compared with the conventional method so it can give an idea of the effect of using soft-switching MPPT SEPIC converter.

* Corresponding author: lukiseptya@gmail.com

2 Design System

In this study work has been used 6 PV modules with the capacity of each PV module is 135 WP. The PV module specifications are shown in Table 1.

Table 1. PV module specifications

Maximum Power (Pm)	135 WP
Open-circuit Voltage (Voc)	61,3 V
Short-circuit current (Isc)	3,41 A
Voltage at point of maximum power (Vmp)	47 V
Current at point of maximum power (Imp)	2,88 A
Modul efficiency	9 %
Number of Cell	45
Temperature coefficient-open-circuit voltage (αVoc)	-0,3
Temperature coefficient-short-circuit current (αIsc)	0,07
Temperature coefficient-power (αPm)	-0,24

Based on the PV capacity has been used, then designed soft-switching MPPT SEPIC converter.

2.1. Design of SEPIC converter

In this work has been designed SEPIC converter with a capacity of 1000 Watt power. SEPIC converter can produce output voltage which can be higher or lower than input but without changing polarity. The inductor current and capacitor voltage limits are removed to investigate current and voltage fluctuations. The inductor current is assumed to work continuously on this analysis. The average inductor voltage is zero and the average capacitor current is zero for seady-state operation.
Kirchhoff's Voltage Law (KVL) on circuit *Vs, L1, C1 and L2*

$$-V_s + V_{L1} + V_{C1} - V_{L2} = 0 \quad (1)$$

If it follows the assumption of the average voltage of the inductor, then the voltage on the capacitor *C1* is

$$V_{C1} = V_s \quad (2)$$

When the switch is closed, the diode will be off and the circuit is shown in Figure 2.b. The transverse voltage *L1* for the *DT* interfal is

$$V_{L1} = V_s \quad (3)$$

When the switch is open, the diode will be on and the circuit is shown in figure 2.c. KVL on the outer path is

$$-V_s + V_{L1} + V_{C1} + V_O = 0 \quad (4)$$

The assumption of capacitor voltage *C1* remains constant at the mean voltage *Vs*

$$V_{L1} = -V_O \quad (5)$$

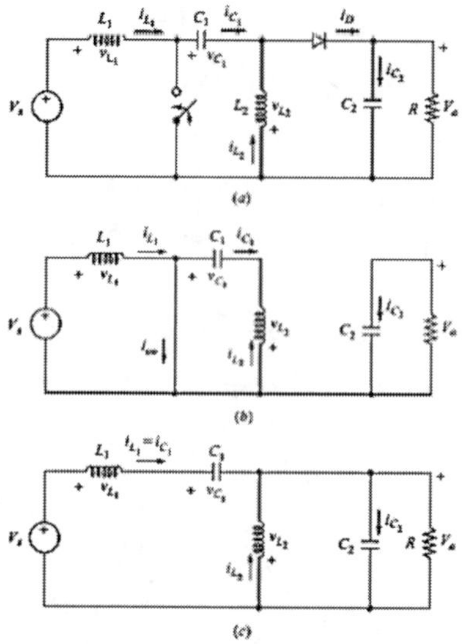

Fig. 2. (a) SEPIC circuit; (b) SEPIC circuit when switch is closed and diode off;
(c) SEPIC circuit when switch open and diode on

For intervals *(1 - D) T*. Since the average inductor voltage is zero for periodic operation (steady state operation), then the equation becomes,

$$(V_S(DT) - V_O(1-D)T = 0 \quad (6)$$

Where *D* is the duty cycle ratio of the switch, ie

$$D = \frac{Vo}{Vo+Vs} \quad (7)$$

The result of this equation is similar to buck-boost converter, with the important difference that there is no reversal polarity between the input and output voltages. The ability to have a voltage output greater or less than input without reversal polarity makes this converter suitable for many applications. The average current inductor which is also the current source can be searched by the equation:

$$I_{L1} = I_S = \frac{V_O I_O}{V_S} = \frac{V_O^2}{V_S R} \quad (8)$$

The variation in i_{L1} when the closed switch can be searched from the equation,

$$V_{L1} = V_S = L1\left(\frac{di_{L1}}{dt}\right) = L1\left(\frac{\Delta i_{L1}}{\Delta t}\right) = L1\left(\frac{\Delta i_{L1}}{Dt}\right) \quad (9)$$

Where Δi_L is

$$\Delta I_{L1} = \frac{Vs\,DT}{L1} = \frac{Vs\,D}{L1.f} \quad (10)$$

The average current in each capacitor is zero, so the mean current in *L2* is,

$$I_{L2} = I_O \quad (11)$$

The variation in i_{L2} is determined from the circuit when the switch is closed. Using Kirchhoff's law of KVL, *C1*, and *L2* with the voltage *C1* assumed Vs is constant, giving

$$V_{L2} = V_{C1} = Vs = L2\left(\frac{di_{L2}}{dt}\right) = L2\left(\frac{\Delta i_{L2}}{\Delta t}\right) = L2\left(\frac{\Delta i_{L2}}{DT}\right) \quad (12)$$

Where Δi_{L2} is,

$$\Delta I_{L2} = \frac{Vs\,DT}{L2} = \frac{Vs\,D}{L2.f} \quad (13)$$

The current waveform is shown in Figure 3. Kirchhoff KVL's Law is applied to the drawings. 3 c, assuming no ripple voltage on the capacitor, indicating that the voltage going through the switch when open is *Vs + Vo*. From Figure 3 b, the maximum reverse voltage of the diode bias when the diode off is also *Vs + Vo*.

The output part consisting of diodes, *C2*, and load resistors is the same as in boost converter, so the output of ripple voltage is,

$$\Delta V_O = \Delta Vcs2 = \frac{Vo\,D}{R.C2.f} \quad (14)$$

Then *C2* is,

$$C2 = \frac{D}{R(\Delta Vo/_{Vo})f} \quad (15)$$

The voltage variation in *C1* is determined from the shape of the circuit with the closed switch. The current capacitor i_{C1} is the reciprocal of i_{L2}, which has previously been determined to have an average value of *Io*. From the definition of capacitance and considering the magnitude of charge,

$$\Delta V_{C1} = \frac{\Delta Qc1}{C} = \frac{Io\,\Delta t}{C} = \frac{Io\,DT}{C} \quad (16)$$

Change *Io* with *Vo/R*,

$$\Delta V_{C1} = \frac{Vo\,D}{R\,C1\,f} \quad (17)$$

Then *C1* is,

$$C1 = \frac{D}{R\left(\Delta V_{C1}/_{Vo}\right)f} \quad (18)$$

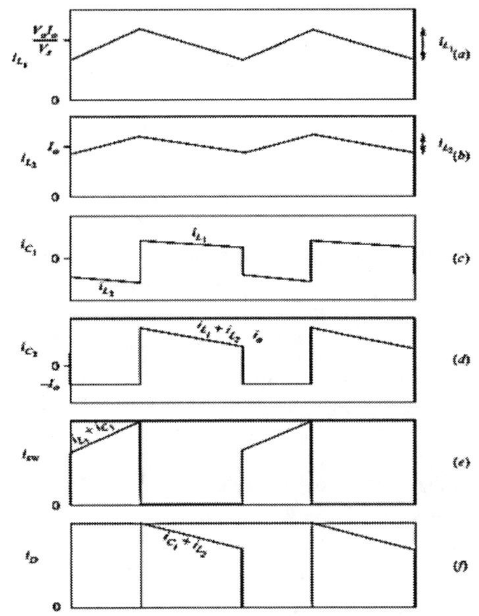

Fig. 3. Current of SEPIC converter. (a) L1; (b) L2; (c) C1; (d) C2; (e) switch; (f) diode

2.2 P&O MPPT

The Perturb and Observe (P&O) algorithm is one of the simplest and easy to implement for MPPT. The design of MPPT requires two parameters to determine the slope of the control input voltage (*Vin*) and control input current (*Iin*)

$$P_{in}(n) = V_{in}(n)\,x I_{in}(n) \quad (19)$$

From these two parameters obtained Power (*Pin*), and voltage (*Vin*), then compared with the previous data reading parameters that is *Pin (n-1)* & *Vin (n-1)*.

The result of comparison is obtained *ΔP* and *ΔV*

$$\Delta V = V_{in}(n) - V_{in}(n-1) \quad (20)$$
$$\Delta P = P_{in}(n) - P_{in}(n-1) \quad (21)$$

And the result of division ΔP and ΔV will be called slope.

$$Slope = \frac{\Delta P}{\Delta V} \quad (22)$$

As in Figure 4, there are 3 types of points that are in 3 positions. To the left of the *dP / dV> 0* peak, the curve curve *dP / dV = 0* and to the right of the peak *dP / dV <0*. On the left side of the MPP the power changes to the voltage change *dP / dV> 0*, while on the right, *dP / dV 0*, it is known that perturbation is done to move the PV working voltage forward MPP control. If dP / dV, then the work point change directs PV away from MPP. Then the P&O algorithm reverse the disturbance direction.

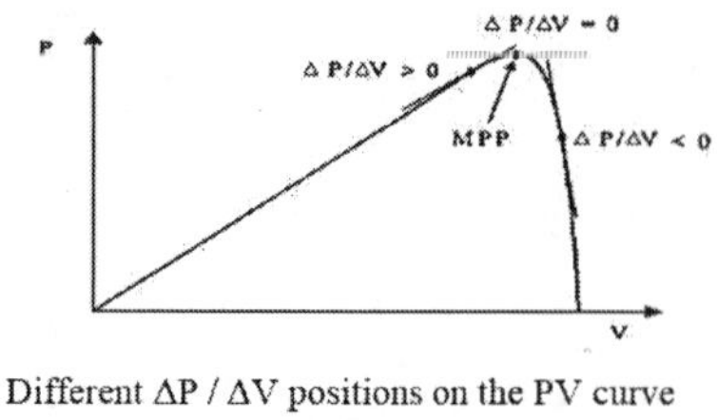

Fig. 4. Different ΔP / ΔV positions on the PV curve

Flowchart of the P&O algorithm is shown in Figure 5. First of all measure or sensing voltage (V_K) and output current (I_K) PV. Then calculate the power with the multiplication of voltage and current of sensing result (P_K). Furthermore, compared to the current power (P_K) is greater than the previous power (P_{K-1}), if yes then compare the current voltage (V_K) is greater than the previous voltage (V_{K-1}), if yes then naiakn Vref by raising duty Cycle on the converter. Conversely if not, then lower the Vref by lowering the duty cycle. If the current power (PK) is not greater than the previous power (P_{K-1}), then compare the current voltage (V_K) is greater than the preceding voltage (V_{K-1}), if yes then naiakn Vref by raising the duty cycle on the converter. Conversely if not, then lower the *Vref* by lowering the duty cycle.

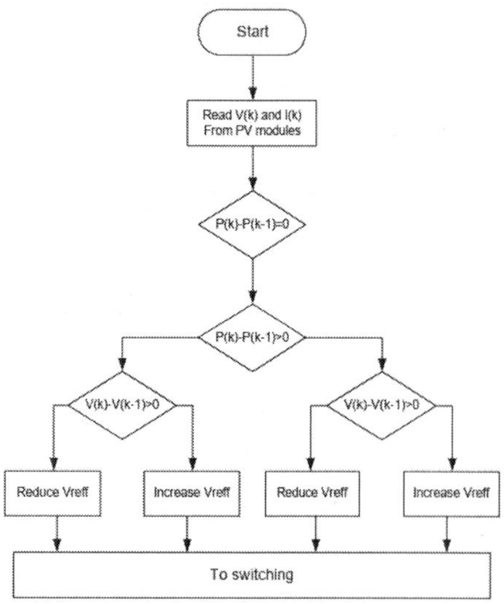

Fig. 5. Flowchart of the P & O algorithm

2.3 Soft-switching

Voltage and current switching do not switch to zero instantaneously at turn-on or turn-off. There is a time duration during switching transitions (ie turn-on and turn-off switching). Power losses during switching exist in overlapping or overlapping regions of current waves and switch voltages during turn-on or turn-off. Since the average power is energy divided by the period, then the higher the switching frequency causes higher switching losses. Sharp and sudden switching transitions are also sources of electromagnetic interference (EMI) noise that can affect the performance of converters and other surrounding electrical equipment. Figure 6 shows the current and voltage graphs of switch *S1* during the switching cycle. At the time of the driver pulse from *S1* off then *S1* turn-off. *S1* takes time t_{off} to completely turn off. During this time, there is an overlap of the current wave and voltage, resulting in a turn-off loss represented by the area below the P_{off} graph. At t_1, the driver pulses on and switch *S1* will turn-on. At this time, the *S1* switch takes t_{on} of time for turn on. During this time, there is an overlap of current wave and voltage from *S1*, so the turn on losses occur.

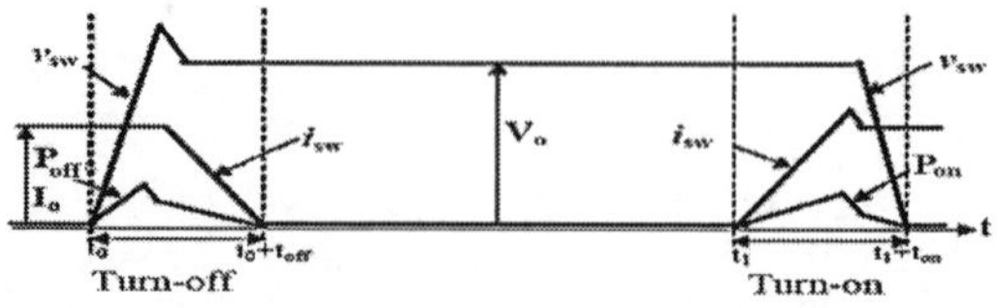

Fig. 6. Power loss during switching turn-on and turn-off due to overlap between switching currents (I_{ds}) and switching voltages (V_{ds})

The switching loss problem of the switching switching operation can be reduced by using soft-switching. The term soft-switching in power electronics refers to the various techniques in which switching transitions are made to be more gradual or not sudden. Switching losses decrease due to power dissipation when switching transitions take place there is no overlap between the voltage and current on the switching.

3 Experimental Result

System testing will be performed on each device to see the data characteristics generated by the equipment and to know, then to be tested as a whole with integrated.

3.1 PV Test

PV testing is performed to determine the characteristics of PV to sunlight that falls on the surface of PV. Testing of 810 WP PV characteristics is shown in figure 7.

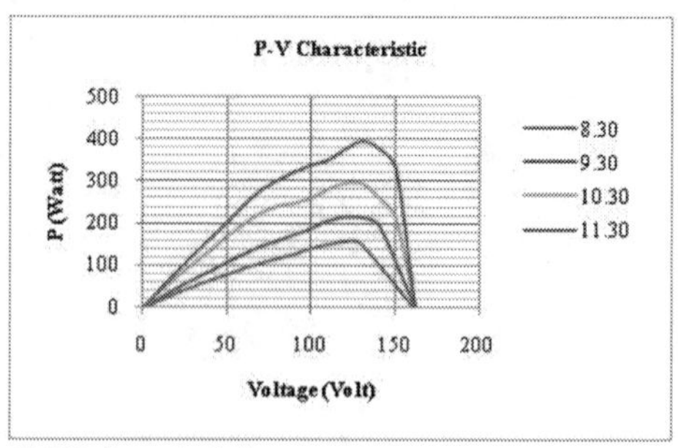

Fig. 7. PV characteristic

3.2 DC-DC SEPIC ConverterTest

This test is aimed to know the function of SEPIC converter as DC-DC converter which is used to increase and decrease voltage. The SEPIC converter test uses a modified resistive load but with a fixed input voltage of

100V and the duty cycle is fixed at 60%. This test is up to a nominal power of 1000 W. Switching waveform shown in Figure 8 is the drain-source waveform MOSFET.

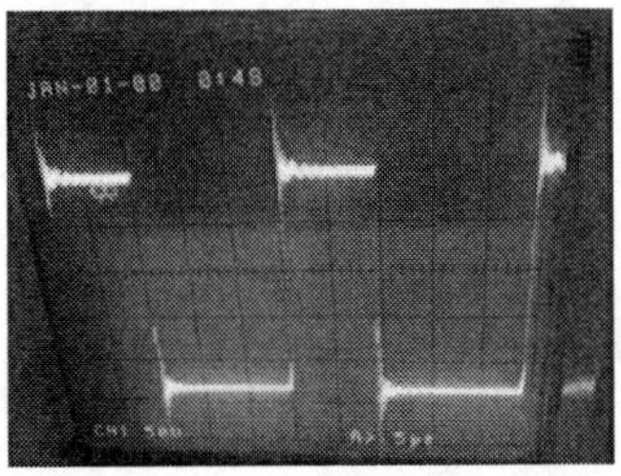

Fig. 8. Drain-Source Signal MOSFET duty cycle 60%

The SEPIC converter test data is shown in Table 2.

Tabel 2. SEPIC Converter Testing

SEPIC							
Duty cycle	**Vin (V)**	**Iin (A)**	**Iout (A)**	**Vo prak (V)**	**Vo teori (V)**	**Error (%)**	**Effisiensi (%)**
60%	100	10	5	160	150	6,67	80,0
	100	8	4,2	153	150	3,33	80,33
	100	5	3	138	150	8	82,8
	100	4,8	2,8	148	150	1,3333	86,33
	100	3,5	2	162	150	8	92,57
	100	2,6	1,4	149	150	0,6667	80,23

From the test results seen the efficiency has been meet the design of the converter is 80% and the error voltage is not large ie error maximum 8%. After performing converter testing, it is expected that the converter is able to work well when integrated with PV fluctuate to find the maximum power of PV.

3.3 Integration Test

At the integration testing stage, the entire system is unified. Maximum Power Point Tracking Testing is integrated with SEPIC Converter as well as previously designed voltage and current sensors. The effect of MPPT P & O on finding the maximum or peak power value of PV 810 WP, testing open loop and close loop. The test is carried out with resistive load at nominal power. In the first test of open loop testing or non-controlled testing, the duty cycle of the SEPIC converter is maintained in all test conditions. While in the second test that is close loop testing or controlled testing, the duty cycle of SEPIC converter varies according to PV peak power condition in all test conditions.

From these tests, there will be differences in PV power output between controlled by soft-switching MPPT-SEPIC P & O and without control. Table 3 and 4 show power output between controlled by soft-switching MPPT-SEPIC P & O and without control. Collecting data between controlled and without MPPT control is done almost simultaneously at the same time and day. This is due to the PV characteristics that have rapid variable data changes, such as current values and PV voltages. This is to find the maximum power value when the solar panel is in a load state when using MPPT controls or not.

Table 3. Data Test Results system without MPPT control

Vin (V)	Iin (A)	Vo (V)	Io (A)	Po (W)	Pin (W)
144,9	1,4	145	0,9	130,5	202,86
153,1	1,4	153,5	0,9	138,15	214,34
146,7	1,4	147	1	147	205,38
156,4	1,4	156,1	0,9	140,49	218,96
150,1	1,4	159,3	0,92	146,556	210,14
153	1,4	159,8	0,92	147,016	214,2
156	1,4	158	0,92	145,36	218,4
154	1,4	160,7	0,92	147,844	215,6
155	1,4	158	0,92	145,36	217
151	1,4	156	0,91	141,96	211,4
150	1,4	156	0,92	143,52	210
150	1,4	154	0,9	138,6	210
141	1,4	143	0,87	124,41	197,4
117	1,5	120	0,82	98,4	175,5

Table 4. Data Test Results system with MPPT control

Vin (V)	Iin (A)	Vo (V)	Io (A)	Po (W)	Pin (W)
120	1,8	135	0,83	112,05	216
122	2,1	141	0,86	121,26	256,2
130	2,3	153,2	0,85	130,22	299
130	2,3	170	0,9	153	299
130	2,6	172	0,93	159,96	338
132	2,6	176	0,94	165,44	343,2
141	2,8	176	0,94	165,44	401,1
142	2,7	183	0,96	175,68	383,4

From the system test results, it can be compared that the MPPT tracking results can be better than tracking when the converter duty cycle is made fixed under all conditions. This proves the program has been properly running and can find the maximum power of PV. The amount of MPPT power increase can be proportionalized in Table 5 Since the power when the duty cycle fix is 40% is very far from the other power, then that compares only MPPT P & O power and 50% duty cycle fix power to see the percentage of MPPT P & O power increase. Figure 9 show graphic of comparation

https://doi.org/10.1051/e3sconf/20184301010

between SEPIC converter with control and without control.

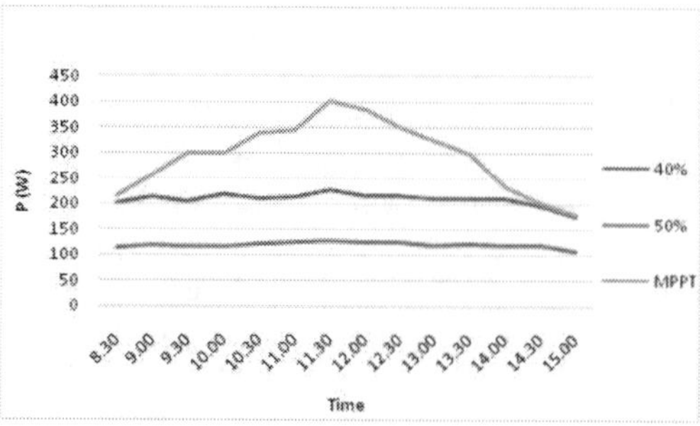

Fig. 9. comparation between SEPIC converter with control and without control.

Table 5. MPPT power increase

Time (+7GMT)	P without MPPT (W)	P with MPPT (W)	Power increase (%)
8.30	202,86	216	6,083333
9.00	214,34	256,2	16,3388
9.30	205,38	299	31,31104
10.00	218,96	299	26,76923
10.30	210,14	338	37,8284
11.00	214,2	343,2	37,58741
11.30	228,4	401,1	44,68085
12.00	215,6	383,4	43,7663
12.30	217	350	38
13.00	211,4	324,3	34,81344
13.30	210	300	30
14.00	210	234	10,25641
14.30	197,4	204	3,235294
15.00	175,5	180	2,5
Average power increase with MPPT			**31,98**

4 Conclusion

Soft-switching MPPT SEPIC converter using P & O algorithm can increase the output power of PV. Based on experimental results that have been done, about 31.98%. Although the method succeeded in tracking the power and able to raise the power to more than 30%, but this power value is still far from the maximum power of PV that has a maximum power of 810WP. Power that can be optimized using this method is only capable of a maximum of 401.1 W that is less than 50% of the maximum power of PV. There are many factors that influence, one of which is the physical state of PV is dirty, the existence of buildings that block the entry of light into the PV and the sensor is less precise in reading the data. So, for the future will be done optimization on it.

References

1. A. Tjahjono, O.A. Qudsi, N.A. Windarko, D.O. Anggriawan, A. Priyadi, M.H. Purnomo, *Electrical Engineering and Informatics (MICEEI), 2014 Makassar International Conference on* 14-19 (2014)
2. J. Ahmed and Z. Salam, IEEE Trans. on Sustainable Energy, **7**, 506-1515 (2016)
3. M. Killi and S. Samanta, in IEEE Transactions on Industrial Electronics, **62**, 5549-5559 (2015)
4. Ma, X. Zhou, Z. Gao and T. Bai, *2017 IEEE International Conference on Mechatronics and Automation (ICMA)* (2017)
5. H. Wang, X. Jin, H. Zhao and Y. Yue, *2017 IEEE International Conference on Mechatronics and Automation (ICMA)* (2017)
6. N. Kumar, I. Hussain, B. Singh and B. K. Panigrahi, IEEE Trans. on Energy Conversion, **32**, 983-992 (2017)
7. N. Barua, A. Dutta, S. Chakma, A. Das and S. S. Chowdhury, *2016 IEEE International WIE Conference on Electrical and Computer Engineering (WIECON-ECE)* (2016)
8. S. Mohanty, B. Subudhi and P. K. Ray, IEEE Trans. on Energy Conversion, **32**, 340-347 (2017)
9. S. W. Lee and H. L. Do, IEEE Trans. on Industrial Electronics, **PP**
10. G. Tibola, E. Lemmen, J. L. Duarte and I. Barbi, IEEE Trans. on Power Electronics, **32**, 9210-9222 (2017)
11. R. Moradpur, H. Ardi and A. Tavakoli, IEEE Trans. on Industrial Electronics, **PP**
12. H. L. Do, IEEE Trans. on Power Electronics, **27**, 2879-2887 (2012)
13. C. L. Shen, Y. E. Wu and M. H. Chen, *2008 IEEE International Conference on Industrial Technology* (2008)
14. A. M. Zacharias and T. R. Devaprakash, *2014 Annual International Conference on Emerging Research Areas: Magnetics, Machines and Drives (AICERA/iCMMD)* (2014)

Design of new ultrasonic transducer with two elements for flow rate measurement using ultrasonic Doppler method

Jevin Tanius Owen[1,*], *Ari* Hamdani[2], *Tomonori* Ihara[3], *Hideharu* Takahashi[2], and *Hiroshige* Kikura[2]

[1]Department of Mechanical Engineering, Tokyo Institute of Technology, 2-12-1, Ookayama, Meguro-ku, Tokyo, Japan
[2]Laboratory for Advanced Nuclear Energy, Tokyo Institute of Technology, 2-12-1, Ookayama, Meguro-ku, Tokyo, Japan
[3]Tokyo University of Marine Science and Technology, 2-1-6 Etchujima, Koto-ku, Tokyo, Japan

Abstract. In this present study, we developed a new two-dimensional velocity profile measurement system by utilizing two-elements ultrasonic transducer with both elements acting as transmitter and receiver. The aim is to obtain two-dimensional velocity information along measurement line and calculate the flow rate. We evaluate the transducer with sound pressure measurement and flow measurement. From the sound pressure measurement, there is one main lobe exist at the center between the elements due to the interference of both waves generated from each element. Therefore two-dimensional velocity measurement is possible. From the flow measurement, we can get two-dimensional velocity profile and calculate the flow rate. We compared the flow rate result with the flow rate reading from the electromagnetic flow meter, and we obtained the error of 4.98% for pipe flow and 10.43% for pipe flow under swirl effect with the measurement range from 8 mm to 50 mm. We used B-spline interpolation with Gaussian fitting to estimate the profile from range 0 mm to 8 mm and finally we were able to obtain full vector profile.

1 Introduction

Ultrasonic Doppler method was first used in medical field. Applying this method, Baker [1] performed the blood flow sensing. This application utilized a one-dimensional technique which has some limitations. For instance, in blood flow sensing, the "real" velocity is extracted from the velocity in ultrasound beam direction by assuming the blood flow parallel to vessel axis. In many areas of clinical interest, however, complex flow patterns, such as helices, shear layers, recirculation zones, and turbulence are present, making assumptions about the flow direction difficult and suspect [2]. Therefore, multi-dimensional velocity measurement is necessary. Peronneau *et al.* [3] proposed single element cross beam system using two transducers as a transceiver (transmitter/receiver). However, this system can only measure two-dimensional velocity at one point, and mechanical adjustment is needed to measure another point. Later, Scabia *et al.* [4] proposed multiple elements cross beam system by using 128 elements array transducer with at least one transmitter and two receivers to measure at one point.

In fluid engineering, Takeda [5] was first to apply ultrasonic Doppler method for measuring one-dimensional velocity profile in pipe flow. This technique then has a recognition as a powerful tool in fluid engineering for its ability of non-intrusiveness and broad range application, including opaque liquid. Wada *et al.* [6] applied this method for flow rate measurement in power plant using multi-lines measurement with cavitation bubbles system and metallic pipe wall. In this field, however, complex flow (*e.g.*, non-axisymmetric flow) exists too. Due to this concern, Hurther and Lemmin [7] developed three-dimensional velocity measurement in open-channel flow using one transmitter and four receivers. Based on this idea, Obayashi *et al.*[8] developed two-dimensional measurement system using only one transceiver and one receiver. They investigated the system accuracy using a rotating cylinder with known rotation speed and distance of measurement line from center. They can measure two-dimensional velocity in a certain depth, but they found the velocity in receiver measurement line with the high error in limited depth. The reason is the echo obtained in the receiver is side scattering echo from transmitter measurement line which has low signal to noise ratio unlike the back scattering echo in transmitter line.

Through the previous cases learning, this present study proposes new measurement system with two elements array transducer to measure two-dimensional velocity profile. Each element acts as a transceiver so that we expect no side scattering occurred. Instead of using crossbeam system, we pursue an interference of both waves generated from each element to produce one main lobe or one measurement volume at the center

* Corresponding author: tanius.j.aa@m.titech.ac.jp

 https://doi.org/10.1051/e3sconf/20184301011

between the elements. With this measurement system configuration, the required number of the pulse generators is only two. Therefore it can reduce the measurement equipment size and electronic cost.

2 Measurement principle

2.1 Single element transducer velocity profiler

In this study, the ultrasonic flow monitor employs ultrasonic Doppler method to provide spatial-temporal velocity information. The ultrasonic transducer transmits a mechanical wave repeatedly, and if a reflector (*e.g.*, bubble or particle) exists along the wave propagation, there will be echoes repeatedly reflected back to the element (see Figure 1). This echo contains information about particle position (x) by the delayed time (T) between wave transmission and reception, while the consecutive echo contains information about reflector velocity (fluid velocity, if their density same) in the wave propagation direction (V_w) by the Doppler frequency (f_d) estimation [9] with equation below.

$$x = \frac{cT}{2} \tag{1}$$

$$V_w = \frac{cf_d}{2f_o} \tag{2}$$

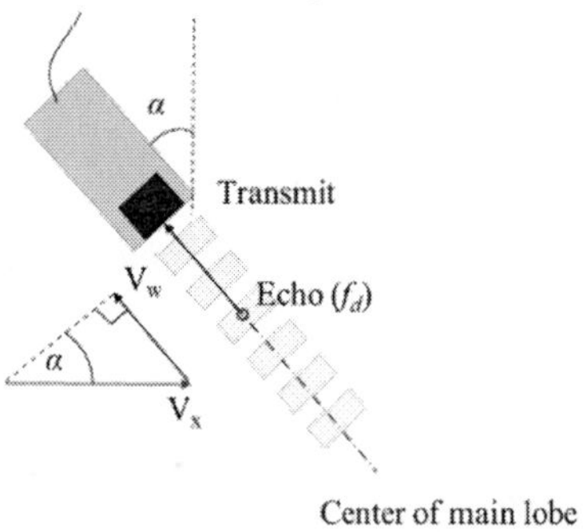

(a) One-dimensional velocity measurement.

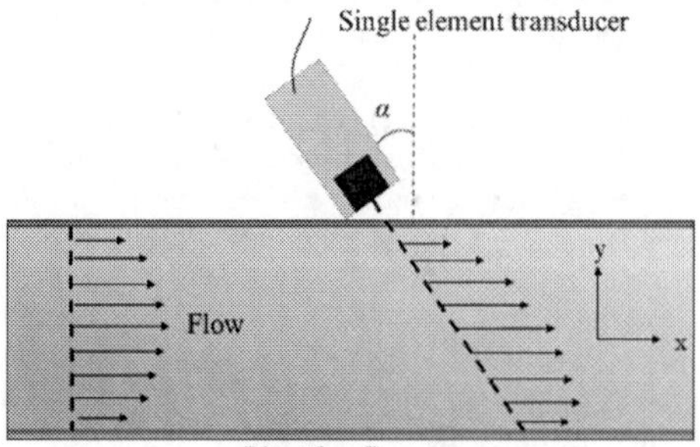

(b) Pipe flow measurement.

Fig.1. Single element transducer profiler.

The variable c and f_o represent sound velocity (*e.g.*, 1480 mm/s in water at 20°C) and transducer basic frequency respectively. In the single element transducer measurement, the transducer is set up with the particular incident angle (α). The velocity of the fluid (V_x) can be calculated by the projection of the velocity in the wave propagation direction (V_w) to the fluid velocity by assuming that the flow is one-dimensional and parallel to the pipe axis. Another assumption is the difference of the velocity profile in the measurement line, and the velocity profile in the flow at one position (straight line) is negligible (suitable in fully developed flow). Finally, the flow profile can be obtained, and the flow rate can be calculated by the integration of the velocity profile of the fluid in the pipe with a certain radius (R). The fluid velocity and flow rate are represented by the equation below.

$$V_x = \frac{V_w}{\sin\alpha} \tag{3}$$

$$Q = \pi \int_{-R}^{R} V_x r dr \tag{4}$$

2.2 Two elements transducer velocity profiler

The mechanism of the measurement using two elements transducer is similar to single element transducer, except the echo is received by both elements. Hence there is two Doppler frequency with a different orientation (see Figure 2), and two-dimensional velocity can be reconstructed as below.

$$V_x = \frac{c}{2f_o}\frac{f_{d1} - f_{d2}}{\sin\theta} \tag{5}$$

$$V_y = \frac{c}{2f_o}\frac{f_{d1} + f_{d2}}{1 + \cos\theta} \tag{6}$$

Using this principle, we can obtain two-dimensional velocity profile without adopting the assumption in single element transducer measurement. In fact, the velocity magnitude (V) and angle (β) of the flow can be calculated with the equation below.

$$V = \sqrt{V_x^2 + V_y^2} \tag{7}$$

$$\beta = \tan^{-1}\left(\frac{V_y}{V_x}\right) \tag{8}$$

Another important thing to be noted, that this measurement occurs in 1 position (the measurement line is perpendicular to the pipe axis), unlike single element transducer. Therefore, when the flow is not fully developed (*e.g.*, transient flow in short entry length or swirling flow), the flow rate (using equation (4)) is better calculated from velocity profile at one particular position.

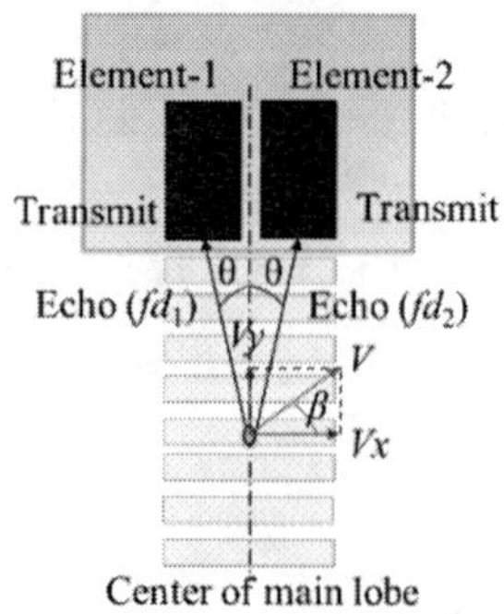

(a) Two-dimensional velocity measurement.

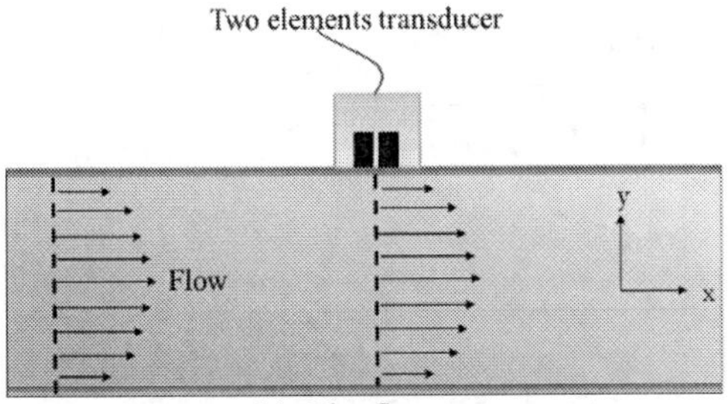

(b) Pipe flow measurement.

Fig.2. Two elements transducer profiler.

3 Design of two elements transducer

3.1 Design consideration

The primary objective of the design is to measure two-dimensional velocity profile. However, it does not simply succeed by using two transducers or two elements array transducer transmitting together. For example, if we use two transducers (*i.e.,* 4MHz, diameter 5 mm) attached to each other and transmitting together, there are two existing main lobe or measurement volume, and two-dimensional velocity measurement is not possible (see Figure 3).

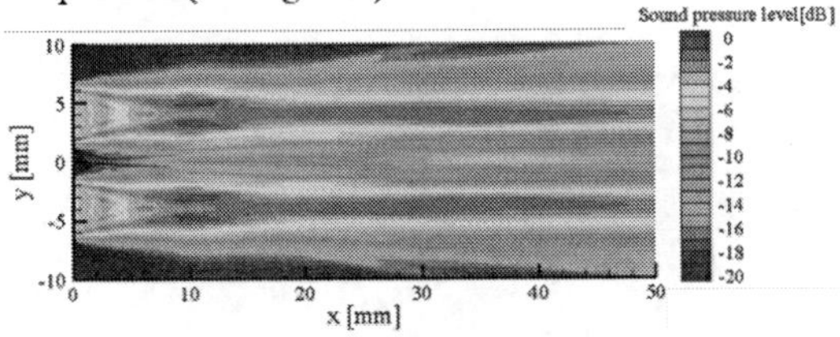

Fig.3. Sound pressure simulation of two transducers.

To ensure the quality of the measurement, we set the design target which is one main lobe in a good intensity and directivity with the acceptable near field length. To achieve this target, we do some parametric study with the parameters that have a significant contribution, *i.e.,* basic frequency of transducers (f_o), element width (a) and inter-element spacing (d). Figure 4 illustrates the transducer design parameters

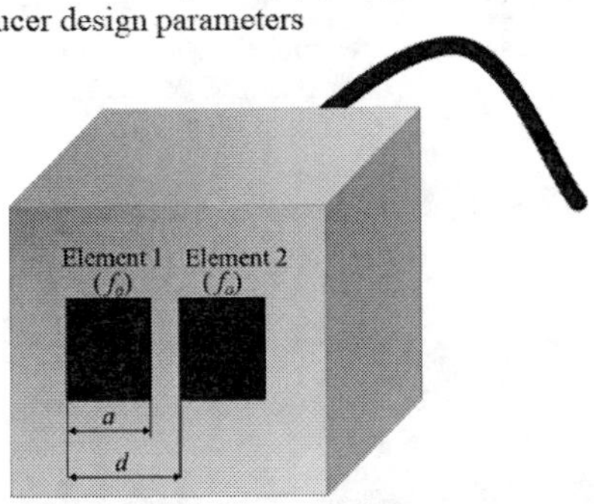

Fig.4. Design parameters of two elements transducer.

This parameters selection will decide the measurement quality or to achieve design targets. To get one main lobe in measurement, we have to choose a small inter-element spacing (d) and consequently a small element width (a) too with sufficient intensity as a constraint. To obtain a good intensity and directivity, we have to choose the high basic frequency (the option is 2MHz, 4MHz, and 8MHz). However, the high basic frequency will result in a long near field zone which is not suitable for measurement. Therefore, we need to compromise the design parameters and select the best configuration for optimum measurement.

3.2 Design selection and simulation

Through the parametric study and sound pressure simulation using software based on *Matlab* code, we finally decided the design parameters for the transducer as in Table.1 below.

Table 1. Transducer design specification.

Transducer detail	Specification
Basic frequency (f_o)	4 MHz
Number of element	2
Element width (a)	2.5 mm
Inter-element spacing (d)	2.7 mm

For basic frequency, we had to compromise intensity and directivity with near field length. Therefore 4 MHz was chosen to attain a good intensity and directivity while having an acceptable near field length. As for the element width and inter-element spacing, we used the value in Table 1 and tried to simulate the sound pressure. The result is the transducer generating one main lobe in the center between both elements, In addition, a good comparison to commercial single element transducers design (4 MHz, 5 mm) is shown (see Figure 5). Based on the simulation data, two-dimensional velocity profile measurement is possible to be carried out. As good as the simulation data, however, we need to confirm it by the experiment of the sound pressure and flow measurement. Finally, we got the design transducer to be manufactured (see Figure 6) by Japan Probe Co, Ltd. and evaluated the design by experiment.

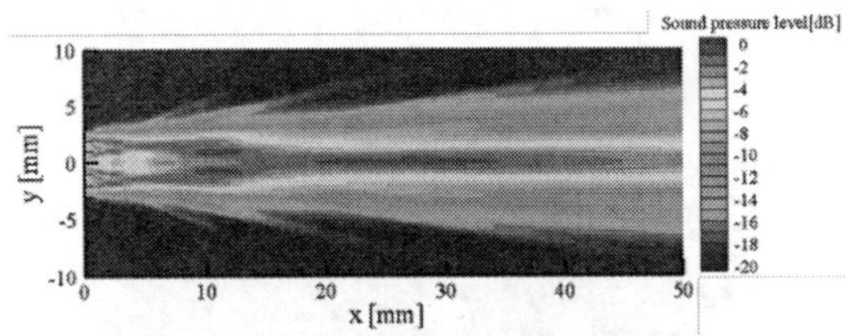

(a) Two-elements transducer 4 MHz.

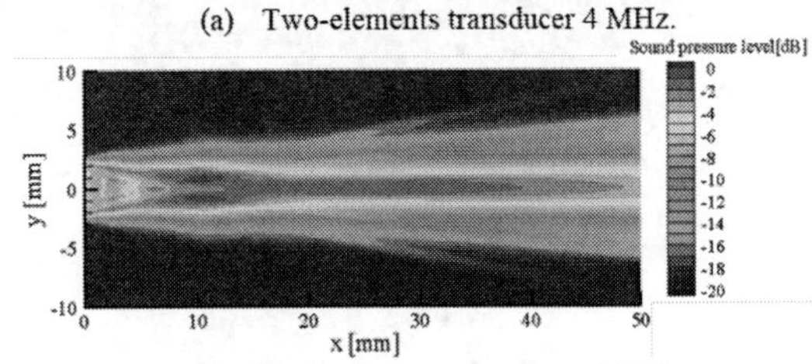

(b) Single-elements transducer 4 MHz.

Fig.5. Sound pressure simulation data.

Fig.6. Two elements transducer.

4 Sound pressure measurement

4.1 Experimental setup and procedures

To confirm the simulation data, sound pressure measurement was conducted in the water tank using two elements transducer transmitting while needle hydrophone is receiving. Figure 7 shows the experimental setup. The transducer is controlled by pulse generator two channel (Japan Probe TIT-10B-USB). The needle hydrophone is mounted at XYZ stage controller which connected to PC with A/D converter. Using this system, the needle hydrophone can be moved, and the sound pressure measurement can be done in particular grid and position (two-dimensional) by controlling the stage controller with the condition in Table 2.

Table 2. Sound pressure measurement condition.

Condition	Detail
Water temperature	23.7°C
Sampling profile	50
Axial distance (x)	50 mm
Half lateral distance (y)	10 mm
Axial resolution (Δx)	1 mm
Lateral resolution (Δy)	0.5 mm

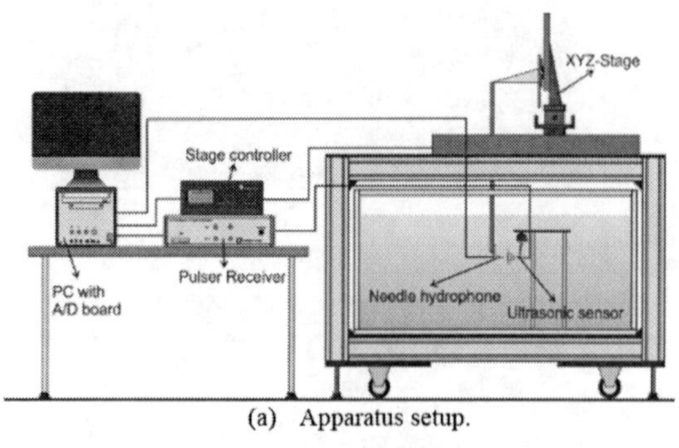

(a) Apparatus setup.

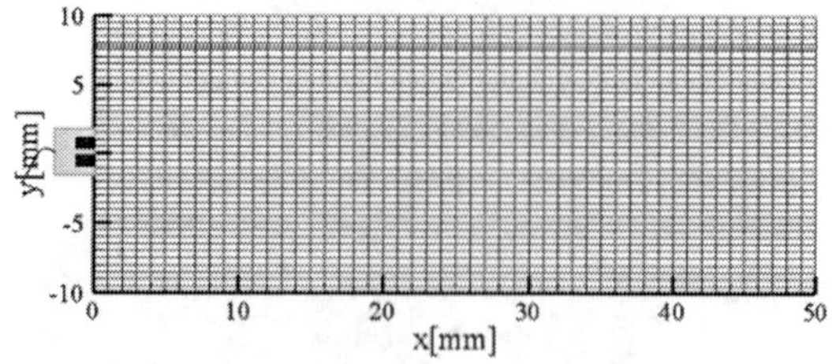

(b) Measurement grid (resolution).

Fig.7. Sound pressure measurement by experiment.

The signal transmitted from the transducer is recorded by needle hydrophone in sound intensity (I), peak to peak voltage. Then the data is plotted on a color graph by using the *Tecplot* software after converted into sound pressure level (SPL) using the equation below.

$$SPL = 20\log_{10}\frac{I}{I_{max}}[dB] \tag{9}$$

4.2 Results and discussion

The sound pressure measurement is being done by using only one element at first to check each element, and then two elements transmitting together. The result is plotted on a color graph and compared to simulation data as below.

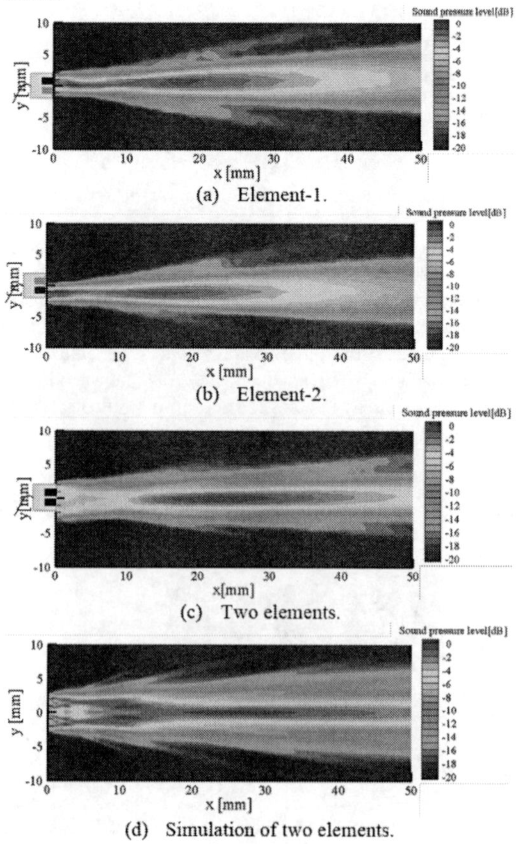

(a) Element-1.

(b) Element-2.

(c) Two elements.

(d) Simulation of two elements.

Fig.8. Sound pressure measurement of two element transducer.

From the result of one element transmitting, we confirm the both elements are working fine. It can be noticed that from one element transmitting, the main lobe is in the center of each element. When the two elements are transmitting together, there is interference of both waves from each element which resulting one main lobe in the center between of the both elements. However, if we notice from the sound pressure of two elements transmitting together in the axial distance (x) up to around 8 mm, the sound pressure shows a lower intensity in the center compared to upper and lower lateral distance (y). In this region, it is suspected that the measurement volume is not one focused in the center. For example, the data is extracted at axial distance (x) of 5 mm and 15 mm (see Figure 9). In the data from the axial distance of 5 mm, we can see that there are two peaks of intensity which indicates the measurement volume is not one focused in the center. On the contrary, the data from the axial distance of 15 mm shows one peak in the center (one measurement volume). Therefore two-dimensional velocity measurement might not be effective in the region with two peaks of intensity (up to 8 mm). There is a difference from the sound pressure data of the experiment and simulation, if we observe the intensity in the center (y=0) along the axial distance (see Figure 10). For example, from the simulation data we can observe the near field effect from the oscillating intensity. This difference might be caused by the parameters (*i.e.*, resolution, number of sample, sampling method, attenuation and *etc.*) between simulation and experiment are not the same. Nevertheless, with the sound pressure data from the experiment, we confirm our design target fulfilled as expected in the simulation data and the design could attain two-dimensional velocity in the flow measurement.

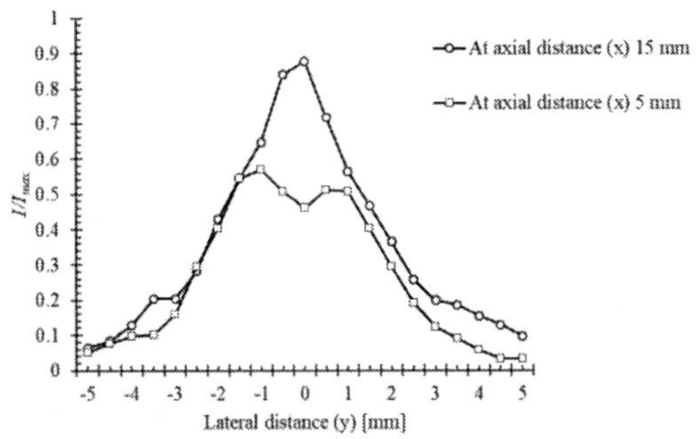

Fig.9. Normalized sound intensity along lateral distance (y).

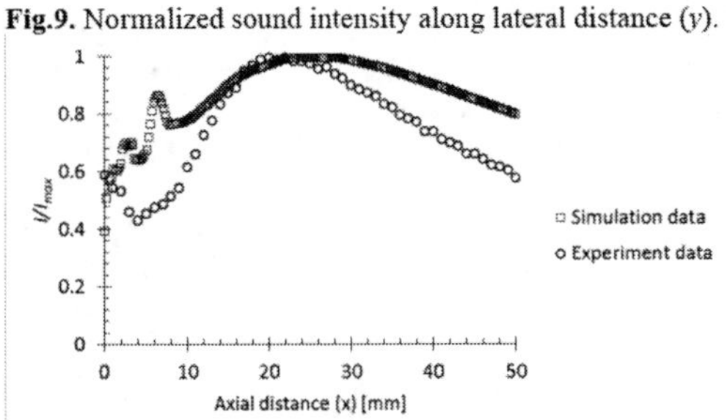

Fig.10. Normalized sound intensity in the center (y=0).

5 Flow measurement

5.1 Experimental setup and procedures

The flow measurement was conducted to confirm the capability of measuring the two-dimensional velocity. The experiment was being carried out in the horizontal acrylic pipe with and without swirling effect using swirl generator (rotating pipe driven by motor). The working fluid is tap water with the temperature controlled by the cooling system (see Figure 11). We used nylon particle (diameter 80 μm) as a reflector which has a similar density as water (1.03 g/cm^3). Therefore it is expected to have the same velocity with water. For the flow rate reading, we used electromagnetic flow meter as the reference and controlled by the ball valve. The transducer is placed in the test section with the pipe thickness of 1 mm and immersed in water as coupling fluid to minimize the attenuation of wave propagation to the working fluid. We used two channel pulse generator (Japan Probe TIT-10B-USB), A/D converter (National Instrument PXI-1033) connected to PC and controlled by developed software. The experiment condition and measurement parameters are presented in Table 3 and Table 4. The measurement data is then plotted as velocity profile (axial and radial) and velocity vector profile. For this present study, we compared only axial velocity for flow rate calculation. Therefore we used single element transducer to get the axial velocity to compare the profile. After that, we calculated the flow rate from each profile and compared them to the flow rate reading of the electromagnetic flow meter.

Table 3. Flow measurement condition.

Condition	Detail
Water temperature	28±1°C
Flow rate (reading)	35 L/min
Reynold number	18000
Pipe inner diameter	50 mm
Motor frequency (in swirling case)	11.9 Hz
Test section from motor	3D
Test section pipe material	Acrylic

Table 4. Measurement parameters.

Parameter	Detail
Basic frequency	4 MHz
Pulse repetition frequency	1 kHz
Channel width	0.74 mm
Channel distance	0.74 mm
Number of cycles	4
Number of repetitions	128
Number of profile	5000
Voltage	120 V
Gain	30 dB

https://doi.org/10.1051/e3sconf/20184301011

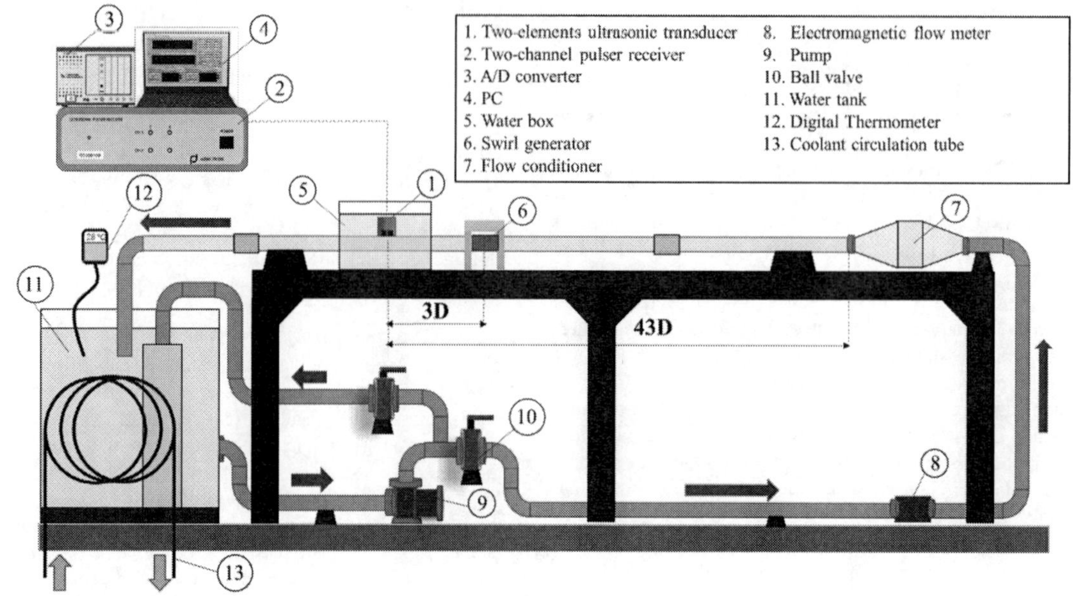

Fig.11. Flow measurement apparatus.

5.2 Results and discussion

5.2.1 Pipe flow without swirl effect

The two-dimensional velocity in the turbulent pipe flow without swirl effect was measured. The result is presented in time averaged velocity profile graph (5000 profiles) with the standard deviation and reconstructed into vector velocity as below.

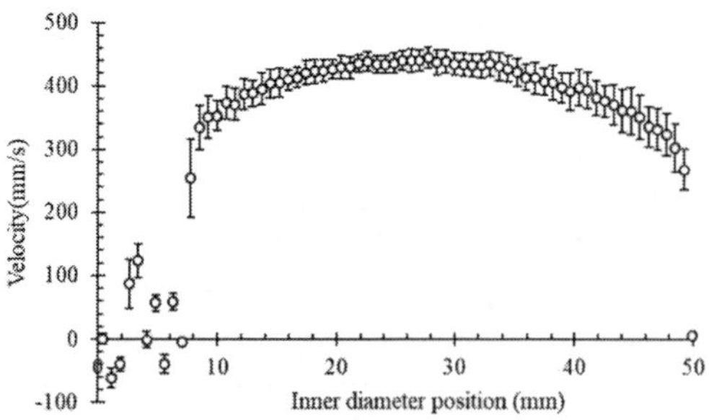

(a) Axial velocity (V_x).

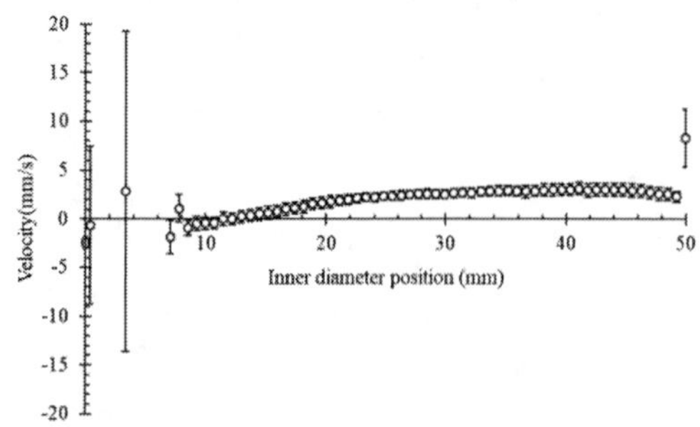

(b) Radial velocity (V_y).

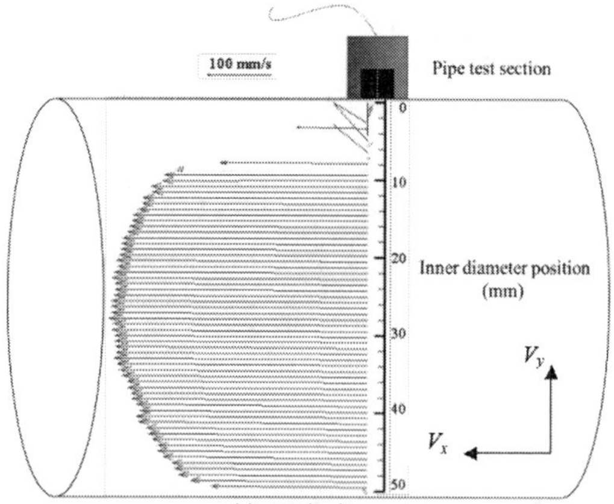

(c) Velocity vector.

Fig.12. Pipe flow measurement using two elements transducer.

From the result, we can see the two-dimensional velocity profile and the vector. The velocity in the axial direction (V_x) has a high standard deviation compared to radial velocity (V_y). The reason is the difference of the denominator from the equation of the axial velocity (equation (5)) and the radial velocity (equation (6)). The denominator in axial velocity is smaller compared to radial velocity. Therefore it will amplify the uncertainty or fluctuation of the profile and increase the standard deviation. In this measurement, the value of radial velocity is very small and has no significant effect on the vector reconstruction. We also can notice that in the region up to around 8 mm, the profile is scattering. If we refer to the sound pressure, it is expected before to be not effective for two-dimensional measurement because the measurement volume does not focus in the center. Moreover from the simulation data, the near field also exist in that location and might reduce the measurement quality. To compare the profile, we use commercial single element transducer (4MHz, 5mm) with an incident

* Corresponding author: tanius.j.aa@m.titech.ac.jp

angle (α) 10-degree to measure in the same flow condition and the result is presented as below.

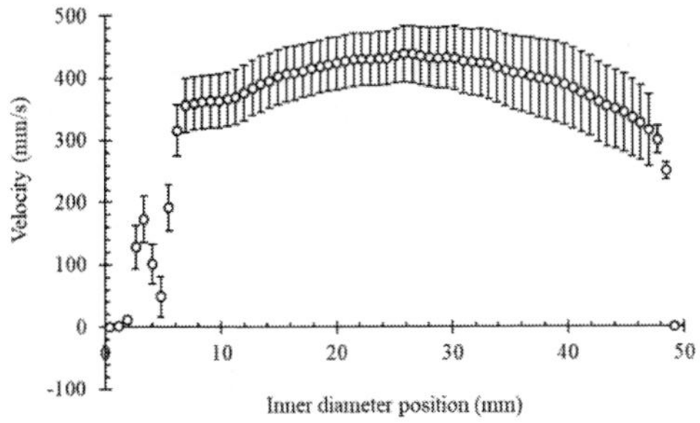

(a) Axial velocity (V_X).

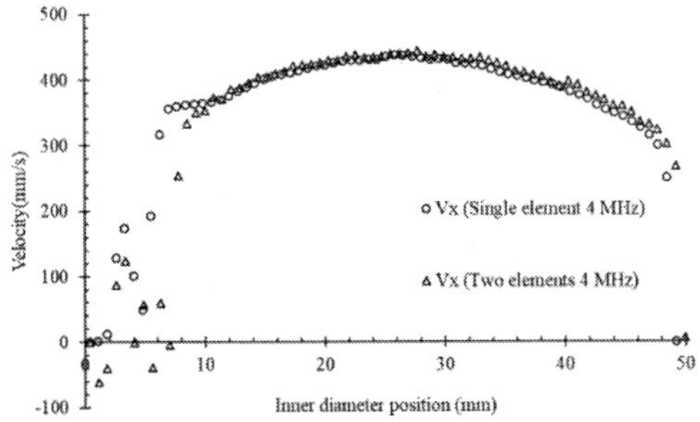

(c) Comparison with two elements transducer.

Fig.13. Pipe flow measurement using single element transducer.

From the profile comparison, it has a quite good agreement, except in the region up to 8 mm. The flow rate can be calculated from integration of the profile (using equation (4)) and compared to the flow rate reading in Table 5. The result is the two elements transducer measurement have larger error than single element transducer measurement. This fact might be caused by the profile scattering in two elements transducer profile at the region up to 8 mm. Therefore the flow rate is lower than single element transducer. As for the standard deviation, the single element transducer has higher average value than the two element transducer. This might be caused by the noise effect of the shear wave. In the interface of solid, the wave will exist in two modes which are a shear wave and longitudinal wave (see Figure 14) entering the liquid. If there are two waves enter the liquid, one of them will become noise in the measurement, and usually, it is the shear wave with the lower transmission rate. This transmission rate can be observed from the transmission coefficient (the ratio of the transmission rate in working fluid to incident beam), and it depends on the incident angle (see Figure 15) [10]. In this case, the two elements transducer and single element transducer will have 0-degree and 10-degree incident angle respectively. In 10-degree incident angle (single element transducer), the noise transmission rate from shear wave will have higher value than 0-degree (two elements transducer). Therefore, in this single element transducer measurement, the noise effect cause the instantaneous velocity fluctuates.

Table 5. Flow measurement quality.

Equipment	Flow rate (l/min)	Error (%)	Average standard deviation of velocity profile(mm/s)
Electromagnetic flow meter (reading)	35	-	-
Two elements transducer	33.26	4.98%	22.20
Single element transducer	33.90	3.14%	45.74

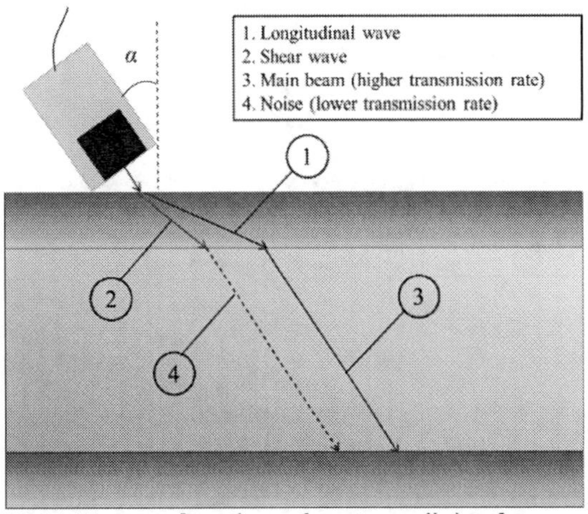

Fig.14. Wave transformation on the water-acrylic interface.

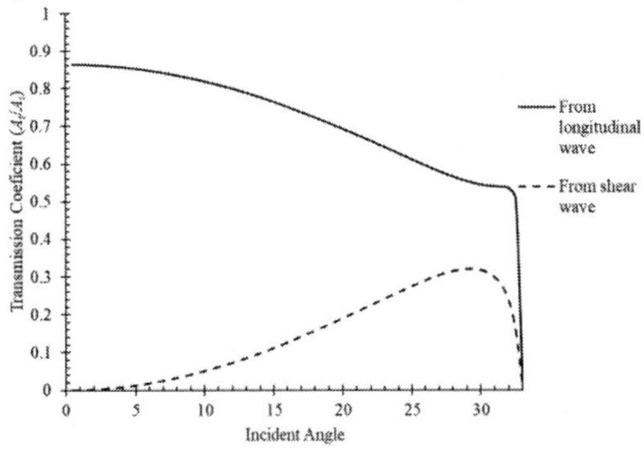

Fig.15. Theoretical transmission coefficient in the water depends on an incident angle with the acrylic-water interface.

5.2.2 Pipe flow with swirl effect

The measurement in this section is by adding swirl effect using a swirling generator in the distance of 3D before the flow reaches test section. The swirling effect is expected to increase the value of radial velocity and produce the not fully developed flow as we can see from the result presented in Figure 16. From the radial velocity profile, the value is higher compared to the pipe flow swirl effect. Therefore, we can see that the radial velocity has the significant effect to the flow direction in the velocity vector unlike in the straight pipe flow. For the comparison (see Figure 17), single element transducer is also used but with two incident angle (10-degree and 45-degree) at this time, to see the effect of incident angle on not fully developed flow.

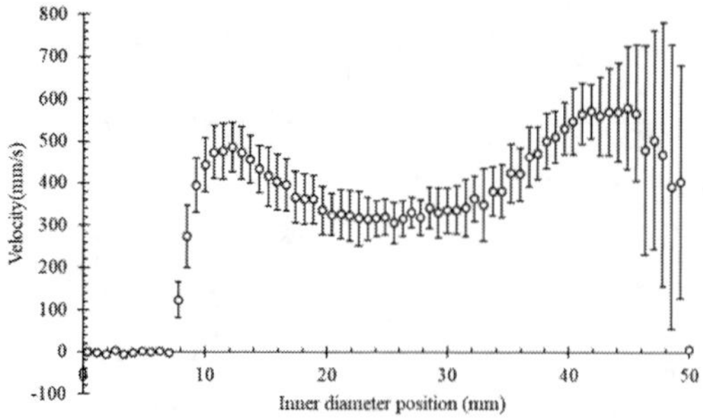

(a) Axial velocity (V_x).

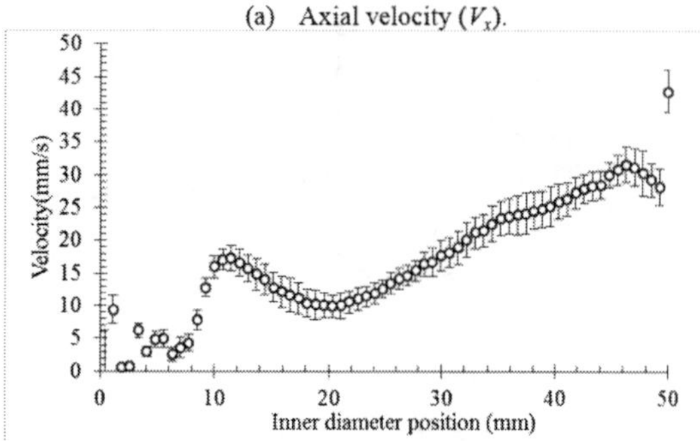

(b) Radial velocity (V_y)

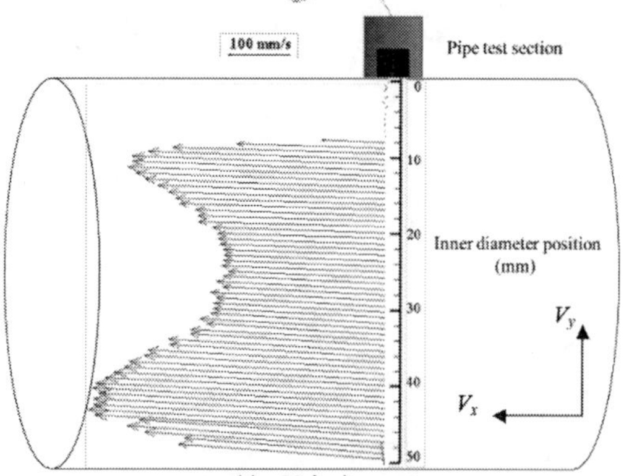

(c) Velocity vector.

Fig.16. Pipe flow measurement with swirl effect using two elements transducer.

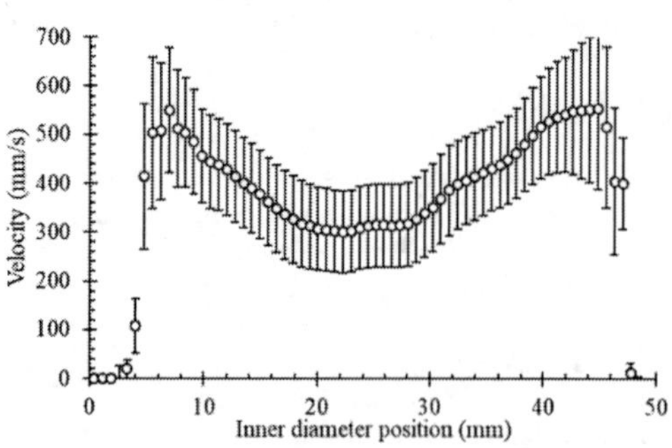

(a) Axial velocity (V_x), 10-degree incident angle.

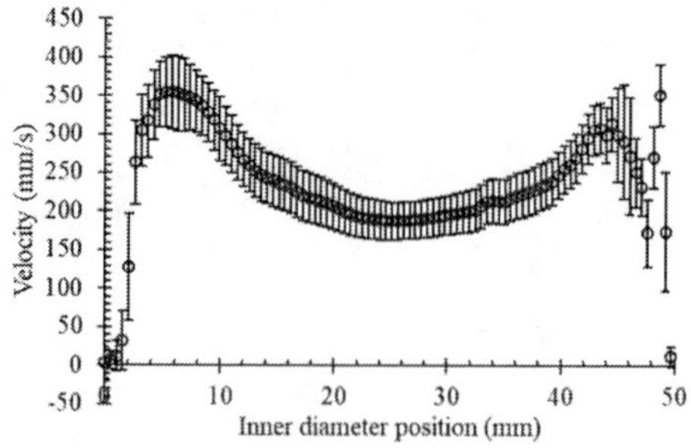

(b) Axial velocity (V_x), 45-degree incident angle.

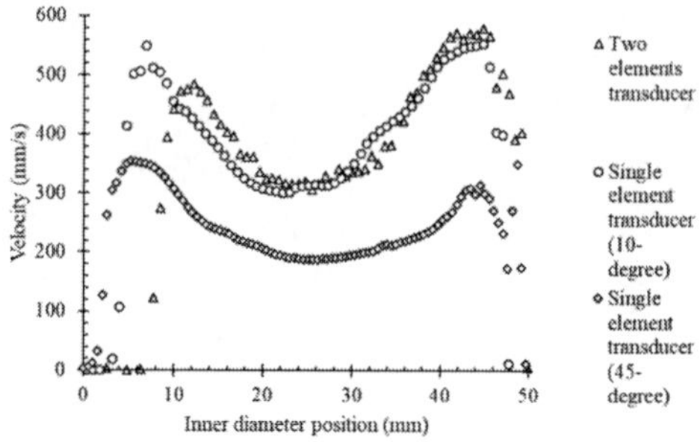

(c) Profile comparison.

Fig.17. Pipe flow measurement with swirl effect using single element transducer.

From the profile comparison, each profile is different to each other. This might be caused by the different incident angle measurement which results for each transducer measurement occurring in not a same position in the fluid while the flow is not fully developed. To evaluate the profile, the flow rate is calculated and compared to the reading flow rate.

Table 6. Flow measurement quality.

Equipment	Flow rate (l/min)	Error (%)	Average standard deviation of velocity profile (mm/s)
Electro-magnetic flow meter (reading)	35	-	-
Two elements transducer	38.65	10.43%	71.57
Single element transducer, 10 degree incident angle	40.05	14.42%	90.02
Single element transducer, 45 degree incident angle	28.91	17.41%	34.16

The overall result for this measurement has lower accuracy than previous measurement (without swirl effect). The reason is the flow under the swirl effect will have non-axisymmetric profile therefore velocity profile in only one measurement line does not fully represent the flow rate. For the profile comparison, the single element 45-degree incident angle have higher error compared to both single element 10-degree incident angle and two elements transducer. The reason is the measured area/position is very wide for not fully developed flow and the velocity profile in the measurement line and the flow profile will be different. Therefore, the calculation of flow rate will result in a higher error even with the small standard deviation. As for the two elements transducer, it has a better error than single element 10-degree incident angle and able to provide two-dimensional velocity profile information.

5.2.3 Two elements transducer profile estimation

One of the counter measure for the limitation in measurement range from two elements transducers is to estimate the profile in up to 8 mm region. The estimation can adopt statistic method and for this study we use B-spline interpolation with Gaussian fitting with the parameters of order 3 and knots 6. The reason is this method can provide a fitting curve with high r-squared value (0.995±0.003). Therefore, the estimation can be done while still keeping the original data (from 8 mm to 50 mm) corresponding to the previous profile (Figure 12 and Figure 16). The interpolation is done by assuming the velocity in the first channel (nearest to the sensor and the wall) as zero. This method is applied for the two-dimensional velocity measurement in pipe flow with and without swirl effect and the flow rate is calculated. The result is presented as below.

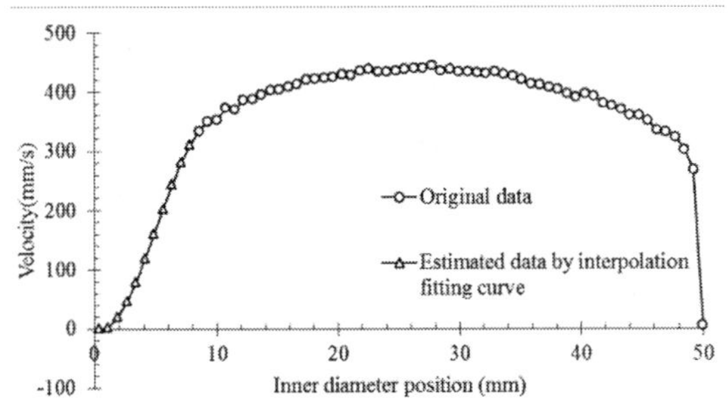

(a) Axial velocity (V_x).

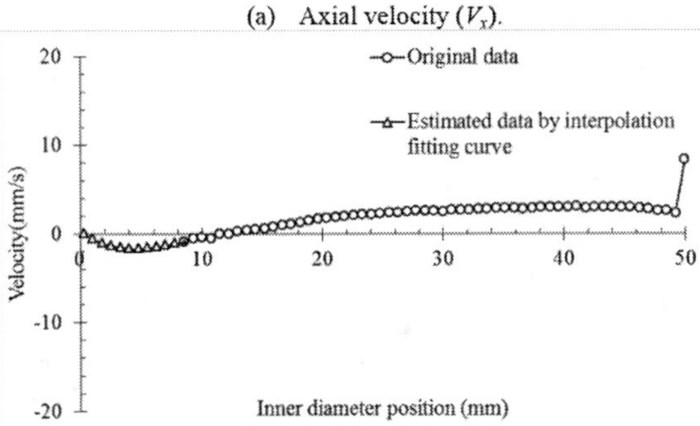

(b) Radial velocity (V_y).

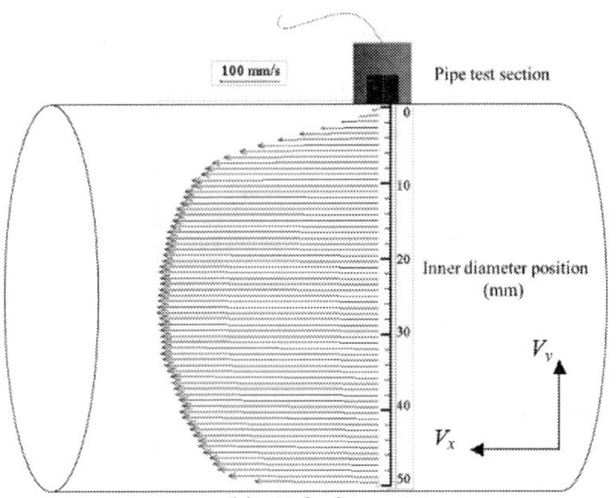

(c) Velocity vector.

Fig.18. Pipe flow measurement using two elements transducer with interpolation curve fitting (original data corresponding to profile in Figure 12)

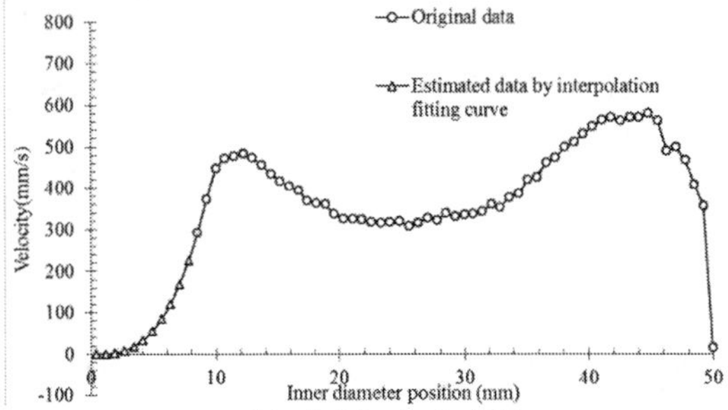

(a) Axial velocity (Vx).

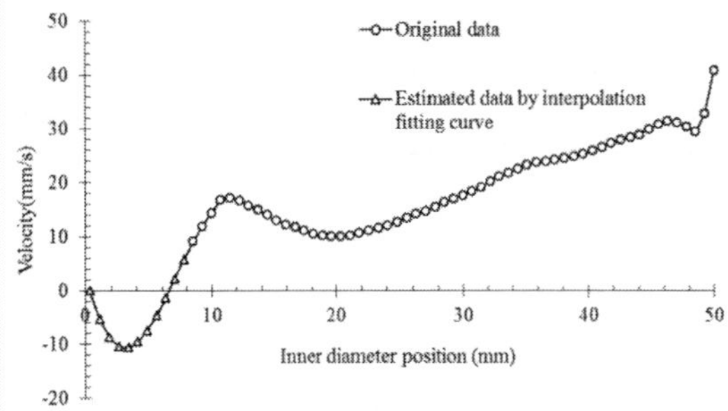

(b) Radial velocity (Vy).

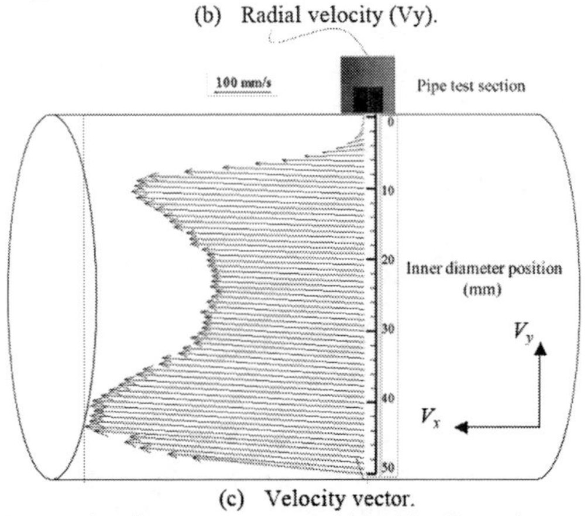

(c) Velocity vector.

Fig.19. Pipe flow measurement with swirl effect using two elements transducer with interpolation curve fitting. (original data corresponding to profile in Figure 16)

Table 7. Flow rate calculation of the profile with interpolation curve fitting.

Equipment	Flow rate (l/min)	Error (%)
Electro-magnetic flow meter (reading)	35	-
Two elements transducer (without swirl effect)	35.39	1.11%
Two elements transducer (with swirl effect)	39.52	12.92%

6 Conclusion

The two-dimensional velocity profile measurement system was developed by using only two elements transducer with both elements acting as a transceiver (transmitter and receiver). The system was designed with some targets and evaluated by sound pressure measurement and flow measurement. From the sound pressure measurement result, we confirm our design targets achieved with the effective measurement range start from 8 mm. From the flow measurement result, we confirm the system capability in obtaining two-dimensional velocity profile, and the flow rate can be calculated.

References

1. D.W. Baker, IEEE Trans. on Sonics and Ultrasonics. **17**, 170-185 (1970)
2. B.L. Dunmire, K.W. Beach, K. Labs, M. Plett, D.E. Standness Jr, Ultrasound in Medicine and Biology, **26**, 1213-1235 (2000)
3. P. Peronneau, W. Sandman, M. Xhaard, Ultrasound in Medicine, 1193-1208 (1977)
4. M. Scabia, M. Calzolai, L. Capineri, L. Masotti, A. Fort, Ultrasound Medicine and Biology, **26**, 121–131 (2000)
5. Y. Takeda. Nuclear Engineering and Design, **126**, 277-284 (1991)
6. S.Wada, H.Kikura, M.Aritomi, M.Mori, Y.Takeda, Journal of Nuclear Science and Technology, **41**, 339-346 (2004)
7. D.Huther and U.Lemmin, Measurement Science and Technology, **9**, 10 (1998)
8. H. Obayashi, Y. Tasaka, S. Kon, Y. Takeda, Flow Measurement and Instrumentation, **19**, 189-195 (2008)
9. C.Kasai, K.Namekawa, A.Koyano, R.Omoto, IEEE Transactions on sonics and ultrasonics, **32**, 458-64 (1985)
10. S.Wada, K.Tezuka, W.Treenuson, N.suzuki, H.Kikura, Science and Technology of Nuclear Installations (2012)

E3S Web of Conferences 43, 01012 (2018) https://doi.org/10.1051/e3sconf/20184301012
ASTECHNOVA 2017

Design of PLC Integrated Process Simulation Software for Steam Generator Using Matlab Simulink

Ikhtiander[1,*], *Soekirno* Santoso[2]

[1]University of Indonesia, Physics Department, Depok 16424, Indonesia
[2]University of Indonesia, Physics Department, Depok 16424, Indonesia

Abstract. This paper describes the work done in order to make Matlab Simulink based steam generator simulator in the simulation of a steam generator. The steam generator under this research is operated with the steam quality of 72%, O2 content is 1.2%, designed steam volume flow is 3600 barrel per day at a maximum and designed fuel gas volume flow is 1300 Thousand Standard Cubic Feet (MSCF) per day at a maximum. The simulator program of the steam generator is separated into individual components consisting of Burner, Radiant, Convection, Exhaust Stack, Feedwater Pump Discharge and Steam Discharge. Within the components, thermodynamics and heat transfer principles such as conduction, convection, radiation and also conservation of mass, momentum, and energy were applied to compute the pressure values, temperature values, and flow rate values of simulated field device based on the command and setpoint from PLC. The validation process has been done with the steam generator is operating in a steady state to the 10 important process parameters of the steam generator. The error percentage calculated from a difference between the simulation result value and the actual value from field data reference divide by actual value from field data reference. The error percentage results are as following : Fuel Gas Orifice Differential Pressure : 2.39%, Fuel Gas Pressure : 1.37%, Fuel Gas Temperature : 5.95%, Fuel Gas Flow Rate : 1.25%, Feedwater Orifice Differential Pressure : 1.94%, Feedwater Pressure : 1.54%, Feedwater Flow Rate : 0.92%, Steam Orifice Differential Pressure 3.26%, Steam Discharge Pressure 1.93% and Steam Quality : 0.05%.

1 Introduction

EOR (Enhanced Oil Recovery) is a production method applied to an oil production field to increase the amount of oilfield production. The EOR serves to increase the production of an oil field when the natural pressure in the oil well, which drives the oil out to the surface has diminished with the age of an oil field, so a method is needed to push the oil out of the well. There are various methods applied in EOR, one of which is a steam flood or thermal injection.

Thermal injection is one of the techniques for increasing the amount of crude oil that can be extracted from an oil field. Steam flooding is one means of introducing heat into the reservoir by injecting steam into the well. In order to produce steam, several steam generators are operated to produce steam with the specific steam flow and steam quality. The steam generator under this research is operated with the steam quality of 72%, O2 content is 1.2%, designed steam volume flow is 3600 barrel per day at a maximum and designed fuel gas volume flow is 1300 Thousand Standard Cubic Feet (MSCF) per day at a maximum.

Oil Field D is an oil field operated by PT C. The steam injection method is the most suitable method for heavy oil type produced by Oil Field D. Heavy oil is a type of oil that has a high viscosity, so it takes steam to reduce the level of the viscosity of the fluid so that the process of removing the oil more easily and efficiently. Figure 1 below provides an illustration of an oil field production system using a steam injection system.

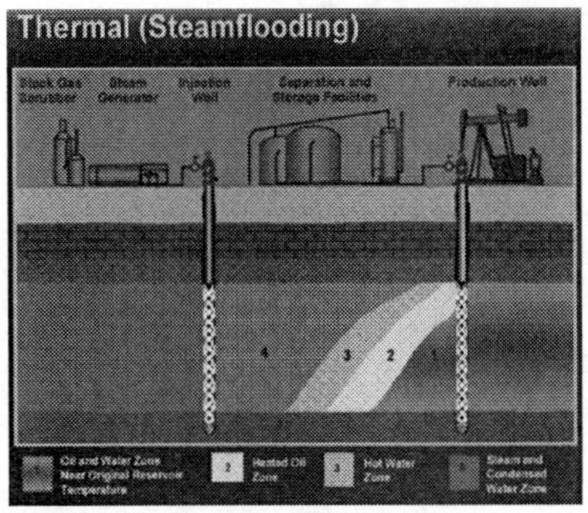

Fig. 1. The EOR system using steam flood.

To produce steam for steam flood process, PT C operates a natural gas-fired steam generator to produce steam with certain steam pressure and steam quality. This steam generator works based on the principles of heat transfer physics, thermodynamic physics and fluid mechanics

* Corresponding author: ikhtiander@ui.ac.id

operating under the control of instrumentation and control systems based on Programmable Logic Controller (PLC).

During the lifecycle of a steam generator unit, necessary activities from an operation, maintenance and optimization will be carried out. Depending on the scope of activities, operators, technicians or engineers may be involved in some of the phases. An operator that is responsible for carrying out operation phase or technicians which are responsible for carrying out maintenance phase shall be competent to carry out activities which they are accountable. Competency only is achieved with the right methods of training activities, the most effective training method is hands-on training which will put trainee into real situations and scenarios. On the other side, engineers that are responsible for carrying out optimization phase will need effective tools for research and experiments activities.

The Matlab Simulink based steam generator simulator software will be proposed as a part of steam generator simulation system that can be used for training purposes or for research purposes. The Steam generator simulation system will be consisting of Allen-Bradley Controllogix Programmable Logic Controller (PLC) and Rockwell FTview ME human-machine interface (HMI) software that they are loaded with and run the existing steam generator control system logic programs. The Allen-Bradley Controllogix PLC acquires data of simulated field device from Simulink based steam generator simulator software, processes the logic and control program and then send the command of simulated field actuator to Simulink based steam generator simulator software through Matlab OPC Toolbox connections.

The simulator program of the steam generator is separated into individual components consisting of Burner, Radiant, Convection, Exhaust Stack, Feedwater Pump Discharge and Steam Discharge. Within the components, thermodynamics and heat transfer principles such as conduction, convection, radiation and also conservation of mass, momentum, and energy were applied to compute the pressure values, temperature values, and flow rate values of simulated field device based on the command and setpoint from PLC.

The objectives of this research are :

1. Design and build a mathematical model of physical processes within the steam generator system.
2. Design and build a computer simulation for steam generator physics process based on the mathematical models resulted.
3. Integrate simulation system with existing steam generator PLC based control systems for learning and research tools.

Limitations of the design of the steam generator simulation are as following :

1. Simulation built with Matlab Simulink programming language and communication between PLC and Simulink is using ethernet network.
2. The inputs of this simulation are the actuator control commands of the PLC while the outputs are the simulated value of the sensors.
3. Sensors to be simulated are limited to the 10 sensors available in the field operation data recording system according to Table 1.
4. The analysis is limited to steady-state conditions and the tolerance range of the simulated results is 10% against the field operation data.

Table 1. List of sensors to be simulated.

No	Physical Quantities	Unit
1	Gas orifice differential pressure	Inch H_2O
2	Gas pressure	PSI
3	Gas Temperature	°F
4	Gas flow rate	MSCFD
5	Water orifice differential pressure	Inch H_2O
6	Water pressure	PSI
7	Water flow rate	BPD
8	Steam orifice differential pressure	Inch H_2O
9	Steam pressure	PSI
10	Steam quality	%

The research method used in this research consists of several stages as follows :

1. Mathematical Modeling.
2. Implementation of mathematical modeling in Matlab Simulink
3. Matlab Simulink programming
4. Testing of modeling programs
   - Testing of input and output between PLC and Simulink.
   - Testing of modeling :
     - Subprocess module (SM) testing
     - Steam generator unit module (UM) testing
5. Analysis and conclusions of research results.

2 Theoretical Review

The steam injection system at Oil Field D used steam with certain temperature and pressure generated by steam generators. The steam generators at Oil Field D is a boiler that uses natural gas fuel. Within the combustion process in the combustion chamber, natural gas will be reacted with air in a certain ratio so as to produce the most efficient combustion process. The heat generated by the combustion process in the combustion chamber will be used to heat the water in the boiler so that the water changes phase to the steam phase. The resulting steam is then injected into the injection well. In Figure 2 there is a functional organization of steam generators operated in Oil Field D, which consists of the main parts that will be based on the modeling to be developed.

 https://doi.org/10.1051/e3sconf/20184301012

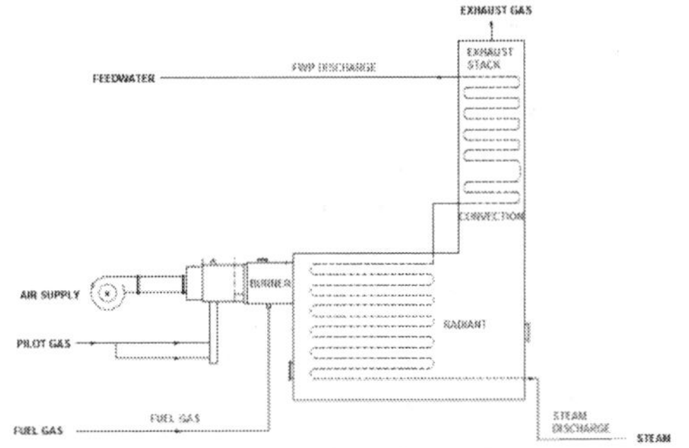

Fig. 2. The functional organization of the steam generator

In [9] Saiful Hidayat simulates a steam generator with modeling consisting of the following main models as presented in Figure 3 below:

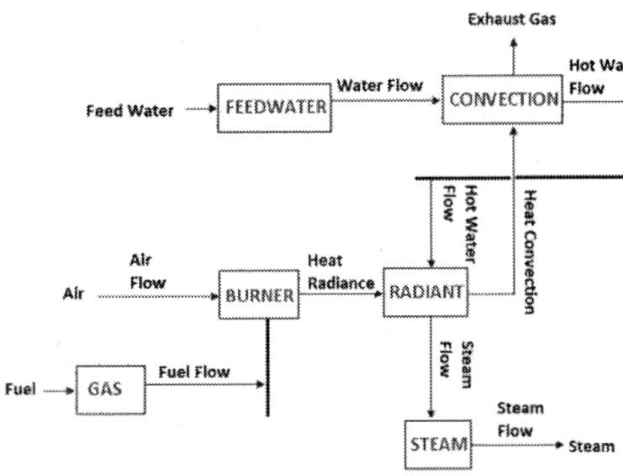

Fig. 3. Steam generator systems modeling.

1. Burner
   In this model, the combustion process between fuel (in this case is natural gas) and oxygen occurred, which is an exothermic process that becomes a source of heat for steam generators.
2. Fuel
   In this model, there is a process of measuring and controlling the gas flow rate regulated by the PLC through a gas flow control valve based on the amount of gas required by the ongoing combustion process. The amount of measured gas flow is determined by the pressure of the gas source, the pressure difference between the input and output of the orifice and the gas temperature.
3. Feedwater
   In this model occurs the process of measuring the flow rate of water by PLC. The amount of measured gas flow is determined by the pressure of the water source, the pressure difference between the input and the output of the orifice measuring element and the water temperature.
4. Convection
   In this model, the heat carried by the exhaust gas is utilized to heat the water stream convectionally before the water is fed into the Steam Generator.
5. Radiant
   In this section, there is a transfer of heat from the process of burner model to the water stream by radiation so that the water temperature increases causing the water to changes the phase to steam phase. Some heat that is not absorbed by water will be carried away by the exhaust gas toward the Convection model.
6. Steam
   In this section, there is a steam quality measurement process by PLC. The quality of steam is the percentage of vapor phase content in the fluid. The amount of steam quality is determined by the steam pressure and pressure differential between the input and output of the orifice measuring element.

In [10] Craig Pauley stated that:

$$J12 = \left(\frac{\left(\sqrt{(1-AB12^4) * BK12}\right)*14.6*E12}{AI12 * AJ12 * B12}\right)^2 \quad (1)$$

where :
J12: Orifice Differential Pressure (Inch H_2O)
AB12: Orifice Beta Ratio
BK12: James / Texaco Effective Mixture ($ft^3/lb.$)
E12: Flow Rate (BPD)
AI12: C -Discharge
AJ12: Orifice Thermal Expansion Coefficient
B12: Orifice Diameter (Inch)

The measurement system of the water flow rate using an orifice plate with PLC as a flow computer presented in Figure 4 below:

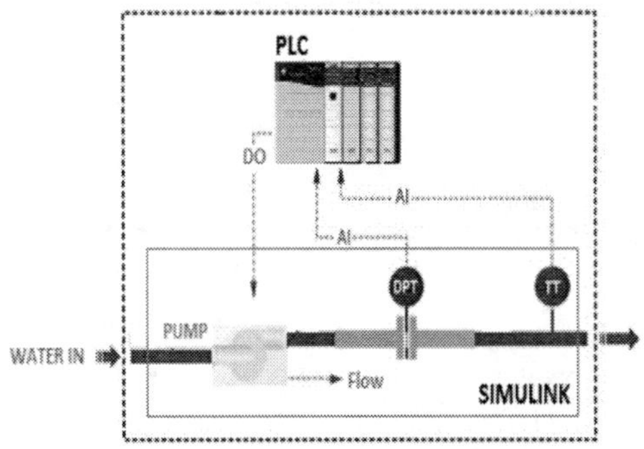

Fig. 4. Basic water flows measurement system using orifice plate and PLC.

In [10] Craig Pauley stated that:

$$BL12 = \frac{(((I12)^{H12}*(BE12-BD12))+BD12)*((E12*14.6)^2*(1-AA12^4))}{128828.59*B12^4*AI12^4*AJ12^4}$$
(2)

Where :
BL12: Orifice Differential Pressure (inch H_2O)

https://doi.org/10.1051/e3sconf/20184301012

B12: Orifice Diameter (inch)
E12: Flow Rate (BPD)
H12: 2-phase Exponent
AA12: Beta Ratio (%)
AI12: C -Discharge
AO12: Pressure (psia)
BD12: Vt (ft^3 / lb_m)
BE12: Vg (ft^3 / lb_m)
BK12: Isentropic Exponent of Wet Vapor k
I12: Steam Quality (%)

The measurement system of steam flow rate using orifice plate with PLC as flow computer presented in Figure 5 below:

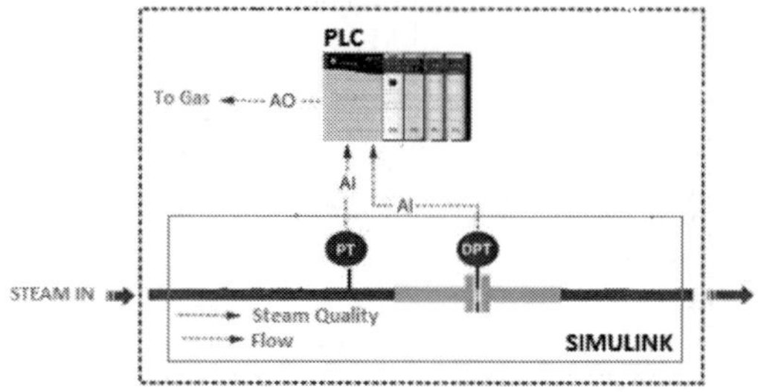

Fig. 5. Basic steam flow measurement system using orifice plate and PLC

In [12] and [23] it is explained that for hydrocarbon gas, the flow calculation follows the following equation:

$$M_{dpv} = \frac{\left(\frac{Qv}{(7709.61 * CdFT * Ev * Y1 * dfo^2)}\right)^2 * (Fgr * Fzf * (MTemp) + 459.67)}{(PF1 * Fzs)}$$

(3)

Where :
M_{dpv}: Differential Pressure (inch H_2O)
Qv: Volumetric Flow (SCFH)
CdFT: Orifice Plate Coefficient of Discharge
Ev: Velocity of Approach Factor
Y1: Expansion Factor
dfo: Orifice Plate Outer Diameter at flowing conditions (inch)
Pf1: Gas Mixture Upstream Pressure at Flowing Conditions (psia)
Fzs: Compressibility Factor of Gas at Standard Conditions
Fgr: Real Gas Relative Density
Fzf: Compressibility at Flowing Conditions
MTemp: Gas Temperature (°F)

The process of measuring the flow of fuel gas using orifice plate with PLC as flow computer using the system presented in Figure 6 below:

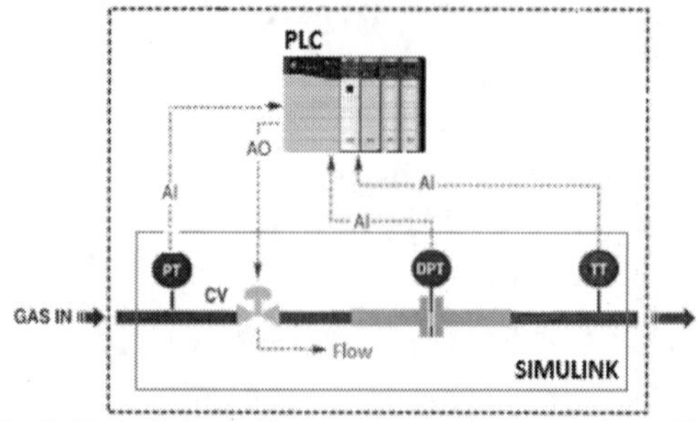

Fig. 6. Basic fuel gas flow measurement system using orifice plate and PLC

3 Design of Simulation System

The system architecture of integrated PLC steam generator simulation is shown in Figure 7 while the connection between PLC and computer is shown in Figure 8 and Figure 9. This simulation system consists of two hardware parts as following :

- Personal Computer
  It is running Matlab Simulink software, PLC programming software, and HMI software. The computer will be receiving commands from PLC and then send the sensor simulations to PLC through ethernet communication.
- PLC
  It is controlling the process of steam generators simulated by Simulink through ethernet communication.

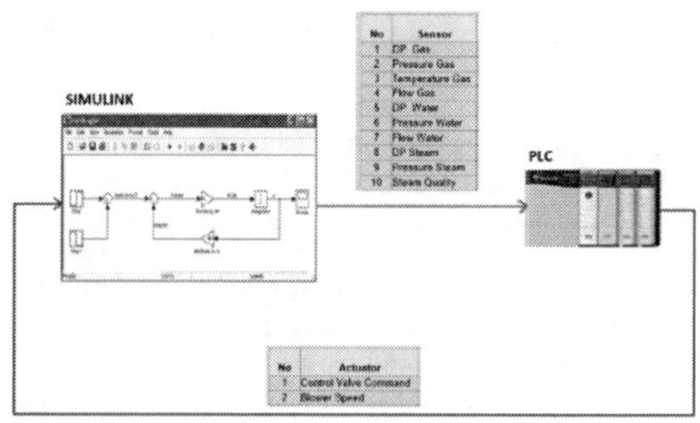

Fig. 7. PLC integrated steam generator system simulation architecture.

 https://doi.org/10.1051/e3sconf/20184301012

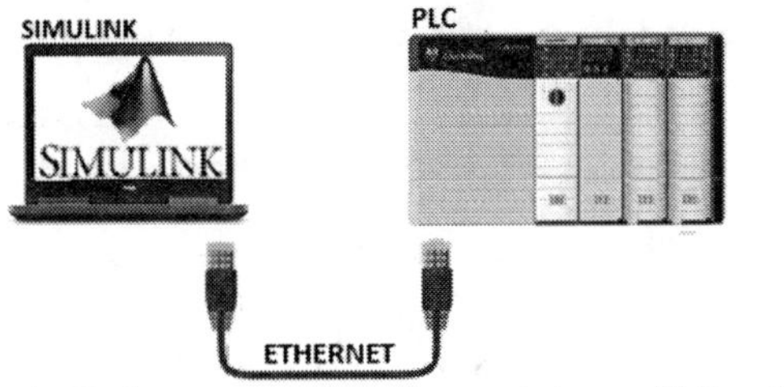

Fig. 8. Hardware connection diagram between PLC and Simulink.

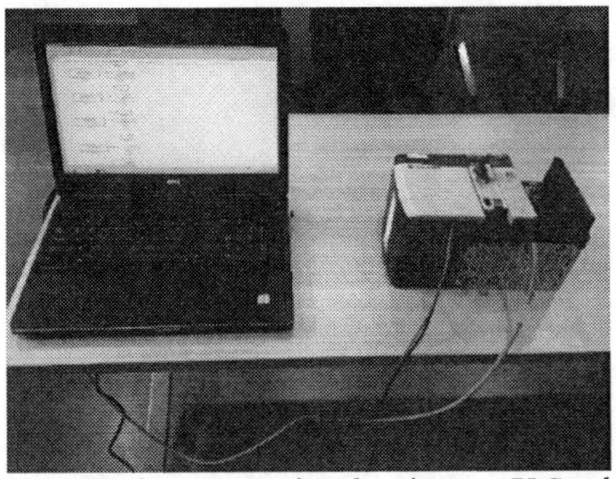

Fig. 9. Hardware connection photo between PLC and Simulink.

The modeling method used is based on the modeling structure of the process control system from the ANSI / ISA-88.01-2010 standard Batch Control Part 1: Models and terminology [17], adapted according to the requirements in modeling the physical processes within the steam generator.

All functions or data are wrapped in objects or classes [17]. Objects interact with other objects in a class to form a new object in a higher class. The advantage of this modeling method is to ease the understanding of the physics process and easy to be taught because it is structured and also ease the implementation of making the simulation program because the program can be built from several function modules which can be reused.

The proposed modeling design is presented in Figure 10 below, where the steam generator unit 14 consists of six process modules, six subprocess modules, and five control modules:

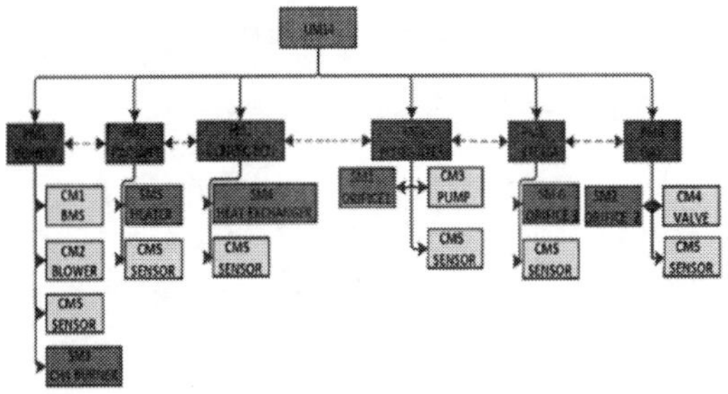

Fig. 10. Modelling of unit 14 steam generator Area 1 at Oil Field D

Simulink simulation software will be communicated to the PLC through a gateway between Simulink OPC Toolbox and RSLinx software as shown in Figure 11 that describe details of software connection between Simulink software and PLC.

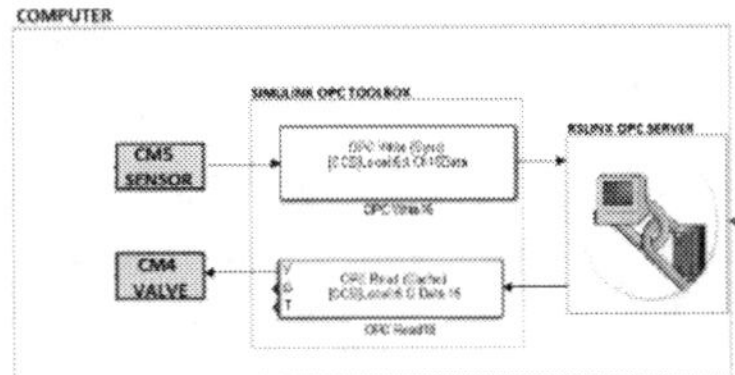

Fig. 11. Software connection diagram between PLC and Simulink.

4 Static Simulation System Testing

Static simulation system testing for subprocess module level is done by running the simulation for individual subprocess modules in Matlab Simulink environment then the result compared with reference from field operation data.

For Orifice_1 subprocess module, the water flow rate in barrel per day (BPD) is an independent variable with a fixed value in accordance with the actual pumping capacity which the value varies between 3580.5 BPD to 3585 BPD. The test results are presented in Figures 12 and Table 2.

Table 2. Table of Orifice_1 subprocess module testing

Reference Data From Field						Simulation Data		
Id	Water Flow (BPD)	Pressure (psi)	Differential Pressure from Reference (Inch H_2O)	Steam Quality (%)	Temperature (°F)	Differential Pressure from Simulation (Inch H_2O)	Error	Error Percentage
493040	3582.88	1089.55	172.98	71.91	152.00	174.83	1.85	1.07
493051	3580.60	1087.85	172.64	72.11	152.00	174.61	1.97	1.14
571406	3582.49	1087.08	173.18	72.35	152.00	174.79	1.61	0.93
571466	3585.01	1086.94	172.99	72.03	152.00	175.04	2.05	1.18
572047	3584.57	1087.85	173.07	71.96	152.00	174.99	1.92	1.11
572093	3584.57	1085.62	172.98	72.16	152.00	174.99	2.01	1.16
572143	3584.64	1086.97	172.98	72.03	152.00	175.00	2.02	1.17
575326	3582.94	1087.60	172.77	72.12	152.00	174.83	2.06	1.19
575380	3584.35	1086.09	172.98	71.62	152.00	174.97	1.99	1.15
575418	3583.36	1085.24	172.88	72.10	152.00	174.88	2.00	1.15
								1.13

https://doi.org/10.1051/e3sconf/20184301012

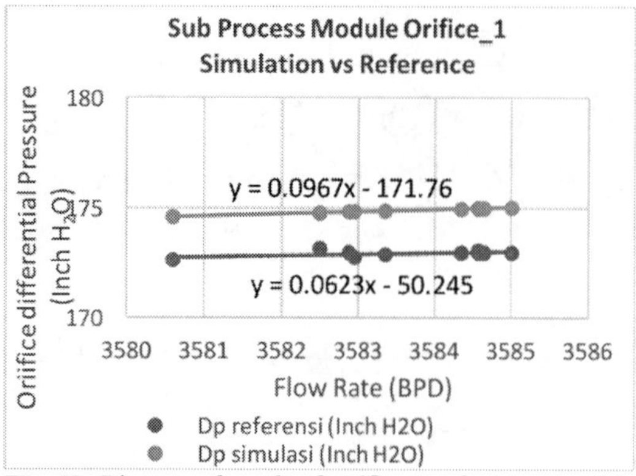

Fig. 12. Diagram of result of Orifice_1 subprocess module testing

Based on the test results presented in the figure and the table above the average of the percentage of error against the field operation data is 1.13%, so it is still within the maximum error range of 10%.

For Orifice_2 subprocess module, the gas flow rate is an independent variable which the value varies from 132.39 MSCFD to 1039.90 MSCFD. Test results are presented in Figure 13 and Table 3.

Table 3. Table of Orifice_2 subprocess of module testing

Reference Data From Field					Simulation Data		
Id	Pressure (psi)	Temperature (°F)	Gas Flow (MSCFD)	Differential Pressure From Reference (Inch H_2O)	Differential Pressure From Simulation (Inch H_2O)	Error	Error Percentage
262388	40.12	95.44	132.39	3.34	3.72	0.38	11.26
262521	39.91	95.66	282.76	16.05	17.20	1.15	7.18
262563	39.94	95.32	388.37	29.28	32.50	3.22	11.01
262607	39.93	94.87	519.94	52.49	58.35	5.86	11.17
262655	39.93	94.40	649.84	83.30	91.19	7.89	9.47
262715	39.91	93.89	779.41	122.11	131.26	9.15	7.50
262778	39.73	93.41	906.99	170.52	178.31	7.79	4.57
262983	39.66	92.74	1039.90	230.56	234.61	4.05	1.76
							7.99

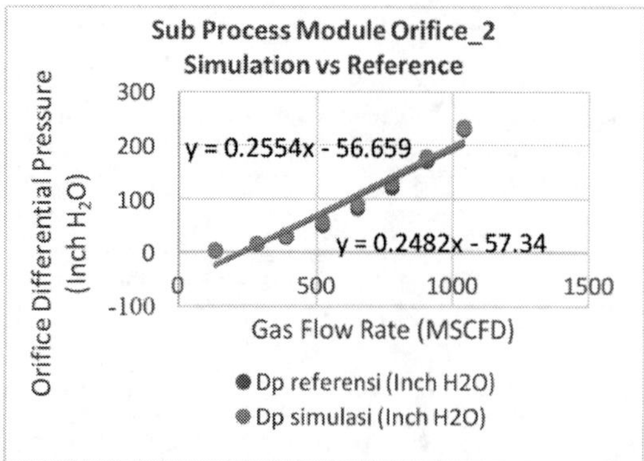

Fig. 13. Diagram of the result of Orifice_2 subprocess module testing

Based on the test results presented in the figure and table above the average of the percentage of error is 7.99% against the field operation data, so it is still within the maximum error range of 10%.

For Orifice_6 subprocess module, the steam quality is an independent variable which the value varies from 20.01% to 72.08%. Test results are presented in Figure 14 and Table 4.

Table 4. Table of Orifice_3 subprocess module testing

Reference Data From Field					Simulation Data		
Id	Pressure (psi)	Steam Flow (BPD)	Steam Quality (%)	Differential Pressure From Reference (Inch H_2O)	Differential Pressure From Simulation (Inch H_2O)	Error	Error Percentage
263210	540.86	3723.31	20.01	270.33	279.25	8.92	3.30
263274	553.64	3679.08	30.19	440.75	449.95	9.20	2.09
263383	564.81	3627.52	40.06	616.19	623.39	7.20	1.17
263810	568.40	3586.88	50.02	834.82	813.59	21.23	2.54
265194	578.94	3539.84	60.05	1039.08	1027.20	11.88	1.14
265963	591.07	3504.18	70.00	1267.60	1239.24	28.36	2.24
266627	592.01	3490.16	72.08	1307.57	1288.57	19.00	1.45
							1.99

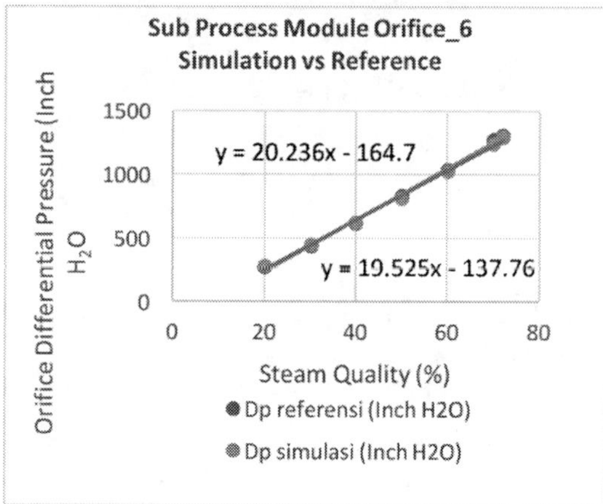

Fig. 14. Diagram of the result of Orifice_6 subprocess module testing

Based on the test results presented in the above figures and tables, the average percentage error generated is 1.99% against the field operation data, so it still within the maximum error range of error of 10%.

To start static simulation system testing for unit module level, standard operation procedures for steam generators are implemented through HMI and PLC so that the steam generator unit is in normal steady state working condition. Under this condition, observations and data recording of the simulated sensors are made, the expected result is that each value recorded by the PLC should be within a certain error range.

From ten (10) static tests that conducted sequentially, one of the results presented in Table 5, the average percentage error of simulation results are presented in Table 6.

https://doi.org/10.1051/e3sconf/20184301012

Table 5. Unit module testing table with reference to dataset #266476

No	Sensor Description	Simulation Result	Reference Value (id 266476)	Error	Percentage
1	DP Gas (Inch H_2O)	327.84	326.43	1.41	0.43
2	Pressure Gas (psi)	40.00	39.58	0.42	1.05
3	Temperature Gas (°F)	77.05	85.68	8.63	10.07
4	Flow Gas (mscfd)	1231.81	1214.29	17.52	1.44
5	DP Water (Inch H_2O)	176.51	164.68	11.83	7.18
6	Pressure Water (psi)	1090.00	1064.35	25.65	2.41
7	Flow Water (bpd)	3617.51	3493.55	123.96	3.55
8	DP Steam (Inch H_2O)	1403.11	1308.12	94.99	7.26
9	Pressure Steam (psi)	595.29	592.29	3.00	0.51
10	Steam Quality (%)	72.00	72.02	0.02	0.03

Table 6. Table of the average result of unit module unit error test results

No	Sensor Description	Error Average (%)
1	DP Gas (Inch H_2O)	2.39
2	Pressure Gas (psi)	1.37
3	Temperature Gas (°F)	5.95
4	Flow Gas (mscfd)	1.25
5	DP Water (Inch H_2O)	1.94
6	Pressure Water (psi)	1.54
7	Flow Water (bpd)	0.92
8	DP Steam (Inch H_2O)	3.26
9	Pressure Steam (psi)	1.93
10	Steam Quality (%)	0.05

Based on the test results, the average percentage of error are as following : differential pressure of fuel gas orifice : 2.39%, fuel gas pressure: 1.37%, fuel gas temperature: 5.95%, fuel gas flow rate: 1.25 %, differential pressure of feedwater: 1.94%, feedwater pressure: 1.54%, feedwater flow rate: 0.92%, differential pressure of steam orifice: 3.26%, steam discharge pressure: 1.93% and steam quality: 0.05%.

5 Dynamic Simulation System Testing

For dynamic simulation system testing, standard operation procedures for steam generators are implemented through HMI and PLC so that the startup process of steam generator will be executed from standby conditions until normal steady state working condition achieved. During this process, observations and data recording of the simulated sensors are made.

Simulation data and reference operational data for steam quality and steam quality setpoint will be presented in form of X-Y chart against elapsed time which will represent the response of steam generator system to PLC control. The two graphics will then be compared to obtain a comparison of dynamic responses between the simulation results and field operational data. The parameters to be compared to two graphics are as following :

- Rise Time
  Rise time is the time required to rise from 10% to 90% of steady-state value.
- Overshoot
  Overshoot is an excess or deficiency of value against the final value, expressed in terms of the final value.
- Settling Time
  Settling time is the time required to reach 5 percent difference with the steady-state value.
- Steady State Error
  Steady state error is the final difference between process value and set point at steady state condition.

Dynamic test results for operational data are presented in Figure 15, based on the curve the following control performance parameters can be calculated :

- Rise Time is approximately 40 minutes.
- Overshoot approximated 5%.
- Settling Time approximated 120 minutes.
- Steady State Error varies within 5% range at maximum.

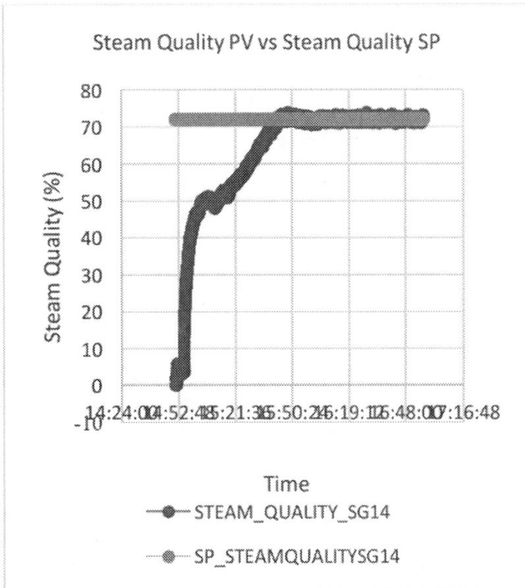

Fig. 15. Diagram of the control system curve based on field operational data.

Dynamic test results for the simulation results are presented in Figure 16, based on the curve the following control performance parameters can be calculated :

- Rise Time is approximated 25 minutes
- Overshoot 0%
- Settling Time approximated 90 minutes
- Steady State Error 0.

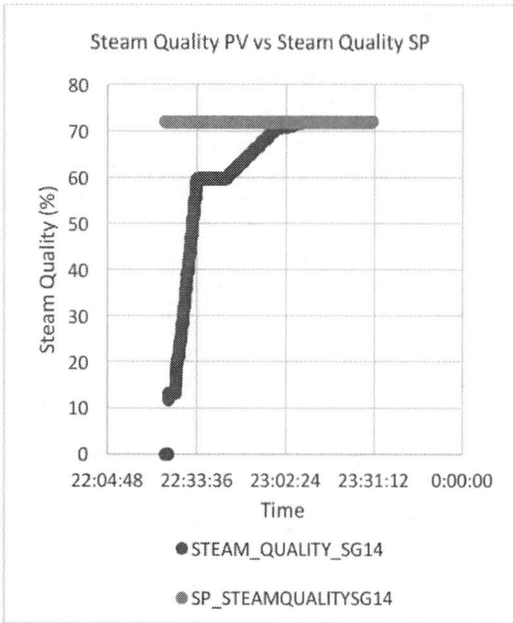

Fig. 16. Diagram of the control system curve based on simulation results

6 Conclusion

Based on the data and analysis of the research on "Design of Integrated PLC Process Simulation Software Using Matlab Simulink", the conclusions that can be taken are as following:

1. The physical process of steam generator can be mathematically modeled based on the thermodynamic principles, heat transfer and fluid mechanics. Errors (a difference between simulation results and field operational data) resulted from subprocess module level testing for water flow, gas flow, and steam flow are 1.13%, 7.99%, and 1.99% respectively.
2. The models then will be used to build a computer simulation of the physical process of the steam generator. Error (difference between simulation result and operational data) resulted from unit module level testing for DP Gas, Pressure Gas, Temperature Gas, Flow Gas, DP Water, Pressure Water, Flow Water, DP Steam, Pressure Steam, Steam Quality are 2.39 %, 1.37%, 5.95%, 1.25%, 1.94%, 1.54%, 0.92%, 3.26%, 1.93%, 0.05% respectively.
3. Models are implemented on Simulink and integrated to the PLC through an Ethernet data communication system to construct complete simulation system for steam generator process control. Communication test results (data sent compared to received data) gives a 0% error rate.

References

1. P. Díaz, Susana, 1st Meeting of EcosimPro Users (2001)
2. B. Elmegaard, Ph.D. Thesis ET–PhD 99–02, Technical University of Denmark, Department of Energy (1999)
3. D. Surianu, Flavius, Intech
4. S. Anderson, Submitted in Partial Fulfillment of the Requirements for the Degree of Master's of the Mechanical Engineering Program, YOUNGSTOWN STATE UNIVERSITY (2008)
5. H. Gan, J. Zhang, H. Zeng, Intl. J. Adv. in Computing Tech. (IJACT) **4** (2012)
6. P.S. Bartells, J.S. Gauthier, Inst. Soc. America (2001)
7. Maican, Camelia, Vinatoru, Matei, Canureci, Gabriela, Iancu, Eugen, WSEAS TRANS. SYSTEMS and CONTROL **3** (2008)
8. Haller, Michel, Konersmann, Lars, Haberl, Robert, Dröscher, Angela, Frank, Elimar, *Eleventh International IBPSA Conference* (2009)
9. S. Hidayat, PT Chevron Pacific Indonesia (2015)
10. C. Pauley, Chevron Corporation (2006)
11. Y. Subejo, PT Chevron Pacific Indonesia (2009)
12. American Gas Association (AGA), Orifice Metering of Natural Gas and Other Related Hydrocarbon Fluids Part 1 General Equations and Uncertainty Guidelines, American Gas Association (AGA) (1990)
13. American Gas Association (AGA), Compressibility Factors of Natural Gas and Other Related Hydrocarbon Gases, American Gas Association (AGA) 2003
14. Aganovic, S. Amar, Department of Energy and Process Engineering Norwegian University of Science and Technology (2013)
15. C. P. Rodrigues, M. A. Lansarin, A. R. Secchi, T. F. Mendes, Engenharia Térmica (Thermal Engineering) **4** (2005)
16. American National Standard, ANSI/ISA–77.20–1993 Fossil Fuel Power Plant Simulators – Functional Requirements, American National Standard (1993)
17. American National Standard, ANSI/ISA-88.01-2010 Batch Control Part 1: Models and terminology, American National Standard (2010)
18. Honeywell International Inc., R RM7800E,G,L,M; RM7840E,G,L,M 7800 SERIES Relay Modules installation instructions, Honeywell International Inc. (2010)
19. PT Chevron Pacific Indonesia, Operations & Maintenance Certification Steam generator, PT Chevron Pacific Indonesia (2008)
20. P.K. Nag, *Power Plant Engineering* (2008)
21. H. Chu, Dept. of Environmental Engineering National Cheng Kung University
22. Steam Station Control Retrofit Project Area 1 Duri - Riau, Piping & Instrument Diagram Steam Generator Duri area - 1 Block 5S, Steam Station Control Retrofit Project Area 1 Duri – Riau (2009)

23. E.D. Neilands, W.E. Davis, Chevron Gas Flow Rate Calc (API MPMS 14.3), Chevron Corporation (2007)

Determination LMWL Wonosobo Area by Using Nuclear Technology, Case Study: Hydrogeology Study for Aqua Danone CSR Program

Agus Budhie Wijatna[1], *Heru* Hendrayana[2], *Fitrotun* Aliyah*[1], *Azwar* Satriya Muhammad[3], *Bungkus* Pratikno[4]

[1]Department of Nuclear Engineering and Engineering Physics, Universitas Gadjah Mada, Yogyakarta, Indonesia
[2]Department of Geology Engineering, Universitas Gadjah Mada, Yogyakarta, Indonesia
[3]PT Tirta Investama Danone Aqua, Jakarta, Indonesia
[4]National Nuclear Energy Agency of Indonesia, Jakarta, Indonesia

Abstract. Corporate Social Responsibility (CSR) program is one of the management concepts whereby the company integrates social and environmental aspect to support their business as the strategy to achieve the company goals for balancing between economic, social, and environmental imperatives. PT Tirta Investama Danone Aqua, which located in Wonosobo has a core business in supplying natural drinking water. To fulfil its responsibility in conserving nature and groundwater reserves, PT Tirta Investama Danone Aqua has a CSR program called Aqua Lestari. The determination of the receiving area of the CSR program should review and analyzed in order the CSR programs is not misplaced. PT Tirta Investama wants the CSR program to be given to the groundwater recharge areas for maintaining groundwater reserves sustainability. Therefore, a comprehensive study concluded to determine the groundwater recharge area studied from the aspects of geology, hydrogeology, and hydro isotope. The application of nuclear technology especially for stable isotope tracer can be used to know the groundwater recharge area, and modeling the flow of subsurface groundwater. Local Meteoric Water Line (LMWL) is used for analysis of groundwater recharge area. The equation of LMWL Wonosobo is δD, ‰ = 8,102 * $\delta^{18}O$ + 14,21‰.

1 Introduction

CSR or Corporate Social Responsibility is a corporate program which gives something to the beneficiaries in order to reach the balancing of economic aspect, social, and environment. PT Tirta Investama Danone Aqua, which located in Wonosobo, Central Java has a core business in supplying natural drinking water in the form of bottled water product. To fulfill its responsibility in conserving nature and groundwater reserves, PT Tirta Investama Danone Aqua has a CSR program called Aqua Lestari. The determination of the receiving area of the CSR program should be reviewed and analyzed in order the CSR programs is not misplaced. PT Tirta Investama wants the CSR program to be given to the groundwater recharge areas in order to maintain groundwater reserves sustainability. Therefore, a comprehensive study concluded to determine the groundwater recharge area studied from the aspects of geology, hydrogeology, and hydro isotope. Nuclear technology application through stable isotope analysis in groundwater compositions has proven useful as geochemical tools for exploration and management of groundwater resources [1]. The hydro isotope technology can measure the spring water of PT. Tirta Investama Danone Aqua comes from the groundwater recharge area which is obtained from a study of geology and hydrogeology.

1.1 Site Description

PT. Tirta Investama Danone Aqua has natural springs located in Wonosobo area, precisely in the mountains of Dieng plateau, Central Java, Indonesia. Most of the research areas are administratively located in Wonosobo District, Central Java Province. Astronomically, the study area is between the UTM coordinates Zone 49S lines 375400 - 390078 meters and 9186333 - 9203520 meters or 109° 52'17 "- 110° 0'15" East Longitude and 7°12'16 "- 7°21'35 "south latitude of 251.55 km^2 or about 25.6% of the total area of Wonosobo regency. The research area is also included in Temanggung Regency in the northeast and Banjarnegara District in the northwest.

Wonosobo regency is one of the city in Central Java with an area of 984.68 km^2. It is located between 7°11 'and 7°36' South Latitude, 109°43 'and 110°4' East

* Corresponding author: fitrotun.aliyah@mail.ugm.ac.id

Longitude. The area mostly mountainous with a range from 250 – 2250 meters above sea level (m.asl).

1.2 Local Meteoric Water Line (LMWL)

In hydrological cycle, the rainfall of precipitation will be a surface water or infiltrate to the ground and become groundwater resource. The surface water is susceptible to be evaporation which degree of evaporation will change the isotopic composition of surface water. On the other side, the infiltrated water is shielded by the reservoir rock from evaporation thus preserving the isotopic composition of the source rainwater. The groundwater consists of two types of circulating if the ground which are shallow groundwater and deeply-circulating groundwater. Shallow groundwater has the short circulating with near exit through locally-fed groundwater system. In another side, the deeply-circulating groundwater system has exit points in regionally-fed spring. The isotopic composition of the type (deeply circulating groundwater) is typically calculated using the rainwater data. The derived isotopic values, in term of Deuterium (^2H) and Oxygen-18 (^{18}O) are collectively called as the Meteoric Water Index (MWI) [1] or Meteoric Water Line [2]. Craig [3] first defined that the Global Meteoric Water Line (GMWL) used to $\delta^2H = 8\ \delta^{18}O + 10$ an ordinary least squares regression. The primary source δ^2H and $\delta^{18}O$ data for GMWL is Global Network of Isotope in Precipitation (GNIP) globally [3]. For the local precipitation data is called Local Meteoric Water Line (LMWL).

Local Meteoric Water Line (LMWL) is a linear correlation of δ^2H and $\delta^{18}O$ contents which observed in the local area (Wonosobo precipitation). LMWL can be used to evaluate the origin of waters. For example: if the values of a pair of isotopes ratio determine from groundwater samples plot in the range of the Meteoric Water Line (MWL), it can be concluded that groundwater was recharged by local precipitations (originate from meteoric water).

Local Meteoric Water Line (LMWL) is often established from analysis of monthly rainwater samples in an area. The LMWL is used to a variety of hydrological application, commonly to determine the relationship of surface or groundwater to a potential precipitation source or to determine the degree of evaporation enrichment of the water [2].

The analysis of groundwater recharge area can't be separated from the calculation of LMWL because LMWL will be a basis whether the sample water is from rainwater/ surface water or groundwater.

The aims of the determination of LMWL equation is as a reference to know any hydraulic interconnection between groundwater samples from PT. Aqua Danone spring water and surrounding springs water. LMWL equation can be used as a reference in natural isotope research to study of groundwater dynamic in hydrology cycle and conduct mapping of groundwater recharge area.

2 METHOD

This study proposed that the samples examined were rainwater samples taken from different locations of elevation between Jembangan volcanoes, Dieng volcanoes, and Sundoro volcanoes. Samples of rain water were taken during the rainy season in five different locations with different elevations representing different characteristics of rainfall and the rainwater isotope fractionation process in the west slopes of the Sundoro volcano. Samples at different elevations are expected to know how the influence of elevation on changes in isotopic composition in rainwater. In each elevation were taken three times of sampling when high rainfall, medium rainfall, and low rainfall, so that the different effects of rainfall on isotopic compositions are easily observed.

Analysis of isotopic $^2H/^1H$ ($D/^1H$) and $^{18}O/^{16}O$ isotope measurements were performed at the Earth and Environmental Laboratory, PATIR-BATAN, using a mass spectrometer. To minimize the occurrence of evaporation and contamination, the bottle cap of the sample is isolated by water tape insulation and fixing the time interval of the sampling to the laboratory analysis process no later than 5 days after sampling.

Measurement and analysis of the $^2H/^1H$ and $^{18}O/^{16}O$ isotope ratios are used to assess the effect of elevation and rainfall on the isotopic composition of the rainwater. Subsequently, the equations resulting from the study of the influence of elevation on the isotope composition ($\delta^{18}O$, δD) of rainfall that falls on the West Slope of Sundoro volcanoes are called the Local Meteoric Water Line (LMWL)-Wonosobo equation. The origin of groundwater can be estimated from the equations formed on the influence of elevation on ^{18}O isotope ratio.

The value of $D/^1H$ and $^{18}O/^{16}O$ isotope ratio from rain water samples is expressed relative to SMOW and written with the notation (δD‰) and ($\delta^{18}O$‰) using equation [4,5]:

$$\delta = \frac{Rsmpl - Rstd}{Rstd} x1000\,‰ \quad (1)$$

$$\delta D.‰ = \frac{\left[\left(D/_H\right)smpl - \left(D/_H\right)smow\right]}{\left(D/_H\right)smow} x1000 \quad (2)$$

$$\delta^{18}O.‰ = \frac{\left[\left(^{18}O/_{^{16}O}\right)smpl - \left(^{18}O/_{^{16}O}\right)smow\right]}{\left(^{18}O/_{^{16}O}\right)smow} x1000\ ‰ \quad (3)$$

With;

R_{smpl} = isotope ratio ($D/^1H$ or $^{18}O/^{16}O$) of water sample

R_{std} = isotope ratio ($D/^1H$ or $^{18}O/^{16}O$) sea water

δ = isotope ratio (D or ^{18}O) of water sample relative to the SMOW, ‰

δD = isotope ratio $D/^1H$ water sample relative to SMOW, ‰

$\delta^{18}O$ = isotope ratio $^{18}O/^{16}O$ water sample relative to SMOW, ‰

Standard mean ocean water (SMOW) is a standard for calibrating the composition of isotope ratio in the water sample. SMOW is from International Atomic Energy Agency (IAEA).

Based on the consideration that the largest of evaporation process occurs in the ocean, then the seawater is served as reference or SMOW. The isotope ratio of Deuterium ($D/^1H$) and Oksigen-18 standard ($^{18}O/^{16}O$) is 0.00015576 and 0.0020052 (**Table. 1**).

Based on this convention then most water (other than sea-water) has an isotope ratio δD and $\delta^{18}O$ <0 ‰ (negative value) or based on Equations 2 and 3 above will apply [6]:

a) if R_{smpl}> R_{std} then the value of δ_{smpl}> 0, means that the D or ^{18}O in the water sample is richer (enriched) than SMOW,
b) if R_{smpl} <R_{std} then the value of δ_{smpl} <0, means that the D or ^{18}O level in the water sample is poorer (depleted) than SMOW, and
c) if $R_{smpl} = R_{std}$, then $\delta_{smpl} = 0$, means that the D or ^{18}O in water = SMOW or also can be interpreted that water sample comes from sea water.

Rainwater sample has the ratio of δD and $\delta^{18}O$ depleted than ratio δD and $\delta^{18}O$ of sea water due to the higher of elevation, the slower of temperature, vapor will be condensed again. Rainwater that falls on this higher elevation has a lower δD and $\delta^{18}O$ isotope ratio. This phenomenon causes groundwater from higher elevation has δD and $\delta^{18}O$ smaller than groundwater from the lower elevation.

Table 1. Isotope ratio of SMOW [6]

	Ratio, H/L (=R_{std})	H/L	% H	% L
(1)	(2)	(3)	(4)	(5)
Standard Mean Ocean Water (SMOW)	$D/^1H$	0.0001557	0.015574	99.98442
	$^{17}O/^{16}O$	0.0003799	0.0379	99.76206
	$^{18}O/^{17}O$	0.0020052	0.002005	99.76206

Notes: H = Heavy isotopes, and L = Light isotopes

Besides the elevation influence, δD and $\delta^{18}O$ value also depend on rainfall. The higher the rainfall, δD and $\delta^{18}O$ ratio of rainwater are getting depleted and the lower the rainfall, the more enriched. Due to the rainfall, pressure and air temperature, and air humidity always fluctuated, so the value of δD and $\delta^{18}O$ in every sample location should be calculated with this equation [4,5]:

$$\delta^{18}O,\ \text{‰} = \frac{\sum_{i=1}^{n} P_i \delta_i{}^{18}O}{\sum_{i=1}^{n} P_i} \quad \ldots\ldots\ldots\ldots\ldots (4)$$

$$\delta^{2}H,\ \text{‰} = \frac{\sum_{i=1}^{n} P_i \delta_i D}{\sum_{i=1}^{n} P_i} \quad \ldots\ldots\ldots\ldots\ldots (5)$$

with:

$\delta^{18}O$ = average isotope ratio $^{18}O/^{16}O$ relative to SMOW in rainwater, ‰

δD = average isotope ratio $D/^1H$ relative to SMOW in rainwater, ‰

$\delta_i{}^{18}O$ = isotope ration $^{18}O/^{16}O$ relative to SMOW in rainfall to(i), ‰

$\delta_i D$ = isotope ration $D/^1H$ relative to SMOW in rainwater to (i), ‰

P_i = amount of rainfall between the sample (i-1) to (i), mm per month

The value of δD and $\delta^{18}O$ in some rainwater samples is linear and follow the equation 6. The research showed that for un evaporated rainwater, the 6 depends on geographical factors, particularly with regard to temperature, humidity, thickness, and precipitation. From rainwater samples taken globally through 91 rainfed stations worldwide, d = 10, so it can be written as the Global Meteoric Water Line (GMWL). [4,5,6] :

$$\delta D,\ \text{‰} = a \times \delta^{18}O + d\ \text{‰} \quad \ldots\ldots\ldots\ldots\ldots (6)$$

and for analysis of elevation influence to $\delta^{18}O$ can be obtained from plotting graph of $\delta^{18}O$ ratio vs elevation, that will allow this linear equation [7];

$$\delta^{18}O\ \text{‰} = \alpha * \text{elevation} - \beta \quad \ldots\ldots\ldots\ldots\ldots (7),$$

or

$$E = c * \delta^{18}O\text{‰} + d\ \text{‰}, \quad \ldots\ldots\ldots\ldots\ldots (8)$$

Based on equation (8), if we have the ratio of δ18O groundwater sample data, we can estimate the elevation of origin or groundwater recharge area.

This study used 5 (five) rainwater samples that have taken from Sindoro volcanoes with different significant elevation. Each location is taken samples for three times at significant different of rainfall i.e at January, February, and March 2017, so that the different effects of rainfall on the isotopic compositions are easily observed. In order to use rainfall data from Meteorology, Climatology and Geophysics Agency (MCGA), the location of rainwater sampling is strived to be within the coverage radius of BMKG nearest rainfall monitoring station. Some of the factors considered in water sampling include:

a) The sampling location has significant elevation differences so that the effect of elevation differences on the isotopic composition is easily observed,
b) The sampling location is still within the radius of the coverage area of the rainfall monitoring station so that the rainfall data in MCGA station can be used as secondary data to take into account the effect of rainfall on the isotope composition,
c) The accessibility of the sampling sites so that the rainwater samples can be taken easily, and
d) Time of rainwater sampling is 3 × which conducted in the month where the rainfall has significant difference so that the difference of rainfall effect to the isotope composition is easily observed.

3 Result and Discussion

By considering determination of rainwater samples location, then the coordinate data of rainwater samples in Wonosobo are shown in **Table 2**. The coordinate of rainwater samples is plotted on location map (see **Fig. 1**).

 https://doi.org/10.1051/e3sconf/20184301013

Fig. 1. Location map of rainwater samples

E-1AH is Banaran, Kayugiyang village, Garung, E-2AH is Dsn. Kalilang, Kayugiyang village, Garung. E-3AH is Glebengan village, Garung, E-4AH is Blederan village, Mojo Tengah, and E-5AH is Kejiwan village.

Table 2. Coordinates of rainwater samples

code	*Latitude*	
	LS	BT
(2)	(4)	(5)
E-1AH	07^0 17.978"	109^0 57.328"
E-2AH	07^0 17.978"	109^0 56.987"
E-3AH	07^0 18.630"	109^0 56.086"
E-4AH	07^0 18.727"	109^0 54.823"
E-5AH	07^0 20.931"	109^0 54.019"

Rainwater samples are collected in 120 ml bottle (**Fig. 2**) then analyzed with Liquid Water Stable Isotope Analyzer (LWSIA-LGR DLT-100).

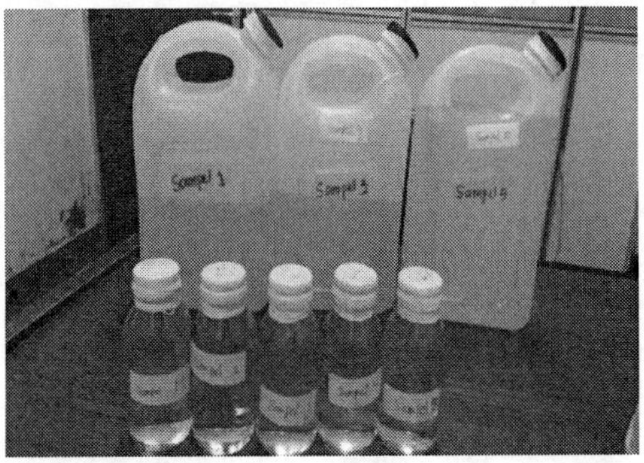

Fig. 2. Rainwater samples

The calculation of rainwater isotope ratio relative to SMOW is shown in **Table 3** and **Table 4**.

Due to the monthly rainfall is fluctuating and theoretically, this phenomenon affects the isotope ratios of δD and $\delta^{18}O$. This is evident from the measurement of the ratios of rainwater isotopes taken at the same elevation but sampling in different months (see **Table 3** and **Table 4**). Theoretically, the higher the rainfall the more depleted of isotope ratio of δD and $\delta^{18}O$ and vice versa the lower the rainfall the more enriched of isotope ratio of δD and $\delta^{18}O$. Therefore, the isotope ratios of δD and $\delta^{18}O$ of water samples in **Table 3** and **Table 4** are then compared using Eq. 4 and Eq. 5 in order to know the effect of the average monthly rainfall value in each elevation.

Table 3. Isotope ratio of δD in rainwater samples relative to SMOW

No	Location code	Elevation, m.asl	δD, ‰		
			Jan 2017	Feb 2017	Mar 2017
(1)	(2)	(3)	(4)	(5)	(6)
1	E-1AH	1474	−51.88	−46.76	−49.85
2	E-2AH	1365	−49.99	−45.81	−47.85
3	E-3AH	1126	−46.05	−41.26	−43.12
4	E-4AH	935	−41.69	−38.73	−42.18
5	E-5AH	744	−40.75	−31.12	−40.08

Source: primary data

Table 4. Isotope ratio of $\delta^{18}O$ in rainwater samples relative to SMOW

No	Location code	Elevation, m.asl	$\delta^{18}O$, ‰		
			Jan 2017	Feb 2017	Mar 2017
(1)	(2)	(3)	(4)	(5)	(6)
1	E-1AH	1474	−8.91	−6.52	−8.12
2	E-2AH	1365	−8.69	−6.24	−8.01
3	E-3AH	1126	−8.15	−6.01	−7.12
4	E-4AH	935	−7.79	−5.71	−6.83
5	E-5AH	744	−7.41	−5.41	−6.29

Source: primary data

In this study, the rainfall data at the sampling site is taken from the result of rainfall record in 2005-2009 at 26-B Monitoring Station, 24-A Monitoring Station, and 24-F Monitoring Station located in Kali Serayu River Basin and surrounding. Because rainwater sampling in each location is done in January, February, and March 2017, the rainfall data used is rainfall data in January, February, and March which is sterilized from MCGA rainfall data in 2005-2009, as presented in **Table 5**.

From **Table 3** and **Table 4** shows that the value of precipitation or rainfall affects the isotopic ratios contained in rainwater. At the same location, the higher rainfall will be the greater dilution of the deuterium and oxygen-18 heavy isotopes in the rain, thus the ratio of δD and $\delta^{18}O$ is also getting depleted. For example, in Banaran Village, Kayugiyang Village, Garung District (E-1AH) which is the highest rainwater sampling location (1474 m.asl), the highest rainfall occurred in January, i.e. 514.20 mm per month, has an isotope ratio in the depleted rain water, i.e. δD = - 51.88 ‰ and $\delta^{18}O$ = - 8.91 ‰, whereas the lowest rainfall occurred in February, i.e. 480.20 mm per month; at the same elevation (1474 m.asl) the ratio of isotopes in rainwater is more enriched, i.e. δD = -46.76 ‰ and $\delta^{18}O$ = -6.52 ‰. The same phenomenon also occurs at other sampling sites, the higher the rainfall the more depleted the isotope ratios. Therefore, to obtain a representative isotope ratios value representing the actual value at a certain elevation, the effect of rainfall variation on the isotope ratios must be calculated using Equations 4 and 5. The results of the calculated values of the isotope ratios of δD and $\delta^{18}O$ in rainwater samples taking into account the variations in rainfall during sampling are presented in **Table 6**.

Table 5. Average monthly rainfall data in 5 sampling sites.

Location code	Elevation (m.asl)	rainfall (P_i), mm		
		Jan'17	Feb'17	Mar'17
(2)	(3)	(4)	(5)	(6)
E-1AH	1474	514.20	480.20	485.40
E-2AH	1365			
E-3AH	1126	456.80	408.20	474.40
E-4AH	935			
E-5AH	744	598.40	643.20	594.00

Table 6. Isotope ratio data of rainwater samples as the funtion of elevation

No.	Location Code	Elevation (m.asl)	Average isotope compotition	
			δD, ‰	$\delta^{18}O$, ‰
(1)	(2)	(3)	(4)	(5)
1	E-1AH	1474	−49.55274	−7.8753
2	E-2AH	1365	−47.93234	−7.6719
3	E-3AH	1126	−43.55726	−7.1329
4	E-4AH	935	−40.96462	−6.8160
5	E-5AH	744	−37.15668	−6.3467

Based on the data of deuterium and oxygen-18 isotope ratios in **Table 6**. a graph of $\delta^{18}O$ versus δD was created to obtain the Local Meteoric Water Line (LMWL) curve in Wonosobo District. **Fig. 3** shows the LMWL curves of Wonosobo District and the Global Meteoric Water Line (GMWL) as a comparison.

LMWL Wonosobo equation can be obtained with the least square regression method form data are shown in Figure 3. The LMWL Wonosobo equation is:

$$\delta D, ‰ = 8.102 * \delta^{18}O + 14.21‰ \quad \text{............} (9)$$

with determination coefficient R^2 = 0.998

The LMWL-Wonosobo equation obtained in this study is not much different with Dansgaard research [8], the global meteoric water line (GMWL) that follows Eq. 10.

$$\delta D, ‰ = 8\,\delta^{18}O + 10‰ \quad \text{.................}(10)$$

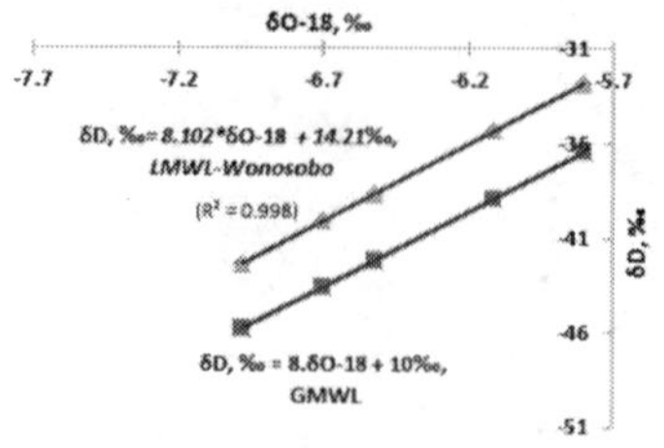

Fig. 3. LMWL Wonosobo and GMWL curve

The difference between LMWL and GMWL is caused by geographical and regional conditions. Geographical differences will have the different temperature, humidity, elevation, and rainfall. The GMWL equation was obtained from the analysis of the average isotope composition of the rainwater samples from various countries in the world that have different climates, whereas LMWL-Wonosobo was obtained based on analysis of isotope rainwater sampling composition taken from the Western Slope of Gunungapi Sundoro located in the tropical area.

The rainwater sample was taken in Kejiwan Hamlet at the lowest elevation (E-5AH; 744 m.asl) from 5 sampling sites has a more enriched isotope ratios, i.e. δD = -37.15668 ‰ and $\delta^{18}O$ = -6.3467 ‰; and on the other hand the rainwater at a high elevation in Banaran Hamlet, Kayugiyang Village, Garung Sub-district (E-1AH: 1474 m.asl) has more depleted isotope ratios, δD = -49.55274 ‰ and $\delta^{18}O$ = -7.8753 ‰ (**Table 6**);

The results of this measurement correspond to the theory that the higher elevation the depleted isotope ratio of δD and $\delta^{18}O$ and vice versa. If the data in Table 6 is made a graph of the isotope ratio $\delta^{18}O$ as an elevation function (E), then the graph will be obtained as shown in Figure 4.

Determining of the isotope ratio δD to the water sample as an elevation function is calculated by entering the value $\delta^{18}O$ into the LMWL-Wonosobo equation. Thus the δD and $\delta^{18}O$ isotope composition of the rainwater at each coordinate point located along the LMWL-Wonosobo line carries the elevation information of the rainwater fall location. If the isotopic composition of a water sample taken from the springs and coordinate wells coincides or close to the LMWL line then it is estimated that the water samples are mostly derived from meteoric water (rainwater) [9]. Furthermore, in this study Equations 9, 11 and 12 will be used as a reference in determining the origin of groundwater samples.

The LMWL was based upon a weighted regression using only rain samples. The weighted LMWL is therefore recommended as a more appropriate hydrological input function. We use the weighted LMWL in the following section to approximate the input or source water composition in various reservoirs by regression to their intersection point with the LMWL [10].

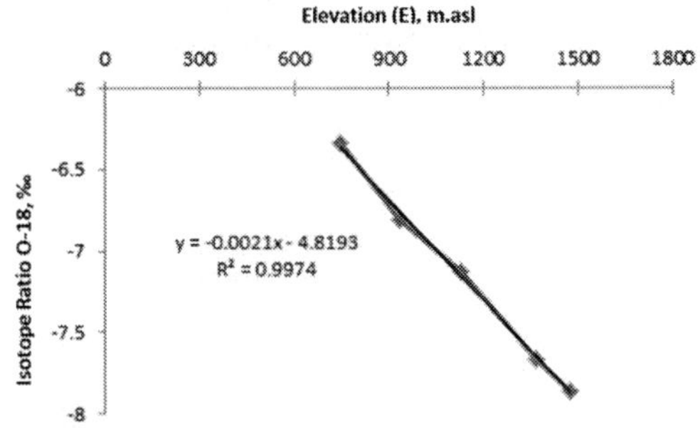

Fig. 4. Elevation versus isotope ratio of $\delta^{18}O$ in rainwater sample

The higher the elevation, the isotope ratio of $\delta^{18}O$ in rainwater is increasingly depleted (Fig. 4), following the following equation:

$\delta^{18}O$ ‰ = -0.0021*Elevation – 4.8193‰, . (11)

With determination coefficient, R^2=0.9974

or, the equation also can be written as follow:

E = -476.190*$\delta^{18}O$‰ – 2294.9047‰, ... (12)

Based on equation (12), if we have the ratio of $\delta^{18}O$ groundwater sample data, we can estimate the elevation of origin or groundwater recharge area. For example, PT. Tirta Investama has Mangli springwater in 720 m.asl which has an isotope ratio $\delta^{18}O$ -7.75 ± 0.2 ‰, with input it to equation 12, the original groundwater elevation as follow:

$E_{\text{Mangli springwater}}$ = -476.190*(-7.75± 0.2) ‰ – 2294.9047‰ .... (13)

$E_{\text{Mangli springwater}}$ = 1395.5 m.asl until .... (14)

$E_{\text{Mangli springwater}}$ = 1509.8 m.asl .... (15)

3 Conclusion

The equation of LMWL Wonosobo is δD, ‰ = 8.102 * $\delta^{18}O$ + 14.21‰ and the estimation of PT Tirta Investama Danone Aqua spring elevation based on elevation equation to the ratio of $\delta^{18}O$ is 1395.5 m.asl - 1509.8 m.asl.

4 Acknowledgments

We would like to thank PT. Tirta Investama Danone Aqua for supporting this research activity, and to the student for assisting of water sample collection. The manuscript was improved by the critical comments of two anonymous reviewers.

References

1. N. Salonga, *World Geothermal Congress* (2015)
2. J. Crawford, C.E. Hughes, S. Lykoudis. J of Hydrology **519**,2331-2340 (2014)
3. IAEA/WMO. *Global Network of Isotopes in Precipitation* (2013)
4. N. Kresic and Z. Stevanovic, Butterworth-Heinemann, **573**, ISBN 978-1-85617-502-9 (2010)
5. I. Clark and P. Fritz, *Lewis Publishers, Boca Raton* (1997)
6. H. Craig. *The Standart for reporting concentrations of deuterium and oxygen-18 in natural water.* Science, 133, 1833-1934, (1961a)
7. A.B. Wijatna, H. Hendrayana, et all, *Report on Hydro-isotope Study in PT Tirta Investama Danone Aqua Plant Wonosobo* (2017)
8. W. Dansgaard, Phys.Lab.11. H. C. Orsted Institure (1964)
9. J.I. Drever, The *Geochemistry of Natural Water,* Pearson Edu.Canada, (1988)
10. T. Baer, S.L. Barbour, J.J. Gibson., J of Hydrology **541**, 1155 – 1164 (2016)

https://doi.org/10.1051/e3sconf/20184301014

Implementation of Maximum Power Point Tracking (MPPT) Technique on Solar Tracking System Based on Adaptive Neuro-Fuzzy Inference System (ANFIS)

Imam Abadi[1*], *Choirul* Imron[2], *Mardlijah*[2], *Ronny* D. Noriyati[1]

[1]Department of Engineering Physic, Institut Teknologi Sepuluh Nopember Surabaya, Indonesia
[2]Department of Mathematic, Institut Teknologi Sepuluh Nopember Surabaya, Indonesia

Abstract. Characteristic I-V of photovoltaic is depended on solar irradiation and operating temperature. Solar irradiation particularly affects the output current where the increasing solar irradiation will tend to increase the output current. Meanwhile, the operating temperature of photovoltaic module affects the output voltage where increasing temperature will reduce the output voltage. There is a point on the I-V curve where photovoltaic modules produce maximum possible output power that is called Maximum Power Point (MPP). A technique to track MPP on the I-V curve is known as Maximum Power Point Tracking (MPPT). In this study, the MPPT has been successfully designed based on Adaptive Neuro-Fuzzy Inference System (ANFIS) and integrated with solar tracking system to improve the conversion efficiency of photovoltaic modules. The designed ANFIS MPPT system consists of current and voltage sensors, buck-boost converter, and Arduino MEGA 2560 microcontroller as a controller. Varying amounts of lamp with 12V 10W rating arranged in series is used as load. Solar tracking system that is equipped with MPPT ANFIS able to increase the output power of photovoltaic modules by 46.198% relative to the fixed system when 3 lamps is used as load.

1 Introduction

The main problem found in the Solar Power Generation System nowadays is low conversion efficiency of photovoltaic module. One way to improve power production from photovoltaic modules is to have it equipped with a solar tracking system. Solar tracking system will keep phovoltaic surface oriented toward sun, allow the module exposed to higher amount of solar irradiation, to produce maximum power [1,2].

Characteristics of a photovoltaic cell is expressed by current versus voltage curve (I-V curve) and power versus voltage curve (P-V) that is influenced by solar irradiation level and temperature of photovoltaic module. There is a particular point on a I-V curve where photovoltaic will produce highest possilble power output called maximum power point (MPP)[3,4]. Process of finding MPP to maximize power extraction is called maximum power point tracking (MPPT). The aim of this study is to develop a MPPT for photovoltaic system equipped with solar tracking system based on *Adaptive Neuro-Fuzzy Inference System* (ANFIS).

1.1 Photovoltaic

Photovoltaic is a semiconductor device that exhibit photovoltaic effect that convert sunlight energy into electrical energy.

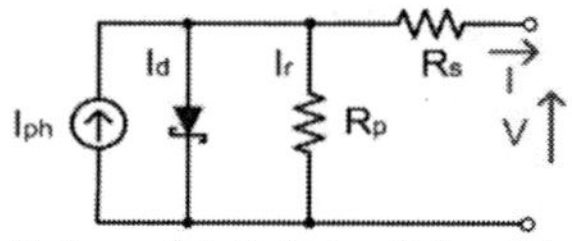

Fig. 1. Single-diode model of photovoltaic cell including parallel and series resistances. [4].

The equation that describe the single-diode model presented in Fig. 1 is

$$I = I_{pv} - I_o\left\{exp\left[\frac{V + IR_s}{aV_t}\right] - 1\right\} - \frac{V + IR_s}{R_p} \quad (1)$$

$$V_t = \frac{N_s kT}{q} \quad (2)$$

where :
I_{pv} is photovoltaic current (A),
I_o is saturation current (A),
q is electron charge ($1{,}602\times10^{-19}$ C),
I is current at the terminal of photovoltaic (A),
V is voltage at the terminal of photovoltaic (V),
V_t is thermal voltage (V),
k is Boltzmann's constant ($1{,}381\times10^{-23}$ J/K),
T is module temperature (K),
a is diode constant,
R_s is equivalent series resistant of photovoltaic *array* (Ohm),
R_p is equivalent parallel resistant of photovoltaic *array* (Ohm),
Ns is series connected photovoltaic cell,

* Corresponding author:imam@ep.its.ac.id

Characteristic curve of photovoltaic is depended on solar irradiation and operating temperature. Solar irradiation particularly affects the output current where increasing solar irradiation will tend to increase the output current. Meanwhile, operating temperature of photovoltaic module affects the output voltage where increasing temperature will reduce the output voltage [5].

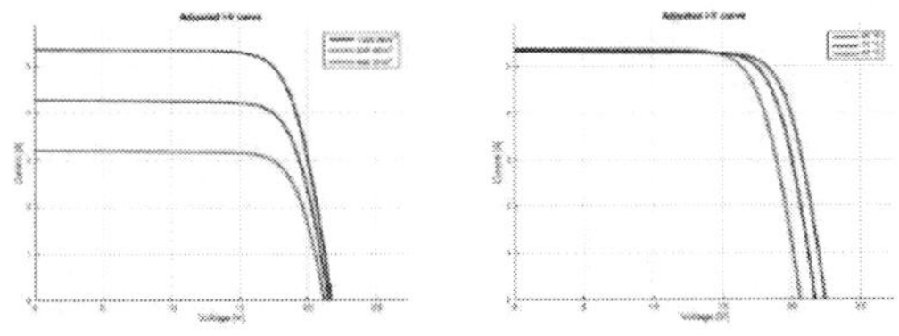

Fig. 1. Effect of solar irradiation and temperature of photovoltaic module on I-V curves[5].

1.2 Maximum Power Point Tracking (MPPT)

Maximum power point tracking is tracking method of MPP to obtain *maximum possible power* from photovoltaic during daylight. The goal of the MPPT is to match the equivalent resistance at the terminal of photovoltaic R_{eq} to the optimal output resistance Ropt that is defined as [5]

$$R_{opt} = \frac{V_{mpp}}{I_{mpp}} \tag{3}$$

When $R_{eq} = R_{opt}$ condition met, the MPP is obtained thus maximum possible power will be produced by photovoltaic modules. Process for matching R_{eq} toward R_{opt} to obtain MPP is illustrated in Fig. 3. Slope of the straight line is the representation of R_{eq} value. The intersection of straight line R_{eq} and I-V curve is operating point of photovoltaic. MPPT will alter this operating point toward MPP. In general, MPPT is consist of a DC-DC converter, a controller, and sensors. Duty cycle of the converter is used as a control variable to change the R_{eq} value [6].

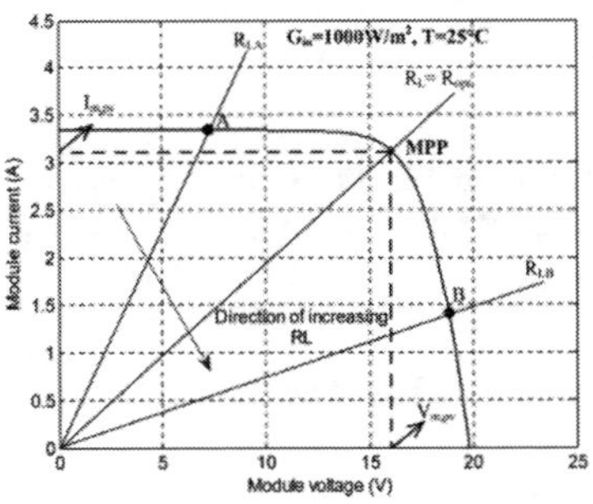

Fig. 2. Ilustration of MPP tracking on the I-V curve[5].

Open-voltage method is a MPPT technique based on observation that maximum power point voltage V_{MPP} has a fixed ratio to open-circuit voltage V_{OC} [6-10] that is defined as

$$\frac{V_{mpp}}{V_{oc}} \cong k \ ; \ k < 1 \tag{4}$$

Where the constant k is found to be between 0.7 – 0.8[11]. Flowchart diagram of open-voltage method is shown on Fig. 5.

1.3 Buck-Boost Converter

B*uck-boost* converter is a type of DC-DC converter that outputs voltage either less or greater than the input voltage. Relationship between input voltage Vi, output voltage Vo, and *duty cycle* D for *buck-boost* converter is stated as:

$$V_o = -\frac{D}{1-D}V_i \tag{5}$$

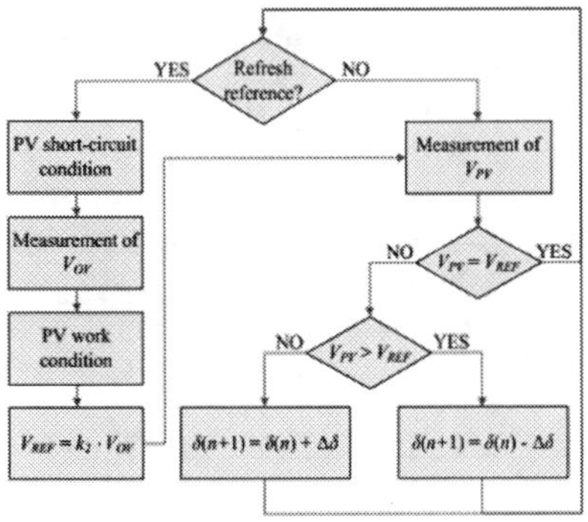

Fig. 5. Flowchart of *open-voltage method*[6].

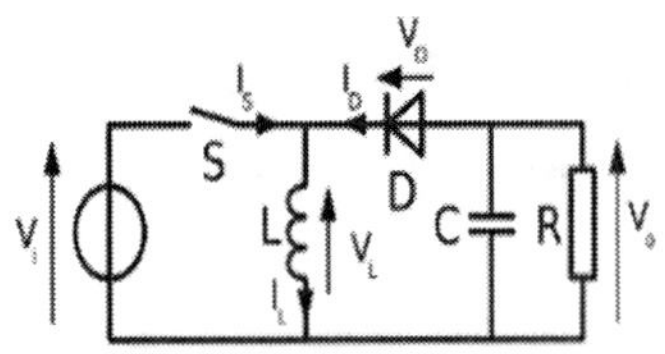

Fig. 6. Buck-boost converter.

1.4 Adaptive Neuro-Fuzzy Inference System (ANFIS)

ANFIS is a kind of adaptive networks that incorporate both Takagi-Sugeno Kang *Fuzzy Inference System (FIS)* and artificial neural network [12]. ANFIS structure is consisted of five layers represent artificial neural network architecture as illustrated in Fig. 7. The square nodes represents an adaptive parts while the circle nodes represents non-adaptive sections. Paramaters of the adaptive nodes will be changed during the training process of ANFIS[9,10].

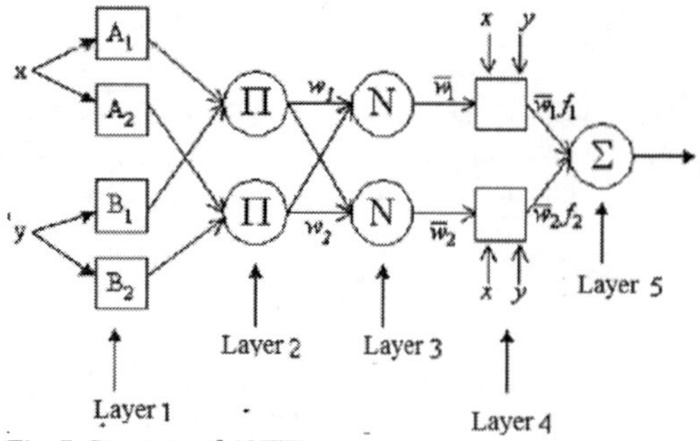

Fig. 7. Structure of ANFIS

2 Metodology

2.1 Block diagram of ANFIS MPPT

ANFIS MPPT developed in this study as displayed in Fig.8 is consist of buck-boost converter, ANFIS MPPT controller, and voltage-current sensor.

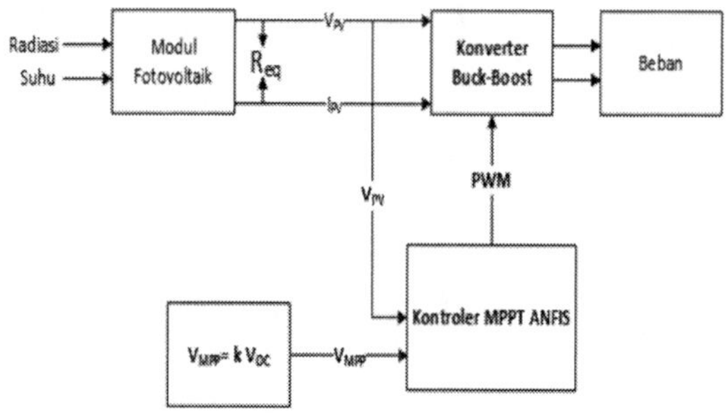

Fig. 8. Block diagram of ANFIS MPPT.

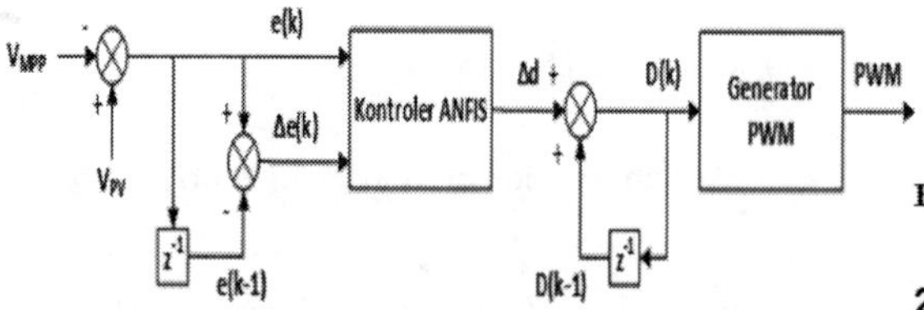

Fig. 9. Block diagram of ANFIS MPPT controller.

The proposed ANFIS MPPT is depicted in Figure 9. Error value e and change of error Δe is taken as the input for ANFIS MPPT controller defined as

$$e(k) = V_{pv}(k) - V_{mpp}(k) \quad (6)$$

$$\Delta e(k) = e(k) - e(k-1) \quad (7)$$

where $e(k)$ and $e(k-1)$ are current and previous error, respectively. Maximum power point voltage V_{MPP} is obtained using *Open-voltage Method.* Output of ANFIS MPPT is change of *duty cycle* $\Delta d(k)$ and duty cycle D value can be written as

$$D(k) = D(k-1) + \Delta d(k) \quad (8)$$

where $D(k)$ is current *duty cycle* and $D(k-1)$ is previous *duty cycle.*

2.2 Design of ANFIS

In this study, ANFIS is designed using MATLAB. The proposed ANFIS consists of five gaussian membership functions for each input as shown in Fig. 10 & 11. Moreover its output is singleton as shown in Fig.12.

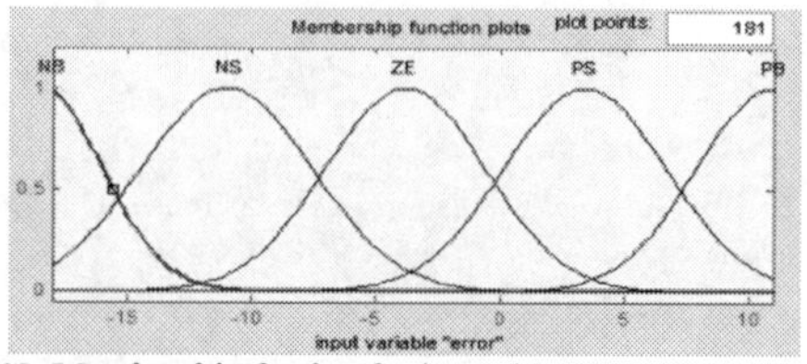

Fig. 10. Membership fuction for input (error)

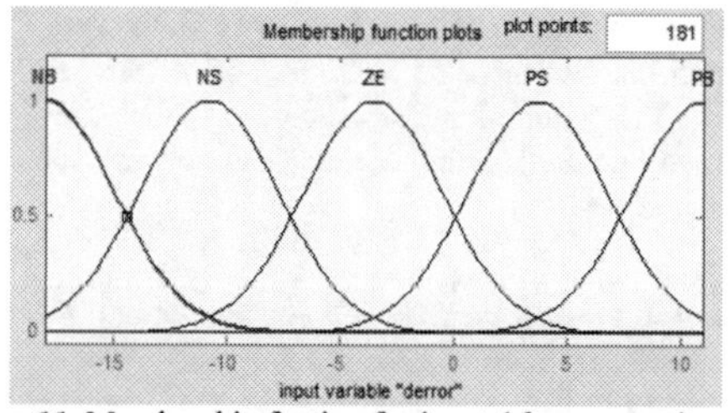

Figure 11. Membership fuction for input (change error)

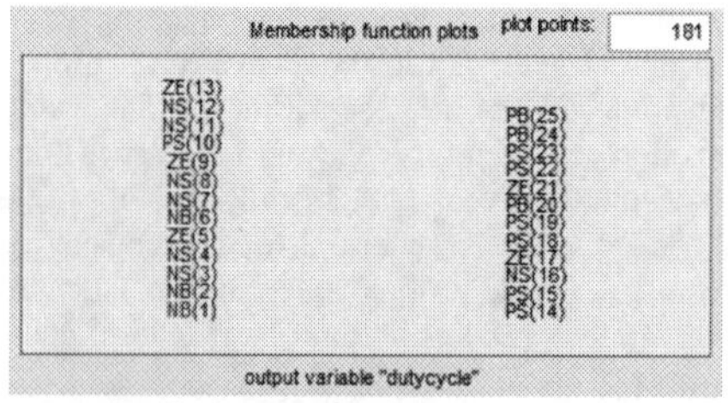

Fig. 12. Membersip function for output.

2.3 Simulation of ANFIS MPPT

Simulation of ANFIS MPPT is performed in PSIM 9.0 and MATLAB/Simulink. *Co-simulation* is carried out by implementing the ANFIS MPPT in *MATLAB/Simulink*, meanwhile photovoltaic module and buck-boost converter is partially run using PSIM 9.0 as seen in Figure 13.

 https://doi.org/10.1051/e3sconf/20184301014

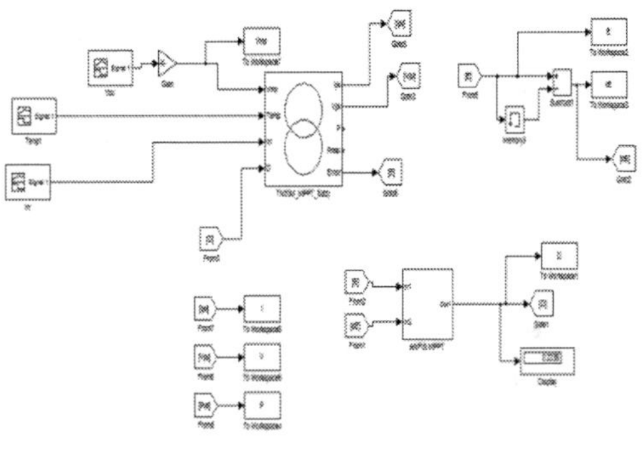

Fig. 13. ANFIS MPPT model in MATLAB/Simulink.

Table 1. Photovoltaic module specification.

Model	TN-20M
Maximum Power at STC (P_{MPP})	20 W
Maximum Power Voltage at STC (V_{MPP})	17,2 V
Maximum Power Current at STC (I_{MPP})	1,16 A
Open Circuit Voltage at STC (V_{OC})	21,5 V
Short Circuit Current at STC (I_{SC})	1,25 A
Temperature Coefficient of V_{OC}	-0,36 %/°C
Temperature Coefficient of I_{SC}	0,05 %/°C
Series Connected Cell per Modul	36

Table 2. buck-boost konverter specification [15].

Parameter	**Quantities**
Vin	9 – 22 V
Vout (nominal)	14 V
Iout	1,5 A
Switching Frequency	25 kHz
Inductor	193 μH
Output Capacitor	470 μF, 50V
Nominal resistive load	8 Ω

2.4 Realization of ANFIS MPPT

ANFIS MPPT hardware developed in this study is shown in Figure 14. In general, prototype is divided into several subsystems for the ease of realization. INA219 is used as current-voltage sensor and Arduino MEGA 2560 is functioned as controller. Specification of the buck-boost converter is shown in Table 2.

Figure 14. ANFIS MPPT realization.

3 Result and discussion

3.1 ANFIS MPPT Simulation Result

ANFIS MPPT is tested and simulated with varying climatic condition. Some parameters involved are shown in Table 3. Value of V_{OC} dan V_{MPP} of photovoltaic module as the effect of varying climatic condition is shown in Table 4.

Table 3. Climatic condition during simulation.

Condition	Temperature (°C)	Irradiation (W/m^2)
1	28	333
2	37	574
3	51	972
4	43	710
5	32	432

Table 4. Value of VOC and VMPP.

Condition	V_{OC} (V)	k	V_{MPP} (V)
1	20,28		15,82
2	20,12		15,69
3	19,59	0,78	15,28
4	19,88		15,51
5	20,22		15,77

This simulation is performed to determine the V_{MPP} tracking performance of ANFIS MPPT. The results as seen in Fig. 15 show that ANFIS MPPT has good performance in tracking V_{MPP} with varying climatic condition. Voltage fluctuation around V_{MPP} is the result of ANFIS controller yield excess control signal Δd for a small value of $e(k)$ and $\Delta e(k)$.

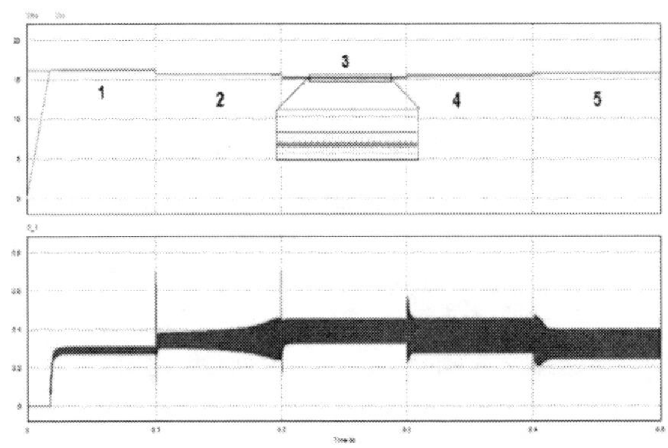

Fig.15. Result of V_{MPP} tracking by MPPT ANFIS (above: voltage; below: *duty cycle*)

 https://doi.org/10.1051/e3sconf/20184301014

Table 5. Comparison of PMPP dan P MPPT

Condition	Parameter	Power (W)
1	P_{MPP}	6.73
	P MPPT	6.61
2	P_{MPP}	11,21
	P MPPT	11,04
3	P_{MPP}	17,51
	P MPPT	17,50
4	P_{MPP}	13,45
	P MPPT	13,33
5	P_{MPP}	8.68
	P MPPT	8.45

Based on Table 5, it can be known that V_{MPP} value predicted by *open-voltage method* is close to actual V_{MPP} proven by P MPPT has a small deviation from actual P_{MPP}.

3.2 ANFIS MPPT Testing on VMPP Tracking

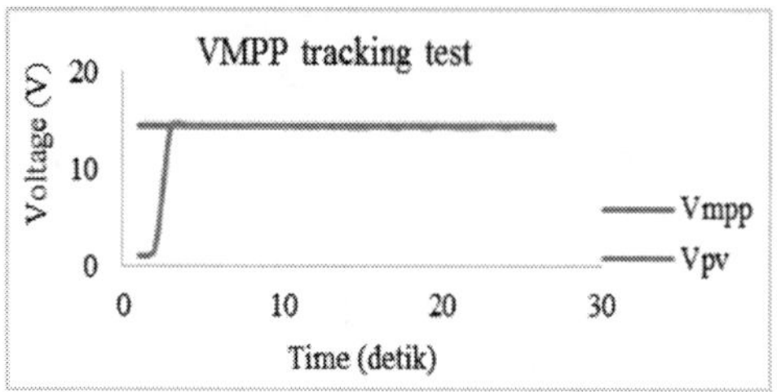

Fig. 16. Result of V_{MPP} tracking using MPPT ANFIS hardware

The result of the experiments show that ANFIS MPPT prototype has good performance on V_{MPP} tracking. The prototype is able to track setpoint with execution time < 5 seconds from short-circuit condition for a given V_{MPP} as displayed in Fig. 16.

3.3 ANFIS MPPT without Solar Tracker System

This experiment is conducted to compare power produced between ANFIS MPPT and non-MPPT system. There are three variabels measured namely Pout non-MPPT, Pin MPPT and Pout MPPT. Pout non-MPPT is power that directly delivered to the load. Pin MPPT is power obtained in the input side of buck-boost converter, moreover Pout MPPT is power obtained in the ouput side of buck-boost converter. Load used for experiment is 12V 10W incandescent lamp. The Experiments are conducted on August 3th, 2017.

3.3.1 Experiment I

This experiment is performed using 2 lamps as load arranged in series for each system at 11.00 – 13.30.

Table 6. Output power comparison for experiment I

No.	MPPT				non-MPPT	Pout Increase (%)
	Duty Cycle	PWM	Pin (W)	Pout (W)	Pout (W)	
1.	0,54	137	**11,69**	**9,08**	**8,5**	6,823529
2.	0,55	138	**12,54**	9,84	8,91	10,43771
3.	0,54	137	**11,65**	9,2	9,07	1,433297
4.	0,55	140	**12,61**	9,6	9,22	4,121475
5.	0,55	139	**12,53**	9,92	8,71	13,89208
6.	0,54	138	**11,94**	9,4	8,55	9,94152
7.	0,54	138	**12,31**	9,97	8,73	14,20389
8.	0,54	138	**12,19**	9,54	8,93	6,830907
9.	0,55	139	**12,53**	10,13	8,9	13,82022
10.	0,54	137	**12,28**	9,75	8,6	13,37209
11.	0,31	79	**2,32**	1,46	0,88	65,90909
12.	0,30	75	**1,99**	1,43	0,86	66,27907
13.	0,42	106	**4,78**	4,63	4,62	0,21645
14.	0,33	83	**2,84**	1,99	1,11	79,27928
15.	0,30	76	**2,3**	1,64	0,7	134,2857
16.	0,40	101	**4,59**	3,68	3,6	2,222222
17.	0,39	98	**4,18**	3,06	2,97	3,030303
18.	0,30	75	**2,04**	1,37	0,54	153,7037
19.	0,44	112	**6,46**	5,72	5,19	10,21195
20.	0,36	90	**3,75**	2,79	1,95	43,07692
21.	0,32	82	**2,8**	2,1	1,91	9,947644
22.	0,34	86	**3,04**	2,29	1,08	112,037
23.	0,32	80	**2,45**	1,83	1,28	42,96875
24.	0,33	83	**2,72**	1,96	1,93	1,554404
25.	0,33	83	**11,69**	**9,08**	**8,5**	6,823529
Average Pout Increase						**34,14997**

Experimental results show power delivered to the input side of buck-boost converter of ANFIS MPPT (Pin) is higher than that delivered to load in non-MPPT system. It indicated that ANFIS MPPT increases produced power from photovoltaic module. However, buck-boost converter used in ANFIS MPPT has efficiency of 70-80% thus power delivered in output side of converter is always lower than input side. According to Table 6, ANFIS MPPT system produced power gain around 34.14997 % relative to the non-MPPT system when 2 lamps were used as load.

3.3.2 Experiment II

This experiment is carried out using 3 lamps as load arranged in series for each system at 13.30 – 16.00.

Table 7. Output power comparison for experiment II.

No.	MPPT				non-MPPT	Pout Increase (%)
	Duty Cycle	PWM	Pin (W)	Pout (W)	Pout (W)	
1.	0,47	119	6,07	4,48	4,13	8,474576271
2.	0,48	123	7,1	5,69	4,22	34,83412322
3.	0,49	125	7,4	5,98	5,15	16,11650485
4.	0,46	117	5,9	4,61	3,75	22,93333333
5.	0,47	120	6,45	5,13	4,88	5,12295082
6.	0,49	123	7,17	5,83	4,45	31,01123596
7.	0,49	125	7,56	6,26	4,6	36,08695652
8.	0,52	132	8,98	7,12	6,68	6,586826347
9.	0,53	133	9,33	7,58	4,65	63,01075269
10.	0,52	133	9,29	7,54	4,63	62,85097192
11.	0,52	133	9,02	7,5	5,58	34,40860215
12.	0,51	130	8,65	7,34	5,57	31,77737882
13.	0,25	63	1,73	1,46	0,91	60,43956044
14.	0,46	116	6,91	5,98	4,32	38,42592593
15.	0,49	124	7,32	6,09	4,25	43,29411765
16.	0,27	67	1,69	1,41	0,56	151,7857143
17.	0,48	121	6,97	5,93	4,29	38,22843823
18.	0,44	112	5,69	4,89	3,85	27,01298701
19.	0,48	122	6,75	5,74	4,33	32,56351039
20.	0,42	106	4,63	4,11	3,51	17,09401709
21.	0,32	80	2,27	1,97	1,76	11,93181818
22.	0,32	80	2,07	1,8	1,08	66,66666667
23.	0,31	77	1,98	1,72	0,97	77,31958763
24.	0,31	78	1,75	1,42	1,05	35,23809524
25.	0,30	76	7,56	6,26	4,6	36,08695652
Average Pout Increase						**39,717277**

Table 8. Output power comparison for experiment III.

No.	MPPT				non-MPPT	Pout Increase (%)
	Duty Cycle	PWM	Pin (W)	Pout (W)	Pout (W)	
1.	0,58	146	10,43	8,76	6,28	39,49
2.	0,57	145	9,97	8,18	6,21	31,72
3.	0,58	146	10,64	8,80	6,11	44,03
4.	0,58	147	10,83	9,35	5,86	59,56
5.	0,58	146	10,61	8,58	6,30	36,19
6.	0,58	147	10,83	9,35	5,62	66,37
7.	0,58	146	10,77	9,49	5,51	72,23
8.	0,57	146	10,60	8,88	4,69	89,34
9.	0,57	145	10,48	8,66	7,07	22,49
10.	0,58	148	11,08	9,47	7,01	35,09
11.	0,58	148	10,93	9,32	7,14	30,53
12.	0,58	148	10,71	9,33	6,80	37,21
13.	0,58	148	10,79	8,83	6,97	26,69
14.	0,56	141	8,72	7,87	6,50	21,08
15.	0,56	143	9,37	8,15	6,71	21,46
16.	0,58	148	11,33	9,49	7,16	32,54
17.	0,59	151	11,73	9,56	6,39	49,61
18.	0,57	145	9,86	7,69	6,01	27,95
19.	0,58	147	10,24	7,78	6,18	25,89
20.	0,59	150	10,87	8,58	6,09	40,89
21.	0,59	149	10,92	8,85	5,79	52,85
22.	0,60	152	11,94	9,54	6,38	49,53
23.	0,60	152	11,76	9,92	6,30	57,46
24.	0,60	152	11,79	9,52	6,26	52,08
25.	0,60	153	12,23	9,84	6,23	57,95
Average Pout Increase						43,21

When using 3 lamps as load, power produced by non-MPPT system (P non-MPPT) is almost equal to input side power of MPPT system (Pin) as shown in Figure 17. MPPT system increases produced power by 39.717277 %. It is assumed that equivalent resistance R_{eq} of 3 lamps used is almost equal to optimal resistance R_{opt} such that it make operating conditon of non-MPPT system is near MPP.

3.3.3 Experiment III

In this case, the experiment provided 4 lamps as load arranged in series for each system. It was conduted at 09.00 – 11.00. Experiment result show that when 4 lamps are used as load, efficiency produced by ANFIS MPPT system is 43.21 % compared to the non-MPPT system. It can be known that operating condition of non-MPPT system has never reached MPP so that it produced less power.

3.4 ANFIS MPPT Based on Solar Tracker System

The experiment was operated using two 20 Wp PV modules as follows: Module 1 was equipped with solar tracking system and Module 2 was a fixed module. Loads used for the project consisted of 12V 10W incandescent lamps.

3.4.1 Experiment I

This experiment was conducted using only Module 1. There were two systems tested in this experiment i.e Solar Tracker with MPPT and without MPPT. *The e*xperiment was conducted on August 4th, 2017 at 09.30; 11.30; and 14.00. In this case, power gain was investigated to know the efficiency improvements.

Table 9. Average power increase of MPPT and non-MPPT system in solar tracker system.

Load	Pout Increase (%)			Pout average (%)
	09.30	11.30	14.00	
2 Lamps	7,81	22,17	5,26	11,75
3 Lamps	3,71	4,35	5,65	4,57
4 Lamps	18,81	10,55	16,42	15,26

Result obtained from this experiment can be seen in Table 9. It shows that ANFIS MPPT can increase power produced from photovoltaic equipped with solar tracking system. Increased average power for various load are 11.75% for 2 lamps, 4,57% for 3 lamps, and 15,26% for 4 lamps.

3.4.2 Experiment II

In this experiment there are two systems tested as demonstrated in Figure 17. First system is module 1 equipped with solar tracking system and ANFIS MPPT. The other one is fixed system and direct-coupled with loads. Load used in this experiment is 3 lamps connected

https://doi.org/10.1051/e3sconf/20184301014

in series. The experiment conducted on August 15th, 2017 at 11.00-13.00. There are three variabels measured namely Pout tracking, Pin tracking and Pout fixed. Pout tracking is power obtained in the input side of buck-boost converter and Pin tracking is power obtained in the ouput side of buck-boost converter. Both of them are measured and determined from the first system. Meanwhile Pout fixed is power that directly delivered to the load in the second system.

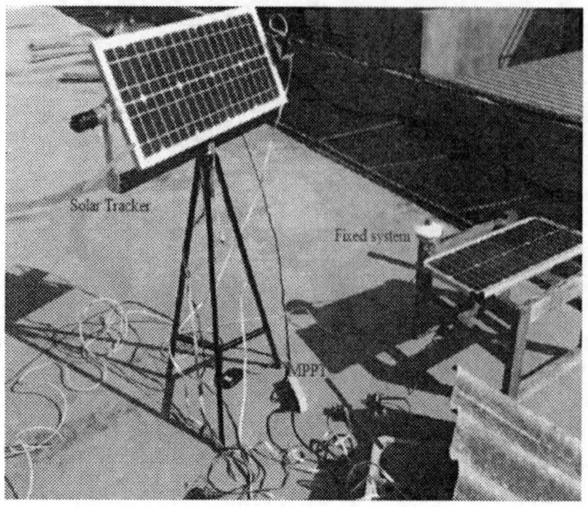

Fig. 17. Setup of Experiment II

Table 10. Power comparison of solar tracker system-ANFIS MPPT dan fixed non-MPPT system.

No.	Solar Tracker - MPPT ANFIS		Fixed	Pout Increase (%)
	Pin (W)	Pout (W)	Pout (W)	
1.	**11,65**	**10,67**	**7,84**	36,09694
2.	**9,31**	8,93	6,38	39,96865
3.	**12,64**	10,54	7,74	36,17571
4.	**14,03**	11,7	7,62	53,54331
5.	**10,47**	9,95	7,52	32,31383
6.	**10,98**	9,21	7,57	21,66446
7.	**13,62**	11,51	7,77	48,13385
8.	**13,32**	12,9	7,8	65,38462
9.	**15,1**	13,2	7,67	72,09909
10.	**12,94**	11,47	7,41	54,79082
11.	**13,57**	11,67	7,35	58,77551
12.	**12,96**	11,16	7,42	50,40431
13.	**3,03**	2,60	2,68	-2,98507
14.	**5,21**	5,2	4,01	29,67581
15.	**9,78**	7,82	5,58	40,14337
16.	**12,4**	11,2	7,71	45,26589
17.	**6,94**	6,55	4,9	67,94872
18.	**9,59**	7,97	4,35	83,21839
19.	**13,26**	11,22	7,65	46,66667
20.	**12,36**	11,44	7,56	51,32275
Average Pout Increase				46,19843

It can be known from Table 10 that photovoltaic was mixed with solar tracking system and ANFIS MPPT (first system) producing 46.19843% more power than fixed system (second system) when 3 lamps were used as load. However, there are conditions where power increase is negative which means power produced by first system is less than the second one. It is assumed that particular climatic condition during experiment was conducted, has made operating condition of second system near the MPP so that it produces power almost equal to the first one. Moreover, buck-boost converter used in ANFIS MPPT of first system has efficiency of 70-80%. It causes that the output load (Pout tracking) is always lower than the input side of converter (Pin Tracking) [13-18].

4 Conclusion

In this study, PV module equipped with ANFIS MPPT and solar tracking system has been proposed to increase PV power production. According to experiment results, the proposed ANFIS MPPT can produce overall PV energy increase relative to the non-MPPT system by 11,75% for 2 lamps as load, 4,57% for 3 lamps load, and 15,26% for 4 lamps load. Moreover, PV equipped with solar tracker system and ANFIS MPPT obtain 46,19843% total power increase compared to fixed PV system for 3 lamps as load.

References

1. A. Veldhuis, A. Reinders, Renewable and Sustainable Energy Reviews **52**, 757–769 (2015)
2. D. Sera, R. Teodorescu, J. Hantschel, M. Knoll, IEEE Trans. Ind. Electron. **55**, 2629–2637 (2008)
3. Imam. A, Ali. M, Adi. S., IREE Intl. Review of Electrical Engineering **10,** 390–398 (2015)
4. M.G. Villalva, J.R. Gazoli, E.R. Filho, IEEE Trans. Power Eelectronics **24**, 1198-1208 (2009)
5. S. Kolsi, H. Samet, M.B. Amar, J. Power and Energy Engineering **2**, 27–34 (2014)
6. A. Dolara, R. Faranda, S. Leva, J. Electromagnetic Analysis and Applications **1**, 152–162 (2009)
7. D. S. Karanjkar, S. Chatterji, S.S. L, A. Kumar, *2014 Recent Advances Engineering and Computational Sciences (RAECS)*, 1-6 (2014)
8. J. Ahmad, *2nd International Conference on Software Technology and Engineering (ICSTE)*, 247-250 (2010)
9. F. Murdianto, O. Panangsang, A. Priyadi, *2015 International Seminar on Intelligent Technology and Its Applications (ISITIA)*, 207-212 (2015)
10. A. Bin-halabi, A. Abdennour, H. Mashaly, Intl. J. Advanced Computer Research **4**, 588–595 (2014)
11. M.A. Eltawil, Z. Zhao, Renewable and Sustainable Energy Reviews **25**, 793-813 (2013)
12. Imam. A, Ali. M, Adi. S. IREMOS Intl. Review on Modeling and Simulation **8**, 640–652 (2015)
13. Imam. A, Adi. S, Ali. M., *BICET Brunei International Conference of Engineering and Technology*, (2014)
14. E. Duran, M. Sidrach-dc-Cordona, J. Galan, J.M. Andujar, *IEEE Power Electronics Specialists Conference*, 2036-2042 (2008)
15. V.H. Pham, Master Thesis (2007)

16. M.H. Tsai, Master Thesis (2012)
17. I.H. Altas, A.M. Sharaf, Intl. J. Information Technology and Intelligent Computing **1** (2007)
18. A. Desraino, *Desain Maximum Power Point Tracking (MPPT) Solar Charge Controller Menggunakan Buck Converter Dengan Metode Logika Fuzzy di Buoy Weather Station.* Institut Teknologi Sepuluh Nopember, Teknik Fisika. Surabaya: Institut Teknologi Sepuluh Nopember, 2016.

Demand Simulation for Water, Food Irrigation, and Energy from Micro Hydro Power Plant in Sungai Bayang, Bayang Utara, Pesisir Selatan West Sumatra

Siti Nurlaila Indriani[1,*], *Ahmad Agus* Setiawan[2], and *Rachmawan* Budiarto[1]

[1]Dept. of Nuclear Engineering and Engineering Physics, Gadjah Mada University, Yogyakarta, Indonesia

Abstract. Clean water or fresh water, food and energy are basic human needs. The three basic needs are dependent to one another. The relationship between the three is called the "The nexus of Water, Energy, and Food". It requires good governance on watershed which will be implemented for example to manage water resources to fulfil demand of clean or drinking water, irrigation in food area and energy sources in hydro power plant. This study conducted analysis and simulation to prepare projection of electricity produced by Micro hydro Power Plant (MHP) It integrates a climate change scenarios to forecast its influence on electricity demand and response of river. In addition, the study also presented projections of influence on irrigated food production scenario in irrigation for rice paddy fields. Projection of The MHP electricity and the water demand including for the food sector is conducted by using the WEAP (Water Evaluation and Planning) software, while electricity demand forecast is conducted by applying the LEAP (Long-Range Energy Alternatives Planning) software. The case studies in this study conducted in river flows Bayang's River. On the river there are three operating MHPs: The Muaro Aie MHP (ity30 kW of installed capac), The Koto Ranah MHP (30 kW) and The Pancuang Taba MHP (40 kW). The LEAP simulation projected electricity demand for Pesisir Selatan until 2025. Demand for South Pesisir Regency up to 2025 is predicted to reach 226.4 GWh with growth of 11.2% per year in BAU scenario, while reach 113.7 GWh with a 5% annual growth in efficiency scenario. The WEAP provided projected electricity production of MHP, basic water needs and irrigation needs for paddy fields in District IV Nagari Bayang Utara until 2025. The MHP electricity production in final year of projection with BAU scenario reaches 0.88 GWh, while with a climate change scenario of 0.63 GWh. The electricity demand fulfilled by MHP is predicted to be 0.39% in the BAU scenario, 0.28% in climate change scenarios, and 0.55% in the electricity savings scenario. Of the three MHP, the MHP Pancuang Taba is the most vulnerable to climate change, while MHP Koto Ranah shows relatively lower fluctuation. The highest staple water requirement is for Pancuang Taba which is 3643.4 thousand m3. The growth of staple water needs until 2025 tends to be constant. and most rice irrigation needs are in agriculture 2 of 976 thousand m3. The growth of irrigation needs of Bayang watershed until 2025 tends to be constant. Most irrigation needs for paddy fields are in irrigation area of "Agriculture 2" reaching 976,000 m3. The growth of irrigation needs in Bayang watershed tends to be constant.

1 Problem Statement

Clean water or fresh water, food and energy are basic human needs. Without all of it, people will not be able to live properly. The three basic needs are also interdependent with each other. In a certain situation, the scarcity of energy will affect water supply. This is because to produce water (with desalination) requires large amounts of energy, for example for desalination, wastewater treatment, water distribution (pumps), and others. Conversely, to produce energy is also needed water, for example MHP (Micro Hydro Power Plant). On the other hand, water scarcity will obviously cause food scarcity. Similarly, food scarcity will cause the standard of human life to decline, thus affecting the quality of work, for example in the energy industry sector. The relationship between these three is called "The nexus of Water, Energy, and Food" (red: water relations, energy / electricity, food). [1]

The rapid growth of the world's population causes the world's water resources to become one of the most

* Corresponding author: sn.indriani@gmail.com

important assets. Water is essential for human consumption and sanitation, for the production of various industrial goods as well as for the production of food and fibre fabrics. Water demand for irrigation is increasing to meet the needs of human food. Care-based design and careful management are essential to achieve water utilization efficiency. [2] The water source of the river is not only to meet basic needs of basic water and food needs in the form of irrigated rice fields and fields, but also used as a source of MHP use. Ultimately this led to the need for governance management of watershed use to be used for basic water needs, food irrigation, and MHP power sources.

Water cannot escape from a water source, one of the water sources is a river. Rivers exist in different regions of the hemisphere. Water from various places flows into the river that will be forwarded to the sea. Watershed is a region on the surface of the earth where there is a water system consisting of the main river and some of its subsidiaries (Ritter 2003). Watershed geometry that affects runoff is the size, shape, slope, orientation, elevation and river density (Shelton 2009). Meanwhile, several physical properties of watersheds that also affect the use of land and land cover, surface infiltration, soil type, permeability, water capacity and the presence or absence of lakes and swamps. River properties such as size, shape, stroke and length affect the capacity of river stores and determine peak discharge times (Shelton 2009). [2]

Climate change is a threat to the earth, because it can affect all aspects of life that will affect the balance of life of the earth. Climate change is the occurrence of changes in atmospheric conditions, such as temperature, and weather that causes an uncertain condition. [3] The rise in temperature will cause the expansion of water masses and sea levels. Climate change will also affect hydrological systems such as watersheds and river flows.

Irrigation has been done by farmers in Indonesia since time immemorial by using streams of creeks. These small irrigation networks are built by using watersheds to irrigate fields and fields that will meet food needs. Food demand is influenced by the ability of the regions to provide food resources, because from the graph, rice imports from time to time increase. So that water, energy, and food needs can be obtained from one resource, that is river water resource.

In the National Energy Policy Presidential Regulation no. 5 of 2006 with the aim to direct efforts to realize the security of supply (energy of supply) to support sustainable development, which is expected by the year 2025 will be achieved target energy elasticity. The elasticity of energy is the ratio between the growth rate of energy consumption and the rate of economic growth. Therefore, the energy conservation program to obtain economic growth, poverty reduction, environmental sustainability, energy security with the basic pattern of energy efficiency supported by the regulation in the field of energy conservation. Water energy is included in the category of Natural Resources that can be renewable. Development of Renewable Power Generators is a priority because renewable energy can be available throughout the ages.

The analysis study of how the river relates to energy generation, about how water resources management, about the nexus of water, energy, food on a macro scale, and on the prediction of electricity needs, general energy needs using Long-range Energy Alternative Planning (LEAP) applications and on the prediction of the production of solar and electric power plants using LEAP.

In a micro-scale review, one or more watersheds covering an area can be managed to be potential water resources to meet the community's need for water resources. Needs that can be managed are basic water needs, water needs for irrigation fields as food sources and energy needs that can be obtained from micro hydro power plants. Furthermore, predictions of electricity demand, basic water sources and water for food irrigation can be obtained. Projection of the demand for electricity, water source and water for irrigation will also be obtained. Similarly, climate change scenarios that will affect the fulfilment of any need will be obtained. So, it is expected that the management of water resources maximally to meet the nexus of water, food, and energy in an area can be done.

Sub-district IV Nagari Bayang Utara is located in Pesisir Selatan Regency, West Sumatera Province. River Basin Batang is located in District IV Nagari Bayang Utara. This research is a follow up study on MHP and the integrated water-energy management scenario in Pesisir Selatan Regency, West Sumatera. This area is chosen because there is one flow that has more than one MHP, also the availability of data, because there are rain gauge stations in the river and in Tarusan River which is adjacent to Bayang River which can be used as data, and then easy access to the location. This location can be said to be quite remote, plus the local people use direct river water to meet water needs, so the watershed is a vital source of life for the surrounding community. Thus, good watershed management is an important issue for this region. [4].

2 Figures and tables

This study projected the MHP electricity production under climate change scenarios and integrated the demand for electricity demand. In addition, projection of food production with irrigation and irrigation field scenarios is predicted with demand for food and basic water needs. MHP projection and food irrigation needs are carried out using WEAP (Water Evaluation and Planning) software, whereas electricity demand prediction is done using LEAP (Long-range Energy Alternatives Planning) software.

The location of this research is Sub district IV Nagari Bayang Utara located in Pesisir Selatan Regency, West Sumatera Province. River Basin Batang is located in District IV Nagari Bayang Utara. This research is a follow up study on MHP and the integrated water-energy management scenario in Pesisir Selatan Regency, West Sumatera. This area is chosen because there is one flow that has more than one MHP, also the availability of data, because there are rain gauge stations in the river and in Tarusan River which is adjacent to Bayang River which can be used as data, and then easy access to the location. This location can be said to be quite remote, plus the local people use direct river water to meet water needs, so the watershed is a vital source of life for the surrounding community. Thus, good watershed management is an important issue for this region. Flowchart research steps can be seen in Figure 1 below.

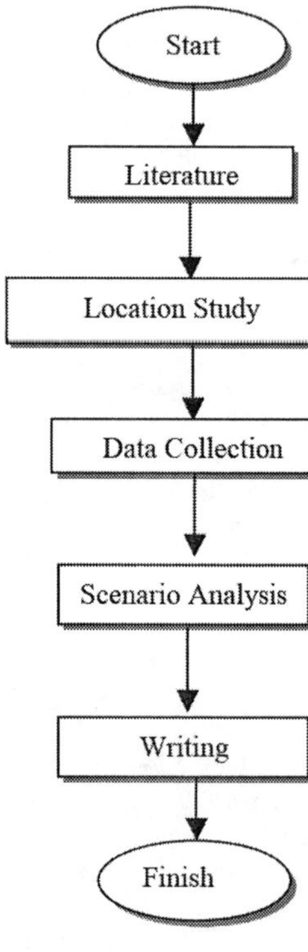

Fig. 1. Research flowchart.

3 Results

3.1 LEAP-WEAP Simulation

Energy demand simulation is done by using LEAP software. The method used in this simulation is based on the end-use model. The equation used as the analysis with the energy intensity variables and the number of consumers as unit activity level.

LEAP modelling steps generally consist of five stages: problem definition, system conceptualization, model representation, model evaluation, and policy analysis.

The simulation of water availability is done by using WEAP software. The method used in this simulation is based on the trend approach. Hydrological modelling is done by incorporating climatological data over the last 20 years to see the trend of climate factors.

Before carrying out the simulation of water availability using WEAP, firstly analyse the mainstay discharge. Calculation of reliable debit is done using Mock Method By following the following stages: Potential Evapotranspiration Calculation (ETo), Calculation of Basic Water Requirement, Calculation of Irrigation Water Requirement.

In this research, WEAP calculation of debit is done by using soil moisture method. This method uses data of rainfall and monthly climatology as input data. The data is then processed to produce the output of evapotranspiration, discharge, and storage of ground water. Water availability calculation scheme by WEAP is shown in Figure 2 below:

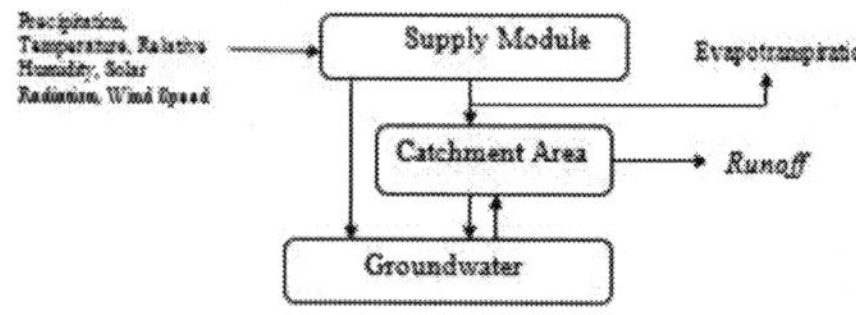

Fig. 2. WEAP calculation scheme.

The implementation phase of water availability simulation with WEAP is as follows:

3.1.1 Basic parameters

The first step in the simulation is to set and determine the basic parameters of the simulation. In the basic parameters, the scope of work is determined on the catchment and the MHP (resource). The next step is to determine the base year of simulation. In this study used as the base year is 2013, based on the data held as the basis for calculation. After that, the final limit of the simulation period is determined in 2025. The last is to determine the units used such as energy units, debit units, and area.

3.1.2 Schematic

The second stage after setting the basic parameters is to create a schematic view of the watershed system or stream to be modelled. To help draw the river and display the position of certain sites such as hydropower, wastewater treatment, groundwater storage, etc., users can enter GIS-based maps as the basic screen in WEAP. In this study, which became the basic screen is a map of

West Sumatra that contained the river and the location of the MHP to be assessed. The next step is to redraw the river and put the MHP according to the layer that has been displayed. Figure 3 Shows the view schematic view in this research.

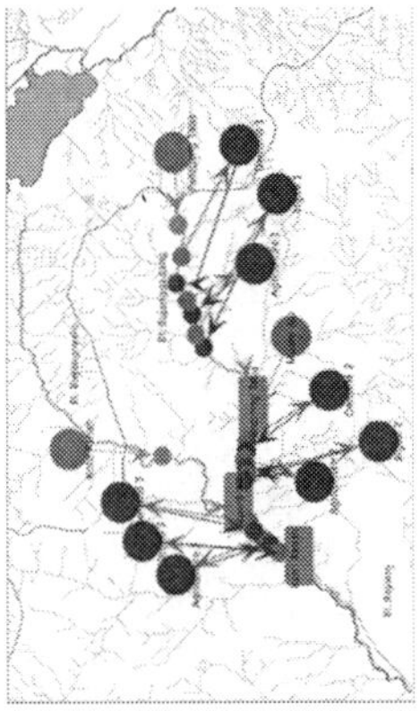

Fig. 3. Schematic view.

3.1.3 Catchments

Catchments are part of a branch that serves to model the watershed to be studied. As the basis of the simulation, first one of the calculation method, based on the data owned. Watershed modelling in this research using Soil Moisture Method.

After determining the calculation method, hereinafter is entering data in current account condition that is base year condition (base year). The base year used is 2013, so the initial input written in the expression section is based on the data of that year.

In this study, the available data is only sufficient to meet the Land use and Climate categories. The data included in the Land use category is the watershed area for each MHP. As for the Climate category, climatological data required are rainfall data, temperature, humidity, wind speed, and solar radiation.

Evapotranspiration in WEAP was calculated using the Penmann-Monteith equation, whereas the discharge was calculated by equation.

3.1.4 Supply Analysis

Supply Analysis is a branch that determines the characteristics of calculating the value of water availability. In this study water availability is calculated to generate electricity at three MHPs. To model the MHP system, technical data i.e. effective head and turbine efficiency data are included to calculate the production of electricity generated factually.

3.1 Input-Output Research

The data used in this study are summarized in Table 1 below:

Table 1. Data input and output on LEAP & WEAP.

Calculation	Input	Assumption	*Software*	Output
Energy Demand	- Energy Intensity - Electricity Consumption - Total Population - Electricity Customers - PDRB		LEAP	Energy Consumtion
Debit	- Temperature - Humidity - Wind Velocity - Sun Radiation	- The annual solar radiation is constant against the earth	WEAP	Evapotranspiration
	- Evapotranspiration - Rainfall - Watershed Area	- Soil moisture coefficient (RRF = 3) - Numbers of groundwater catch (SMC = 200mm) - No land use change	WEAP	Debit
Plant Irrigation	- Harvest crop production - area of harvest - number of cropping cycles per year - rainfall - crop water requirements	- planting cycles per year based on crop cycle theory prevailing in Indonesian society - crop water requirements calculated from Kc	WEAP	Irrigation water volume
Basic Water Requirement	Population	- ideal water requirement 50liter / day / person	WEAP	Volume of staple water
MHP Electricity Production	- Debit - *Head* MHP	- The efficiency of MPH system is considered constant - MHP works 8760 hours / year	LEAP & WEAP	Power

https://doi.org/10.1051/e3sconf/20184301015

There are two stations that record rainfall and climatology data on the Bayang basin. Both stations are located on different sub-watersheds, but can represent climatological conditions that affect the flow of the river (shaft) of Shadow. The location of the station is shown in Table 2.

Table 2. Rainfall monitoring and climatology stations.

Station	Coordinat	Sub-DAS
Danau Diatas	01^{0}04'40,3" LS / 100^{0}46'35,7" BT	Bt. Bujanggadang & Bt. Bayanggadang
Tarusan	01^{0}15'12" LS / 100^{0}29'13" BT	Bt. Bayangnyalo

Rainfall and climatology data used in the calculation is in the form of monthly data for the last 20 years (1992 - 2012). The rainfall chart on both stations is shown in Figure 4.

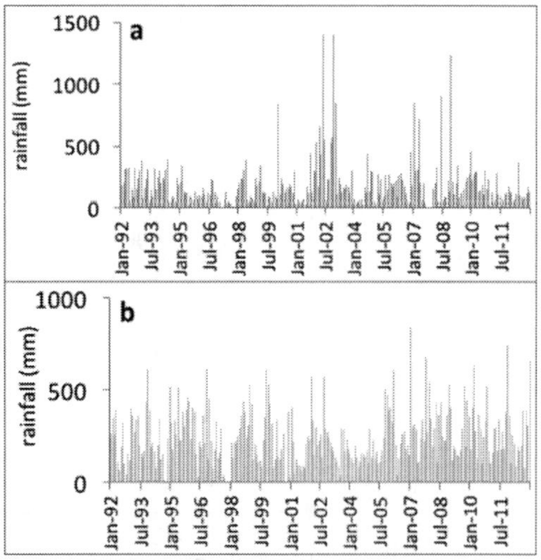

Fig. 4. Rainfall chart on both stations.

The rainfall graph shows the variability of rainfall events per month on both stations. In the Diatas lake station data, there are three times extreme rainfall events exceeding 1000 mm in one month. While at the data station Tarusan, the highest rainfall only reached about 800 mm. Average rainfall per year on both stations is 2328 mm for the Lake Diatas station, and 2998 for the Tarusan station. Other input data are temperature, relative humidity, wind speed and solar radiation.

3.2.1 Business as Usual (BAU) Scenario

BAU scenario is a continuation of the trend that occurred in previous years, both in terms of electricity demand and water availability. In terms of electricity demand, since 2008 there was an average growth of 11.12%. The projected output of electricity demands up to 2025 with a constant growth rate is shown in Figure 5.

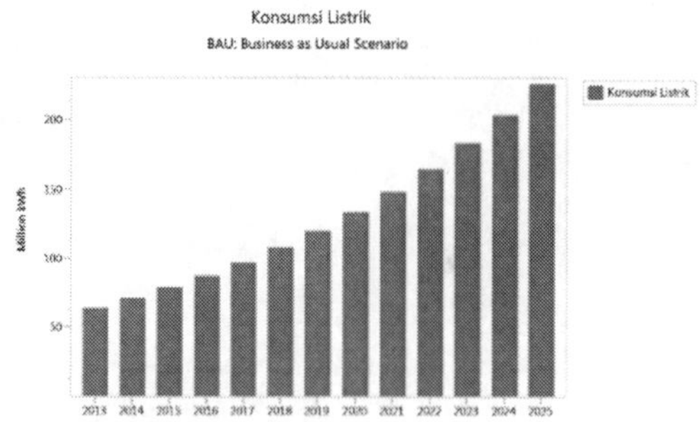

Fig. 5. Output of electricity demand.

In the BAU scenario, there is a very sharp increase in electricity demand side. By 2025, electricity demand is projected to be at 226.4 GWh, whereas in 2013 electricity demand was at 63.3 GWh. This means that within the next 12 years, electricity demand is predicted to grow to 257.67% or almost 2.5 times.

In terms of water availability, the trend approach is used to predict watershed conditions and sub the next 12 years. This trend approach is used in the BAU scenario with the assumption that there is no significant land use change and the MHP remains functioning normally during the projection year.

The electricity production of each MHP is shown in Figure 6, While the overall MHP electricity production until 2015 is shown in Figure 5.28. On the graph shows that MHP Pancuang Taba is relatively more susceptible to fluctuation of debit compared to Muaro Air and Koto Ranah MHP. The fluctuation of the discharge that occurs is influenced by the variation of rainfall in that period.

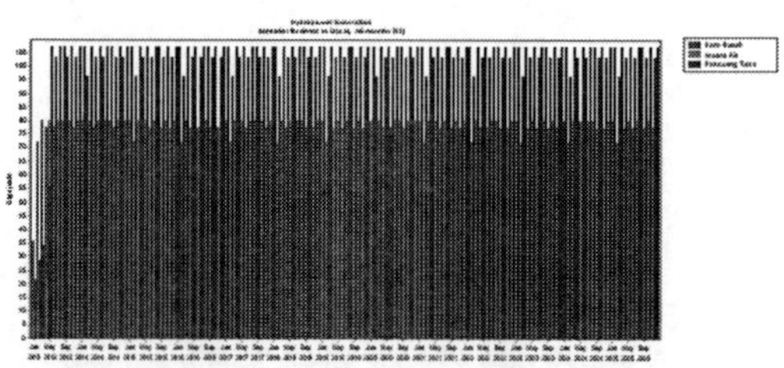

Fig. 6. The electricity production of each MHP.

In the projection of electricity MHP BAU scenario there are several times a significant decrease in electricity production, among others in the period January - May 2014, March - June 2017, and August - December 2020. MHP Pancuang Taba power production decreased drastically and MHP Muaro Water is reduced by 50%. This is due to the occurrence of dry season in the period of no-rainy upstream periods recorded in 1994, 1997, 1999, 2001, 2004 and 2007. The simulation results show the flowing discharge at the Bt. Sub-watershed. Bayanggadang and Bt. Bujanggadang is below the 0.2 m3 / s.

Fluctuating electricity production is also predicted to occur in the year 2024 - 2025. This can happen if the rainfall conditions that occur is a repetition of rainfall in 2011 & 2012 where monthly rainfall changed in the extreme.

Annual electricity production is shown in Figure 7. Decrease in the discharge that occurred in several periods also influence the decrease of electricity production. In this scenario of BAU, the decrease of MHP electricity production is predicted to occur in 2014, 2017, and 2020. The highest electricity production occurs in 2015 which reaches 3,152 GWh, while the lowest electricity production occurs in 2014 of 2,751 GWh. Electricity production in 2025 (the end of the projection) is predicted to be 3.152 GWh.

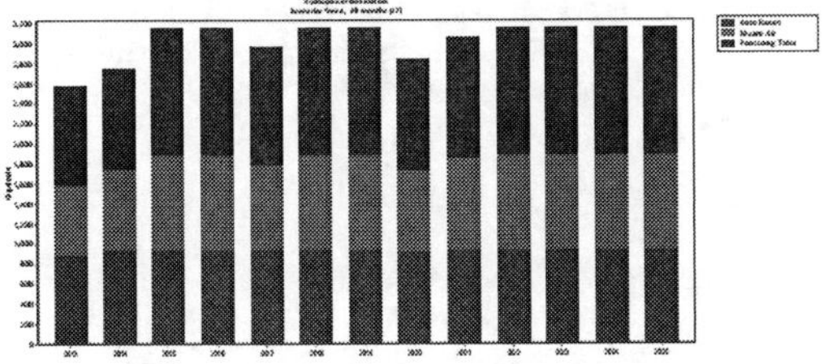

Fig. 7. Annual electricity production.

The projection of electricity generation by MHP is then compared with the projection of electricity demand of Pesisir Selatan regency. So that the electrical production data from WEAP. It can be exported in the form of excel to facilitate the calculation. Electrical production is then divided by electricity demand in Figure 8. From LEAP which can also be exported into excel form and then the result can be seen in graphic form. This comparison will result in the percentage of electricity demand fulfilled as shown in the graph in Figure 8.

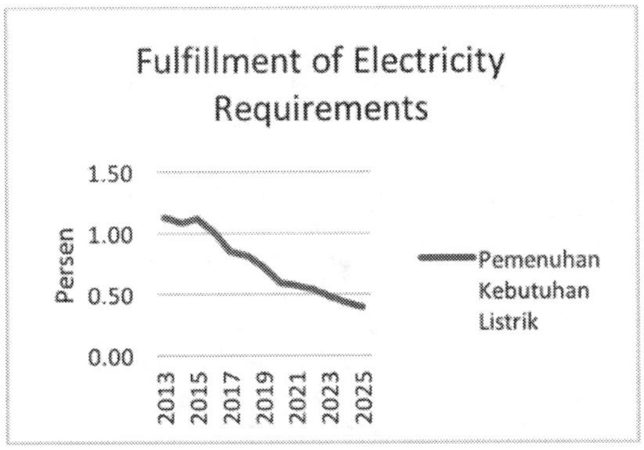

Fig. 8. Fulfilment of electricity requirements.

The percentage of electricity demand fulfilled by MHP shows a downward trend until 2025. This is due to the demand of electricity itself which continues to increase to 226.4 GWh or by 257.67% in the final year of projections. In 2013, MHP is able to generate electricity by 0.72 GWh or 1.13% of total electricity demand for the year. However, by 2025, MHP will only generate 0.875 GWh of electricity or 0.39% of total electricity demand for the year. Calculation of the percentage of electricity consumption can be seen in Table 3.

Table 3. Electricity consumption percentage.

Electricity Consumption	63.3	70.4	78.3	87.1	96.8	107.7	119.7	133.1	148.0	164.6	183.1	203.6	226.4
Electricity Production	0.72	0.76	0.88	0.88	0.83	0.88	0.88	0.79	0.85	0.88	0.88	0.88	0.88
Fulfillment (%)	1.13	1.09	1.12	1.01	0.85	0.81	0.73	0.59	0.58	0.53	0.48	0.43	0.39

In WEAP Schematic, basic water requirement is assumed as Demand 1, Demand 2, and Demand 3. Demand 1 is the staple water requirement of Pancung Taba population, Demand 2 is the main requirement of drinking water of Muaro Aie, whereas Demand 3 is the basic need of drinking water Residents of Koto Ranah area.

To enter the value of the basic water needs, data on the number of residents of Pancung Taba, Muaro Aie and Koto Ranah are multiplied by the amount of water required in one day, 60 liters then multiplied by 365 days, the water needs of the three regions in the period One year. In Figure 9, can be seen input data entered in Demand 1, Demand 2, and Demand 3.

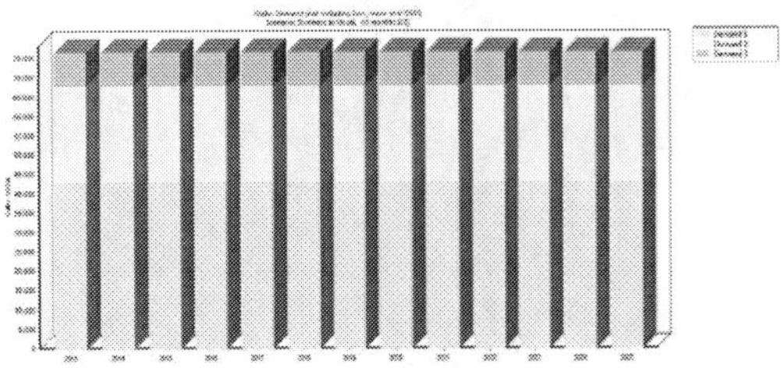

Fig. 9. Data entered in Demand 1, Demand 2, and Demand 3.

Irrigation water demand can be used as a demand that will be calculated for planting water requirement for each plant in m3 / Ha. Data to be included in the form of land area and water volume requirement within one year. Because the required data is the land area of Nagari Pancung Taba, Muaro Aie, and Koto Ranah, so to find the planting area of each nagari obtained from the percentage of the area multiplied by the total planting area of North Bayang District. In this WEAP simulation.

The water requirements of the plants to be simulated are for the needs of rice and *palawija*.

To find the rice cultivation area and the crops of each Nagari is obtained by multiplying the percentage of Nagari area compared to the area of North Bayang then multiplied by the total area of rice harvest and the crops. As for the water requirement on rice according to the Ministry of Agriculture is at least 99.7 m3 / Ha which is then multiplied 3 times the planting time within a year. As for the crops of *palawija* according to the Ministry of Agriculture, requires 2300-2400 m3 / Ha for water planting purposes, which will be taken the middle value between the two is 2350 m3 / Ha. Next will be multiplied also with 3 times the planting time in a year. For rice and *palawija*, it is an almost equal standard in Indonesia if it has average planting time 3 times a year.

In Figure 10 can be seen all the demand required in WEAP simulation, namely the demand of water staple named Demand 1, Demand 2, and Demand 3. Then the demand for rice irrigation is called Agriculture 1, Agriculture 2, and Agriculture 3. Then the demand of the crops Called Irrigation 1, Irrigation 2, and Irrigation 3.

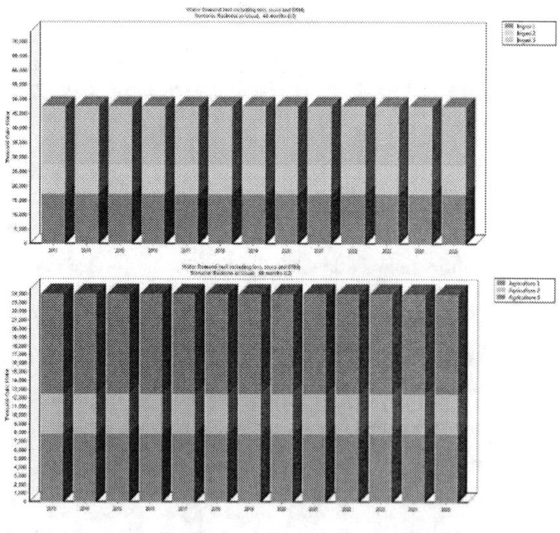

Fig. 10. Demand required in WEAP simulattion.

4 Conclusion

Based on the research that has been done, it can be concluded as follows:

1. Demand for South Pesisir Regency up to 2025 is predicted to reach 226.4 GWh with growth of 11.2% per year in BAU scenario; And 113.7 GWh with a 5% annual growth in efficiency scenario.
2. MHP electricity production in final year of projection (2025) with BAU scenario is 0.88 GWh; While with a climate change scenario of 0.63 GWh.
3. The electricity demand fulfilled in 2025 by MHP is predicted to be only 0.39% in the BAU scenario; 0.28% in climate change scenarios; And 0.55% in the electricity savings scenario.
4. Of the three MHP, MHP Pancuang Taba is the most vulnerable to climate change, while MHP Koto Ranah is relatively more stable.
5. The highest staple water requirement is at Demand 1 ie Pancuang Taba which is 3643.4 thousand m3. The growth of staple water needs until 2025 tends to be constant.
6. Most rice irrigation needs are in Agriculture 2 of 976 thousand m3. The growth of irrigation needs of Bayang watershed until 2025 tends to be constant.
7. The most irrigation needs of rice is in Irrigation 3 of 1726.5 thousand m3. The growth of irrigation needs of Bayas watershed until 2025 tends to be constant.

References

1. The *Water, food, and energy nexus*. 2014. http://www.unwater.org/topics/water-food-and-energy-nexus/en/
2. Linsley, ay. K., Joseph B. Franzini., dan Djoko Sasongko. *Teknik Sumber Daya Air*. Erlangga, Jakarta, 1996.
3. Setiawan, Dodi. *Kajian Pengaruh Perubahan Iklim dan Tata Guna Lahan di Daerah Aliran Sungai (DAS) Citarum Hulu terhadap Pembangkit Listrik Tenaga Air Saguling*. Departemen Geofisika dan Meteorologi, Fakultas Matematika dan Ilmu Pengetahuan Alam, Institut Pertanian Bogor, Bogor, 2013.
4. Napitupupu, Janter. Jurnal *Pembangkit Listrik Tenaga Mini Hidro (PLTM) dalam Pengelolaan Energi Hijau*. Universitas Darma Agung, Medan.
5. *Blueprint Pengelolaan Energi Nasional 2006-2025*. Jakarta, 2006
6. Welsch, M., dkk. *Adding Value with CLEWS – Modelling The Energy System and Its Interpendencies for Mauritius*. Hal. 1-11. September, 2013
7. Anugrah, P. *Proyeksi Pembangkitan Listrik Tenaga Mikro Hidro dengan Skenario Manajemen Air-Energi yang Terintegrasi di Kabupaten Pesisir Selatan, Sumatera Barat*. Skripsi. hal. 1-82, Jurusan Teknik Fisika, Fakultas Teknik, Universitas Gadjah Mada, Yogyakarta, 2014.
8. Handayani, Wuri., dan Indrajaya, Yonky. *Analisis Curah Hujan dan Debit Suib Sub DAS Ngatabaru, Sulawesi Tengah*. Hal. 1-10. Juli, 2011.

https://doi.org/10.1051/e3sconf/20184301015

9. Bazilian, Morgan., dkk. *Considering The Energy, Water, and Food Nexus: Towards an Integrated Modelling Approach.* Hal. 1-11. Oktober, 2011.
10. Wijaya, M.E. *Peningkatan Keamanan Pasokan Jangka Panjang Sistem Perluasan Perencanaan Listrik Jawa-Bali-Madura di Indonesia.* Hal. 1-10. Desember, 2009.

Longitudinal and Shear Wave Velocities Measurement in Compacted Bentonite for Water Content

Shun Kimura[1,*], *Kazumi* Kitayama[1], *Hideharu* Takahashi[1], *Kazushi* Kimoto[2], *Katsuyuki* Kawamura[1], and *Hiroshige* Kikura[1]

[1]Tokyo Institute of Technology, Tokyo, Japan
[2]Okayama University, Okayama, Japan

Abstract. Bentonite is a good candidate of buffer material for disposal repository of high-level radioactive waste. Understanding groundwater behavior in bentonite buffer material is important in order to evaluate the bentonite buffer performance and guarantee long-term safety. Elastic constants of the bentonite buffer material are important parameters for the long-term safety. Water content in buffer material may have an influence on its elastic properties. For this reason, the monitoring system of the water saturation level in compacted bentonite is required. In this study, the ultrasonic velocity measurement method for evaluation of water content in compacted bentonite was proposed. At first, the effect of a degree of saturation in compacted bentonite on the longitudinal and shear wave velocities was investigated experimentally. In addition, the elastic property, bulk modulus, in unsaturated compacted bentonite were evaluated by ultrasonic velocities. As a result, it can be confirmed that ultrasonic velocities can evaluate a degree of saturation and bulk modulus of compacted bentonite.

1 Introduction

High-level radioactive waste (HLW) management is one of the important issues of nuclear power. HLW should be isolated from the human environment for more than 10,000 years because of its high radioactivity and long half-life. Disposal repository of HLW is designed to dispose of the waste in the deep underground with engineered and natural geological barriers as shown in Figure 1. The engineered barriers consist of the vitrified glass, overpack, and buffer material. Buffer material provides stable chemical and physical environment for inner engineered barriers. Bentonite, one of a clay mineral, is considered to be buffer material since it has low water permeability, low diffusion for nuclides, and good swelling property. Bentonite is planned to be compacted and installed in the disposal repository. Compacted bentonite buffer is unsaturated at the beginning of the waste management, and a degree of saturation will increase with time. Replenishment of water is called resaturation. Understanding the resaturation behavior inside compacted bentonite buffer is important because the water in bentonite has a tremendous effect on reactions between bentonite and radionuclides. The elastic properties of compacted bentonite have a dependency on a variation of water content. The elastic properties can be obtained by longitudinal and shear wave velocities. It means the variation of water content will be evaluated as a function of these wave velocities. Ultrasound is one of the elastic waves and is a useful technique for the laboratory experiments. Recently, ultrasonic velocity in compacted bentonite has been investigated for laboratory experiments [1], and seismic imaging technique for nonintrusive monitoring of the barrier in repositories [2] has been studied. For these reasons, in this study, the non-destructive ultrasonic measurement technique for monitoring of water saturation level in compacted bentonite was developed in our laboratory. Longitudinal and shear wave velocities measurement was conducted in order to investigate the relation between the velocities and water content in bentonite. In addition, bulk modulus, one of the elastic constant, was evaluated by the velocities.

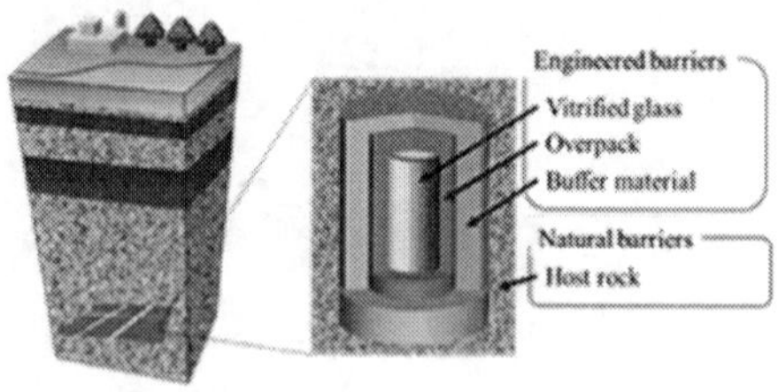

Fig.1. Schematic of disposal repository design of HLW.

2 Experimental methodology

2.1. Sample preparation

* Corresponding author: kimura@us.nr.titech.ac.jp

The sample material of pure smectic bentonite powder (Kunipia-F) was prepared for the compacted specimen. Kunipia-F is a purified bentonite and a content of a smectite, the main component of bentonite, is more than 99%. Table 1 shows the chemical composition of Kunipia-F. The procedure of preparation was as follows. Bentonite was dried out for 24 hours at the temperature of 120°C. Pure water was added to the dried bentonite to adjust water content. Water content W_c is defined as

$$W_c = \frac{M_w}{M_s} \quad (1)$$

Where M_w is the mass of water and M_s is the mass of bentonite. The sample has four levels of water content: ~20%, ~25%, ~30%, and ~35%. Then, the bentonite powder is put into a cylindrical mold, and is compacted from both ends of the mold. Specimens have a diameter of 28 mm and a length of from 10 mm to 16 mm. Dry density of specimens is between 1.00 g/cm^3 to 1.70 g/cm^3. In this study, the water content in compacted bentonite was evaluated by a degree of saturation. A degree of saturation can be calculated as follows:

$$S_r = \frac{W_c \cdot \rho_s}{e \cdot \rho_w} \quad (2)$$

Where e, ρ_w, and ρ_s are a void ratio, density of water content and density of bentonite, respectively. Figure 2 shows the water content and the dry density of bentonite specimens. Red circles, green circles, blue circles and purple circles indicate four levels of bentonite, ~20%, ~25%, ~30%, and ~35%, respectively. The black line represents the fully saturated condition, a degree of saturation of 100%.

Table 1. Chemical composition of Kunipia-F [3]

SiO2	58.0
TiO2	0.2
Al_2O_3	21.9
Fe2O3	1.9
MnO	0.2
MgO	3.4
CaO	0.5
Na2O	3.0
K2O	0.1
H2O	10.8

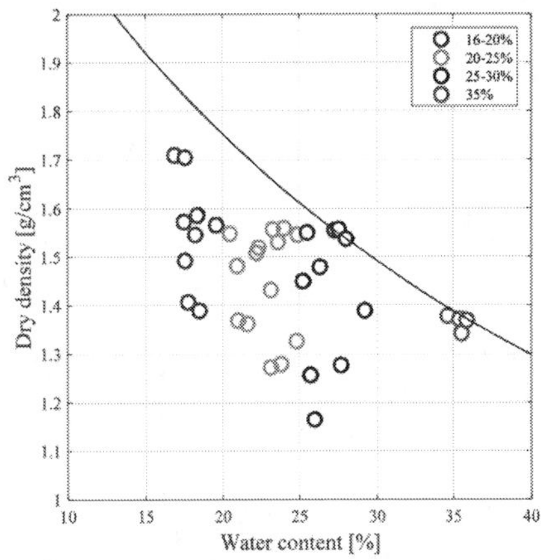

Fig.2. The water content and the dry density of bentonite specimens. Red circles, green circles, blue circles, and purple circles are water content of ~20%, ~25%, ~30%, and ~35%, respectively.

2.2 Experimental setup

Figure 3 shows the experimental apparatus for ultrasonic velocity measurement. The experimental apparatus consists of two ultrasonic transducers for longitudinal wave (B0.5C20N, Japan Probe Co., Ltd.), two ultrasonic transducers for shear wave (1Z10×10SN, Japan Probe Co., Ltd.), a pulser/receiver (JPR-10CN, Japan Probe Co., Ltd.), a computer, and an external amplifier (PR-60A5, Japan Probe Co., Ltd.). The transducer for longitudinal wave has an element with a diameter of 20 mm. The center frequency of the transducer is 500 kHz. The transducer for shear wave has an element with a size of 10×10 mm. The transducer for shear wave has the center frequency of 1 MHz. Transducers were installed in opposite position at both ends of a specimen. The pulser/receiver was used for a transducer control by changing voltage and center frequency of the emitting wave. The frequency of the emitting wave was 500 kHz. Ultrasonic wave was emitted into the specimen from a surface and was received on an opposite surface.

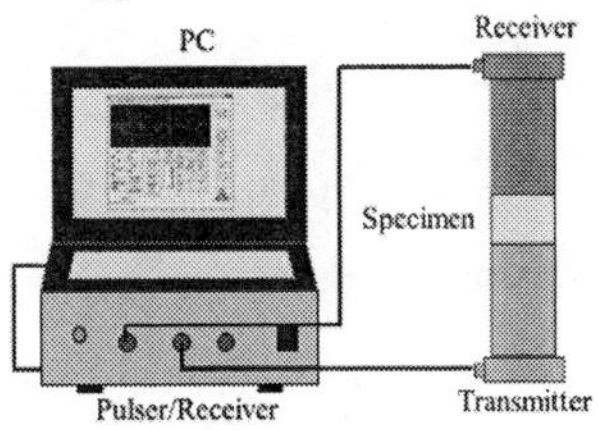

Fig.3. Schematic of the ultrasonic velocity measurement apparatus.

2.3 Group velocity determination

The basic theory of ultrasonic velocity measurement is based on the transit time from the emitting transducer to the receiving transducer. The velocity can be obtained by the propagation length and the transit time in Eq. (3).

$$v = \frac{L}{\Delta t} \quad (3)$$

Where v is the ultrasonic velocity. L is the propagation length of the ultrasonic wave between the emitting transducer and the receiving transducer. Δt is the transit time.

The simplest method is the direct measurement of the first arrival of the received wave [4]. This method identifies the travel time from the time value of the first inflection of the receiving wave. Another typical method is a cross-correlation of emitting and receiving wave [4]. The peak at a time shift of a cross-correlation function indicates the transit time in this method. However, bentonite has a high attenuation of ultrasonic wave with frequency dependency. This attenuation causes the waveform distortions in the time domain. These time domain techniques are not applicable due to the attenuation of the ultrasonic wave in compacted bentonite. Therefore, frequency domain method based on Fourier transform was proposed. A signal in the time domain can be separated into Fourier amplitude and phase spectrum. The Fourier transform of a receiving signal is represented in Eq. (4).

$$R(\omega) = \int_{-\infty}^{+\infty} r(t)e^{-i\omega t}dt = |R(\omega)|e^{-i\phi_r(\omega)} \quad (4)$$

Where ω is angular frequency. $R(\omega)$ is Fourier transform of the receiving signal. $|R(\omega)|$ and $\varphi_r(\omega)$ are Fourier amplitude and the phase spectrum of the signal, respectively.

The attenuation in the time domain is caused by the attenuation of Fourier amplitude or the variation of the phase spectrum. Here, the variation of the phase spectrum in this measurement is not assumed. The differentiation of phase spectrum, the slope of phase spectrum, indicates the group delay. The group delay corresponds to the transit time of the signal. In order to obtain the transit time, the calibration of the initial phase using the signal of reference specimen is required. The polycarbonate specimen was used as the reference specimen. The transit time represents in Eq. (5).

$$t_g(\omega) = \frac{d}{d\omega}(\phi_r - \phi_p) \quad (5)$$

Where $\varphi_p(\omega)$ is the phase spectrum of polycarbonate. t_g is the calibrated transit time. The calculated velocity in this method is called group velocity.

2.4 Elastic properties

Variations of water content may affect elastic properties of compacted bentonite. Elastic properties are characterized by elastic constants. Bulk modulus is one of elastic constants and defines resistance to compression. Therefore, bulk modulus means incompressibility of a material. Bulk modulus can be calculated from longitudinal wave velocity and shear wave velocity.

$$K = \frac{\rho(3V_L^2 - 4V_s^2)}{3} \quad (6)$$

Where K is bulk modulus. ρ is a density of a material. V_L and V_s indicate longitudinal wave velocity and shear wave velocity, respectively. In this ultrasonic velocity measurement, the relation between a degree of saturation and bulk modulus was investigated.

3 Results and discussions

Figure 4 and 5 show examples of the receiving longitudinal wave, Fourier amplitude and phase spectrum of the reference specimen and compacted bentonite, respectively. The original signal was filtered by Hanning window in order to separate the first peak in Fig. 4 (a) and 5 (a). The black line indicates the original receiving wave and the red line indicates the filtered wave. Spike noises appear below 5 μs in receiving wave. These noises caused by emitting ultrasonic wave. The receiving wave cannot be distinguished in this region. Comparing Fourier amplitude in Fig. 4 (b) and Fig. 5 (b), it is can be found that the amplitude in compacted bentonite is attenuated strongly at a frequency of 300 kHz or more. Figure 6 shows an example of the calibrated transit time of longitudinal wave. The calibrated transit time can be calculated from the phase spectrum of reference specimen and compacted bentonite in Fig.4 (c) and Fig.5 (c). The calibrated transit time was obtained by averaging the group delay between 100 kHz to 500 kHz.

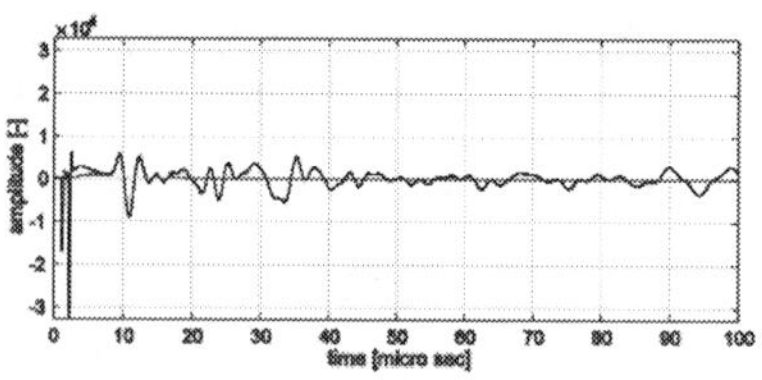

(a) Receiving wave.

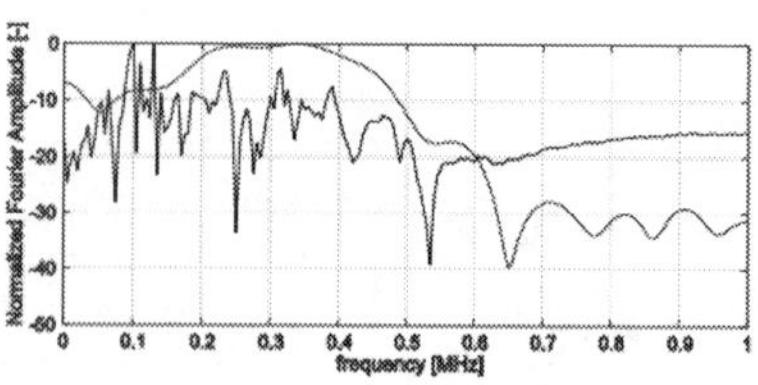

(b) Fourier amplitude.

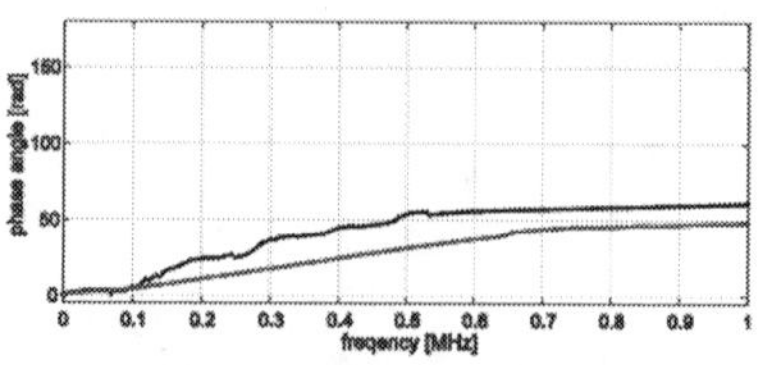

(c) Phase spectrum.

Fig.4. Receiving longitudinal wave and Fourier transform of the reference specimen.

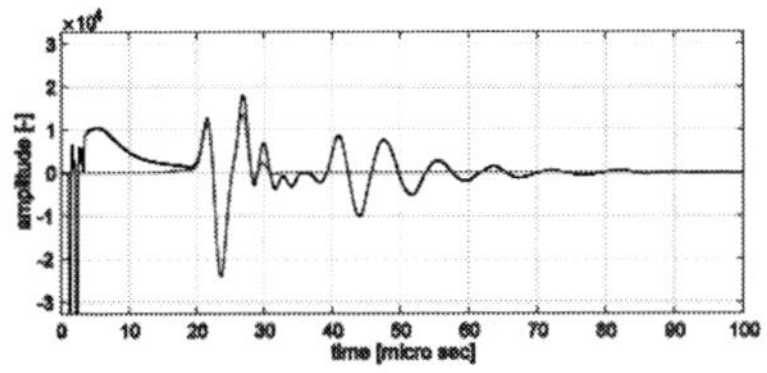

(a) Receiving wave.

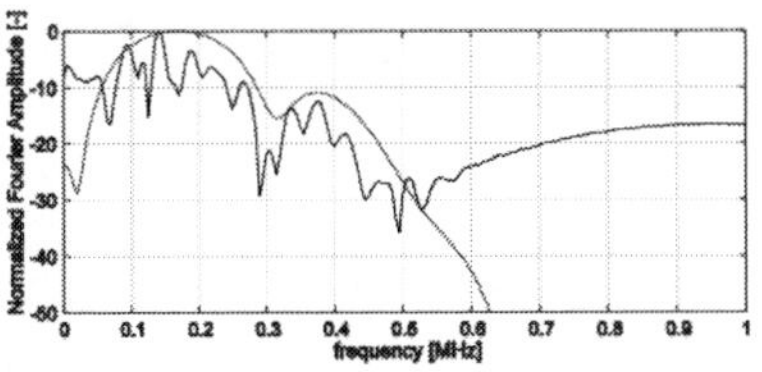

(b) Fourier amplitude.

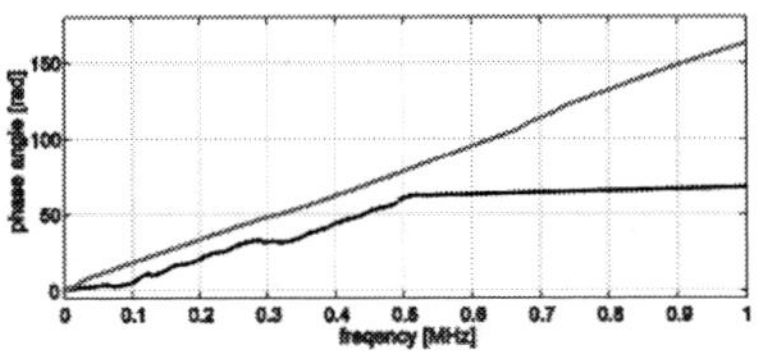

(c) Phase spectrum.

Fig.5. Receiving longitudinal wave and Fourier transform of compacted bentonite.

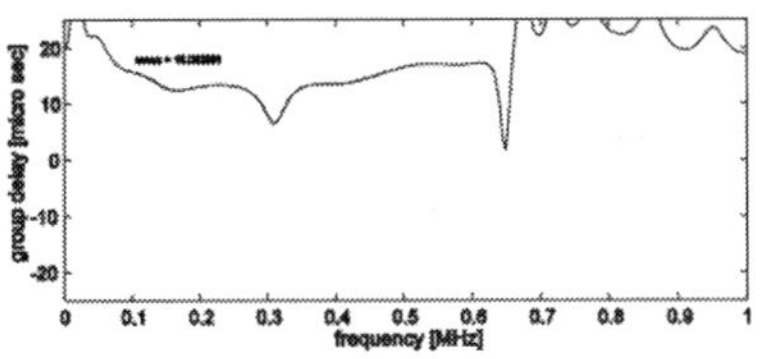

Fig.6. The calibrated transit time of the longitudinal wave.

Figure 7 shows the variation of longitudinal wave velocity with varying a degree of saturation. Red circles, green circles, blue circles, and purple circles are water content of ~20%, ~25%, ~30%, and ~35%, respectively in Fig.7. These colors in Fig.7 correspond to the colors in Fig.2. The longitudinal wave velocity increases with increasing the degree of saturation. It can be found that the velocity at a degree of saturation of 100% is very close to at the longitudinal wave velocity of 1,480 m/s in water. Furthermore, the longitudinal wave velocity at low water content level is higher than high water content level in the same degree of saturation.

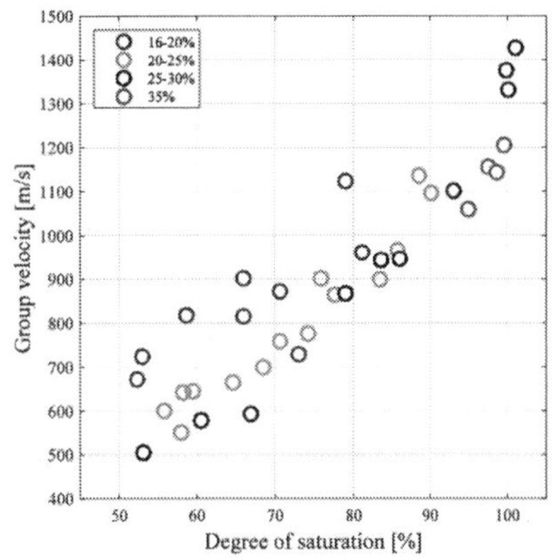

Fig.7. The variation of longitudinal wave velocity with varying a degree of saturation.

Figure 8 and 9 show examples of the receiving shear wave, Fourier amplitude and phase spectrum of the reference specimen and compacted bentonite, respectively. The black line represents the original receiving wave and the red line represents the filtered wave by Hanning window. Comparing with the shear wave of the reference and compacted bentonite, the waveform of compacted bentonite is distorted obviously. The distortion of the wave occurred due to the attenuation at a frequency around 500 kHz as shown in Fig.9 (b). It can be confirmed that the identification of the first arrival of the distorted wave like in Fig.9 (a) is difficult. However, the phase spectrum of shear wave is approximately linear with frequency. Figure 10 shows an example of the calibrated transit time of shear wave. The calibrated transit time of the shear wave was obtained by averaging the group delay between 300 kHz to 500 kHz.

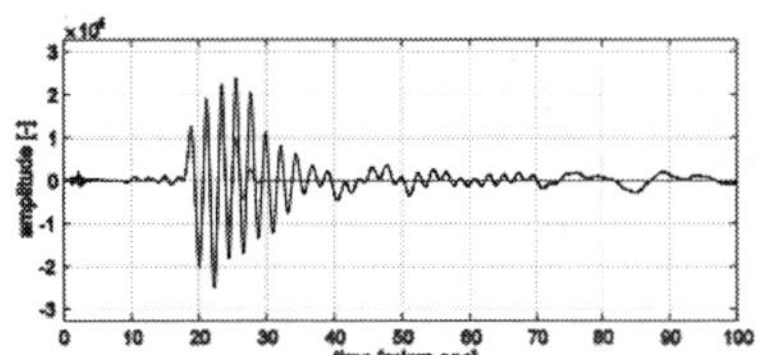

(a) Receiving wave.

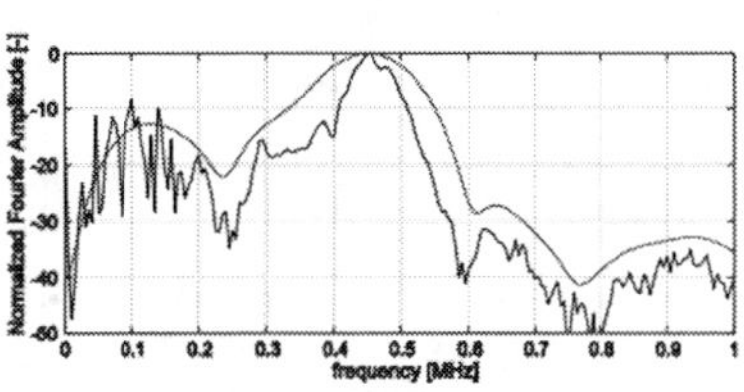

(b) Fourier amplitude.

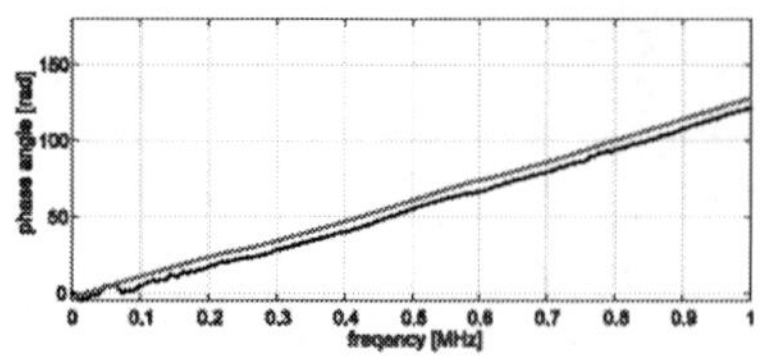

(c) Phase spectrum.

Fig.8. Receiving shear wave and Fourier transform of the reference specimen.

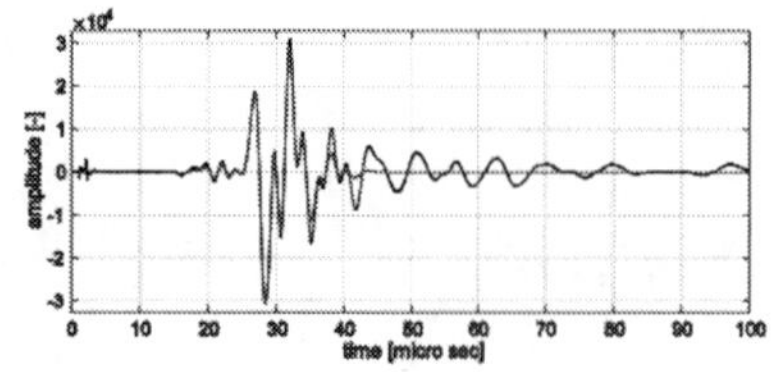

(a) Receiving wave.

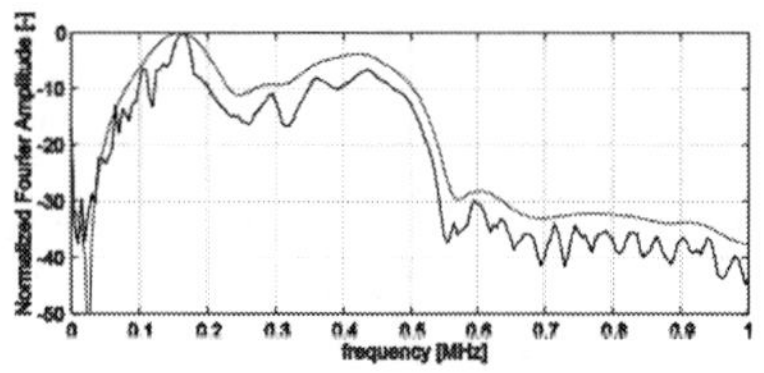

(b) Fourier amplitude.

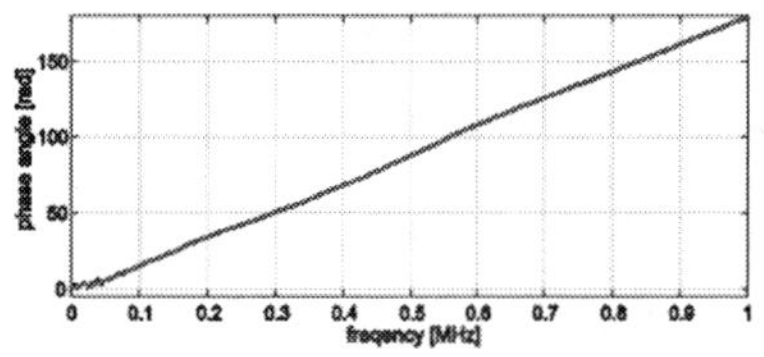

(c) Phase spectrum.

Fig.9. Receiving shear wave and Fourier transform of compacted bentonite.

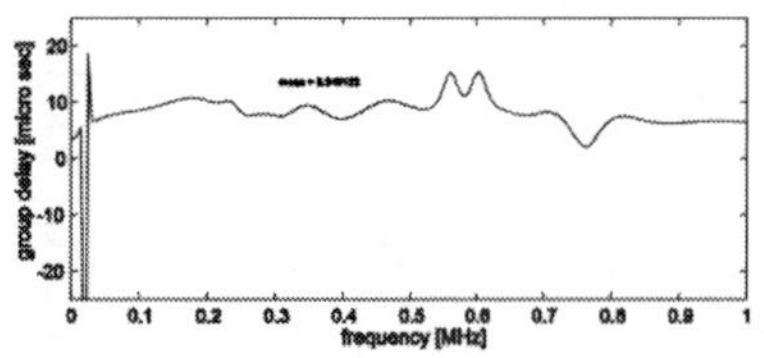

Fig.10. The calibrated transit time of the shear wave.

Figure 11 shows the variation of shear wave velocity with varying a degree of saturation. The shear wave velocity increases with increasing the degree of saturation. The velocity increases by ~200 m/s over 50% to 100% range. It can be seen that the water in compacted bentonite has a significant effect on longitudinal wave velocity than shear wave velocity.

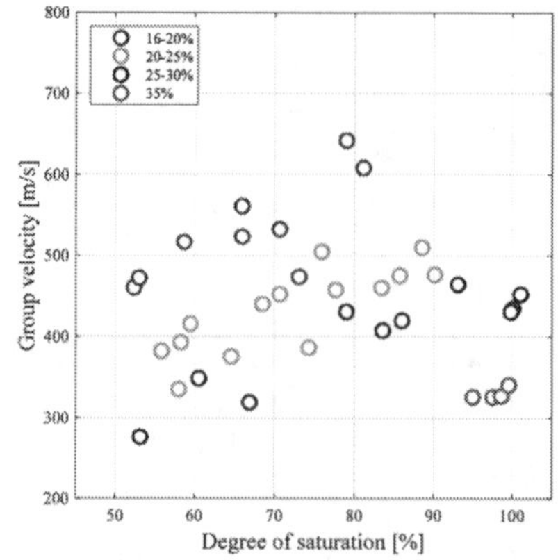

Fig.11. The variation of shear wave velocity with varying a degree of saturation.

The bulk modulus of compacted bentonite was calculated from measured longitudinal and shear wave velocities. Figure 12 shows the variation of the bulk modulus with varying a degree of saturation. The higher the degree of saturation increases the bulk modulus. In addition, the bulk modulus indicates almost the same value for all water content level in the same degree of saturation. Therefore, it is suggested that bulk modulus of compacted bentonite depends only on a degree of saturation. It means a degree of saturation in compacted bentonite can be evaluated by the bulk modulus.

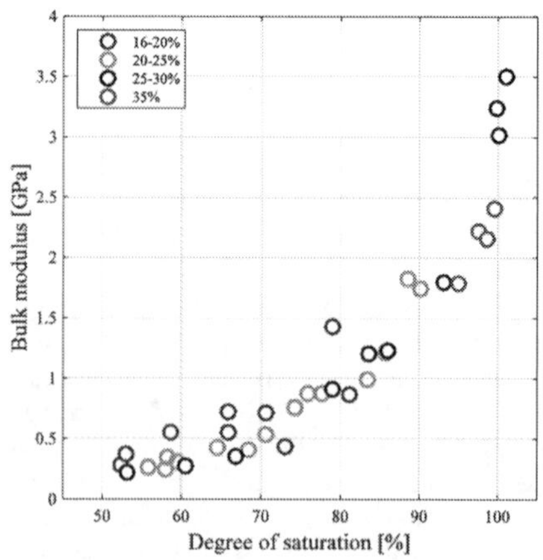

Fig.12. The variation of the bulk modulus with varying a degree of saturation.

4 Conclusions

The non-destructive ultrasonic measurement technique for monitoring of water saturation level in compacted bentonite was proposed. Longitudinal and shear wave velocities measurement for water content investigation was carried out. Frequency domain method was applied to the wave velocity calculation. The relation between longitudinal and shear wave velocities and a degree of saturation was investigated. Then, the bulk modulus of compacted bentonite was evaluated from measured velocities. Consequently, conclusions are as follows:

1. The ultrasonic wave attenuation in compacted bentonite was observed. In particularly, shear wave was distorted strongly. However, the frequency domain method for transit time determination can be applied in the high attenuation material like compacted bentonite. It was confirmed that this method can measure longitudinal and shear wave velocities in compacted bentonite.
2. Longitudinal and shear wave velocities increase with increasing a degree of saturation. The tremendous effect of water in compacted bentonite on longitudinal wave velocity was confirmed. The longitudinal velocity for fully saturated condition is very close to the velocity in water of 1,480 m/s.
3. The bulk modulus of compacted bentonite was evaluated from measured wave velocities. The higher the degree of saturation increases the bulk modulus. The bulk modulus of compacted bentonite depends only on a degree of saturation. It can be seen a degree of saturation in compacted bentonite can be evaluated by the bulk modulus.

References

1. N. Tisato, S. Marelli, J. Geophys. Res.: Solid Earth **118**, pp. 3380-3393 (2013)
2. S. Marelli, E. Manukyan, E. Maurer, S.A. Greenhalgh, A.G Green, Geophysics, **75**, Q21-Q34 (2010)
3. H. Sato, T. Ashida, Y. Kohashi, M. Yui, N. Sasaki, J. Nucl. Sci. Technol., **29**, pp. 873-882 (1992).
4. A. Viana da Fonseca, C. Ferreira, M. Fahey, Geotech. Test. J., **32**, pp. 91-107 (2009).

E3S Web of Conferences **43**, 01017 (2018)
ASTECHNOVA 2017
https://doi.org/10.1051/e3sconf/20184301017

Activity Based Smart Lighting Control For Energy Efficient Building By Neural Network Model

Ivan Hidayat[1], *Faridah*[2*], and *Sentagi Sesotya* Utami[2]

[1]Student of Engineering Physics Department, Gadjah Mada University, Indonesia.
[2]Lecturer of Engineering Physics Departement, Gadjah Mada Unviersity, Indonesia.

Abstract. The goal of this research is to attain a smart control algorithm that can be used in a lighting system based on the activities of a building's occupants, using the neutral network model method. The study case of this research is the activities inside the Asrama Mahasiswa Kinanti UGM building. Control algorithm was built based on qualitative data from the occupants of the building, which were more or less the daily activities of the occupants. The results of the qualitative data will be essential in choosing a sensor and its placement. Several scenarios of activities represented by the combination of sensors' outputs are used as the control system input. The optimum illumination of the lighting system for these scenarios was produced through simulation using DIALux. An artificial neural network model was then developed and used as the smart control algorithm. Input for the neural network is the combination of sensor output and illumination output for each scenarios, given the simulation results. Based on the qualitative data acquired through a survey of the occupants' activities, the design of the lighting control system requires a system that uses occupancy sensors, weight sensors, photoelectric sensors, and photo sensors. The various positions and activities being done by the occupants are represented by the sensors output. A manual remote will be used to adjust the sensors regarding details that cannot be specifically detected. Ongoing specific activities inside the inhabited room gives off a system output. This scenario portrays the lighting conditions of the room, which includes the number of lights that are turned on or turned off. A smart control algorithm was developed using the backpropagation neural network model with 10 neuron inputs, the first hidden layer with 20 neurons, second hidden layer with 20 neurons, whilst the output layer has 5 neurons. The activated function for the first hidden layer is tan-sigmoid, for the second hidden layer is log-sigmoid, and the output layer is using pure linear. The training function uses trainlm. The MSE system's value is 2.72 x 10-8 with a larger R total value, which is 0.99892.

1 Problem Statement

Visual comfort is one of the important factors of a building that needs to be considered. Visual comfort relates on how well a person can work under the lighting condition provided. In order to achieve the desirable visual comfort, most buildings are designed with artificial lighting systems, which consume electricity or energy. The electricity energy for lighting system is approximately 20 – 40% of the total energy consumption of a building [1].

Several attempts have been done to reduce the lighting system energy consumption, which mostly by replacing the light bulbs with highly efficient bulbs such as LED. By still creating the same visual comfort, the lighting system energy consumption can be reduced up to 55% [2].

Building automation technology is another alternative way to reduce the energy consumption. Many innovations of lighting automation system have been done. Target of the design is a system that consumes less energy but still maintains occupant's visual comfort. A system that enables to perform such task by utilizing daylight as the light source. Not only it is energy efficient, it also creates light quality that is suitable to human's visual response [3,4]. Daylight sensor is use to depict the availability of daylighting in a room. Some researches have added occupancy sensor to identify a human presence in a room [5,6]. Based on the sensors' outputs, scenarios of control are created such as on off switch control or dimming, where the goal is to avoid excessive energy consumption of the lighting system.

Human activities are also a factor that can be used as the consideration in developing a lighting control system. Location of occupants and their activities in a building can also be used to define the amount of lighting intensity required for each part of the lighting system. Building automation technology using smart control algorithm enables a lighting system to perform well while considering the human's location and activity factor. A method that is commonly used to develop a smart control algorithm is Artificial Neural Network [6,7].

Kinanthi dormitory is a student's dormitory owned by Universitas Gadjah Mada (UGM) with total occupants of 368 persons. Each room has a capacity of two persons. It is one of the green building icon for UGM. As a green building, the building energy consumption should be efficient. Therefore, lighting automation system in this dormitory is needed to create a

* Corresponding author: faridah@ugm.ac.id

living space that is energy efficient and still supports the students' academic needs.

In this paper, the research aim to create a smart control algorithm for lighting system based on occupant's activity pattern using artificial neural network. Kinanthi dormitory in UGM is used as the case study. Quantitative data were obtained from observation on the occupant's daily activity pattern and used as a consideration in selecting the sensor and its placement. Several control scenarios representing output of the sensors are being ordered and used as the control system inputs. The optimal lighting intensity of the lighting system produced by these scenarios was observed through computer simulation using Dialux. The SNI 03-6197-2000 was used as preferences. The artificial neural network was then developed as the control system algorithm by inserting several combinations of the sensors' outputs and the lighting intensity outputs obtained from each scenarios.

2 Methodology

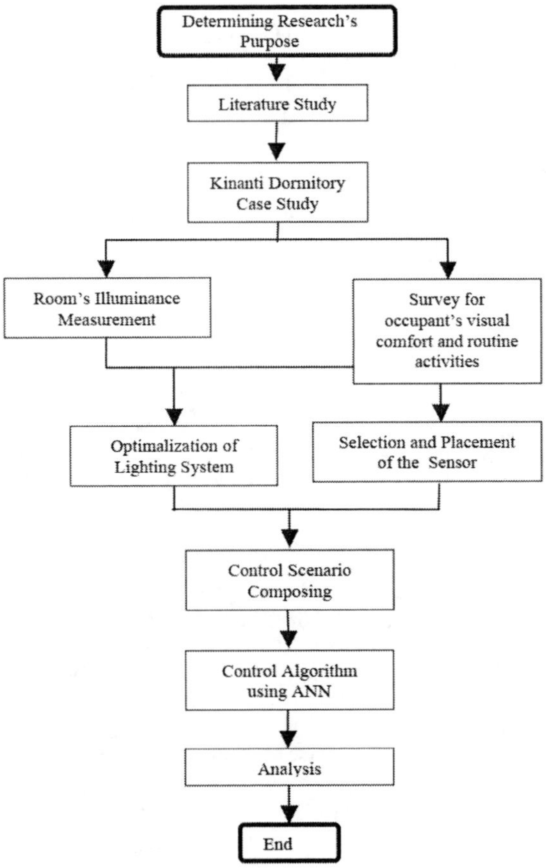

Fig 1. Research Methodology.

The research diagram is shown in Figure 1. The profile of Kinanthi dormitory in UGM used as the case studied was obtained from the interior design drawings as shown in Figure 2.a. However, after the building was built, the interior is slightly different than the design as shown in Figure 2.b.

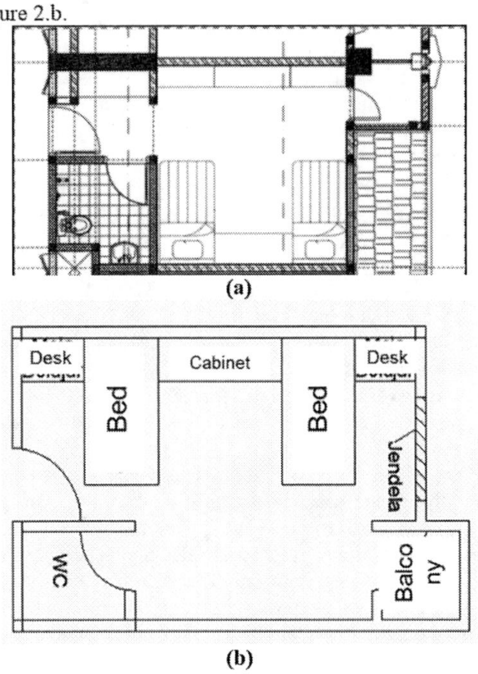

Fig.2. Living area of Kinanthi Dormitory UGM: (a) As built drawing (b) existing.

Table 1. Questionaire's questions about exting visual comfort.

No.	Desired Information	Question
1	Occupants' visual comfort. This information is used to know whether lighting system need to be improved or not.	Do you still meet problems to do activities with lighting in the room? (for example: too dark or blinded by the glare)
2	Daylighting's comfort	Is the sunlight too bright to be used as light source in the afternoon?
3	Certain area in the bedroom that is comfortable to study.	Which area in the bedroom that you feel comfortable for studying?
4	Study's Time	What time do you study everyday?
5	Comfortable lighting to sleep.	How is the lighting that you wanted to sleep?
6	Time spent by the occupant to stay inside the room for each day.	How long do you stay inside the room for each day?

Each space rent in the dormitory provides space for desks for studying, beds, and an indoor bathroom. This research focuses on the functioning of the space(s) for studying and rest activities only. The Survey was conducted to understand the existing visual comfort and provided qualitative data of the activities that are commonly done by the occupants. The survey utilized questionnaire as the instrument with random respondents from the occupants. The questionnaire consists of questions that address expected data as shown in Table 1. Respondents responded to the questionnaire by choosing an answer for each question. This closed questionnaire technique was used to simplify the analysis.

Question no. 1-2 in Table 1 are questions related to the visual comfort level of the existing lighting system. Responses to this questions combined with the illumination measurement result of the room are used to see whether the existing lighting system requires some optimization or not. Meanwhile, question number 3 -5 in Table 1 are questions that addressed the relationship between the occupant's routine activity and other room functions that are actually provided. Responses to these questions are used to select the sensor's type and its placement.

Control scenario is the scenario that might possibly apply for the Kinanthi dormitory lighting system, given the relationship between input and output of the controller. Inputs of the controller output' combinations from the sensors installed. These output' combinations describe types and position of activities and the amount of daylight's availability. Meanwhile, outputs of the controller are actuators' signal that will switch on and off the lighting fixtures of the room, with a combination relevant to the inputs. The illumination distribution from all the combinations action of the actuators was simulated in order to ensure the lighting intensity is suitable for each activity according to SNI 03-6197-2000.

The relations between outputs and inputs that generated the control scenarios are then used to create the Artificial Neural Network model. Briefly, artificial neural network (ANN) is one of algorithm machine that simulates the work of living creature nerves system. The ANN can be used to model a complex relation between input and output of a system in order to find the data pattern.

3 Results

3.1 Illumination existing condition

Figure 3 provides the lighting intensity measurement of the dormitory room. From the measurement results, it shows that during the daytime, the daylighting covers most of the space with an intensity level that complies the SNI standard for studying. However, during night time, the artificial lighting is not providing a sufficient intensity level for studying on the workspace area.

Meanwhile, from the survey result of the visual comfort level, there are several findings as follow. First, 50% of the respondents feel neutral of the lighting quality. This indicates that occupants do not have complains about the darkness or brightness of the room. Thirty percent of the respondents agree that there is a lighting problem in the room and 20% disagree. The second result is that 40% of the respondents are not seeing daylighting as a source light that creates glare. Forty percent feels normal to have daylighting during daytime and 20% agree that sunlight is to glare for a daytime light source.

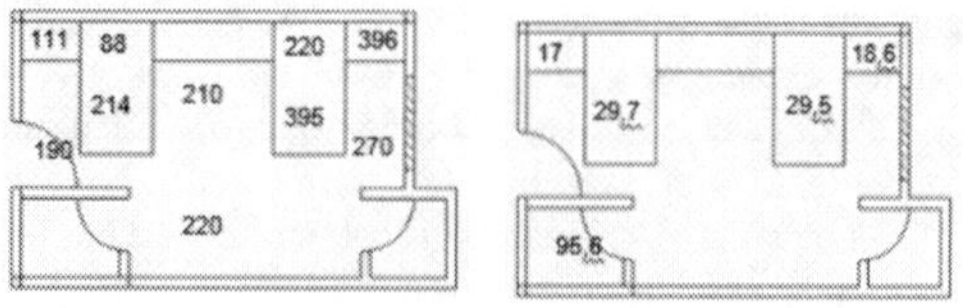

Fig.3. In-site measurement result in the living area (lux) (left) daytime with daylighting, and (right) at night using artificial lighting.

The lighting system optimization is required based on the result of the measurement and visual comfort survey. Furthermore, artificial lighting condition should be improved meanwhile the daylighting is already acceptable. Improvement of the artificial lighting is necessary since the illumination level of the existing condition does not meet the standard even though there were no complaints from the occupants.

3.2 Artificial Lighting Optimization

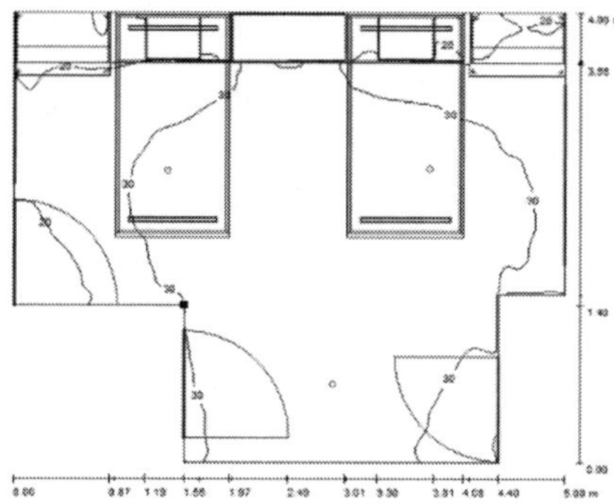

Surface	ρ [%]	E_{av} [lx]	E_{min} [lx]	E_{max} [lx]	u0
Workplane	/	30	7.28	39	0.241
Floor	88	17	0.31	33	0.019
Ceiling	70	17	0.50	41	0.030
Walls (9)	86	19	0.00	76	/

Fig. 4. Simulation of the existing condition.

Results of the simulated lighting condition of the existing room are shown in Figure 4. Similar to the onsite measurement results, the simulation results were also far from the standardized visual comfort mentioned in SNI. Kinanthi dormitory still requires some changes in its artificial lighting given the average illumination of 30 lux, the highest value of 39 lux and the lowest at 7,28 lux.

An aide to the illumination quality, data obtained from in-situ measurement and computer simulation are similar by means the room modeling is close to the actual condition. This model is then used to create an optimal lighting design. Reflection factor selected from materials for the simulation is similar to the real room.

There are several optimizations done in the computer simulation. First, by replacing the LED light bulbs into LED with a higher efficacy. Three lights were replaced using Phillips CorePro LEDspot MV GU10 15W 1521 lm. Figure 5.a shows the simulated illumination distribution after the light bulbs were replaced. The illumination on some area at the room corners are still not sufficient and therefore, it is still lack of lighting on the study desk. A desk lamp is needed, using Philips WL 121V LED5S/830 of 8 watts and 500 lm (see Figure 6). The simulation results of the illumination distribution can be seen in Figure 5b.

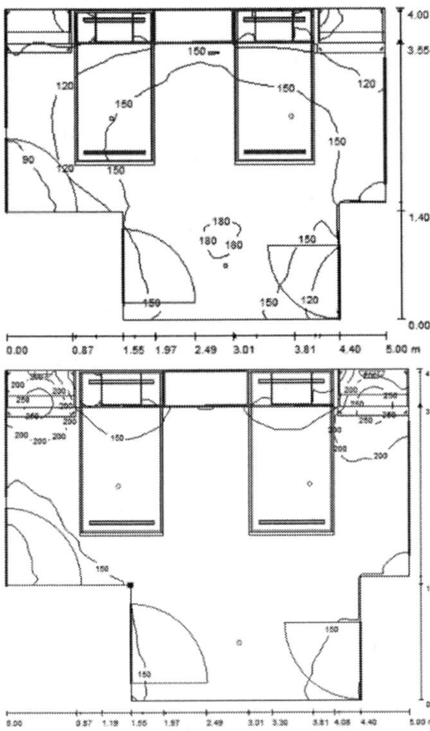

Fig.5. Simulation result of lighting intensity distribution in the living area for optimum condition: (a) without an added fixture, and (b) with an added fixture, on the study desk.

By replacing the light bulb and adding more luminaire, the lighting quality of Kinanthi dormitory room already fulfilled the required visual comfort standard provided by SNI. The average illumination on the work-plane is 172 lux with the lowest at 49 lux and the highest illumination at 288 lux. Another satisfying condition is the maximum lighting load of 3,6 W/m2, which is lower than the (SIN specified maximum lighting load) SNI of 17 W/m2. Table 3 provides the comparison of the existing condition with the artificial lighting optimization of the dorm.

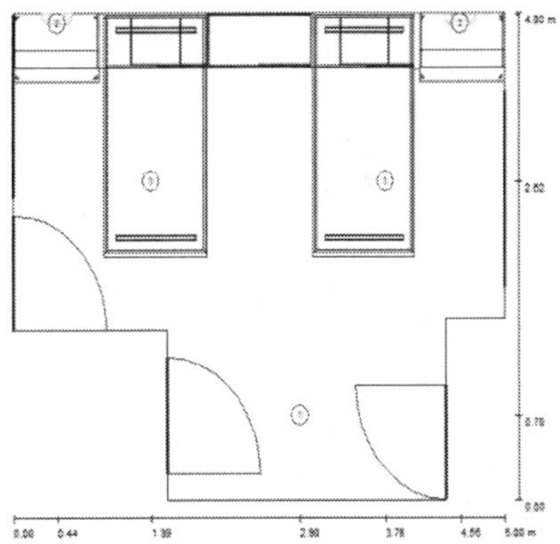

Fig.6. Fixture placement in optimum condition.

Table 3. Comparison between existing condition and optimum artificial lighting in the living area.

Parameter	*Existing Condition*	*Optimum condition*
Average illumance on workplane (lx)	30	172
Energy usage per square area (W/m^2)	1,06	3,06
Total *Luminaire* (lm)	1050	5563
Total Power (W)	18	61

3.3 Sensor and its placement

Several survey results of the common occupant's activity in the Kinanthi Dormitory UGM are as follows,

1. Bed and study desk are areas that are commonly used for studying.
2. The usage pattern of study desk one and two are quite different, which means that each person uses their study desk at the different time.
3. There are respondents wanted their light off during sleeping, dimmed and left on by 50%, 30%, and 20%, respectively.

Based on the survey result, sensors needed for the control system were chosen, where outputs of these sensors were used to provide information on the occupant's activity and their positions in the room for each activity. Sensors used are occupancy sensors, weight sensors, photoelectric sensors and also photosensor. Placement of the sensors is described in Figure 7. Each sensor has their own function. The weight sensor is used to depict whether there is a person on the bed or not. The photoelectric sensor is used to depict whether the person is sitting or not while he was on the bed.

Three types of occupation sensors are used. First is to depict the presence of a person inside the room. Sensor two and three are used to detect whether the person is studying on the desk or not. The photosensor is used to detect the amount of intensity that entered the occupation space. Since most of the daylighting hit the window, therefore, the photosensor is able to find the correlation between the amount of daylight that impinged the window and the amount of intensity level that penetrates into the living area.

Besides the sensors mentioned above, remotes are also used to provide input to the sensor for several stimuli or variable that were not able to be detected specifically. For instance, occupants laying on the bed while reading a book, there is a desire between lighting level for sleeping but not totally shut down. This has created 10 types of inputs to the control system which are, 2 inputs are using weight sensors (W1 and W2), 2 photoelectric sensors (P1 and P2), 3 occupancy sensors (SO1, SO2, and SO3), 1 photosensor, and 2 inputs from the remote (R1 for turning on the light when a person is reading while lying down; R2 for dimming the ambient room light and only leaving on the both desklamp).

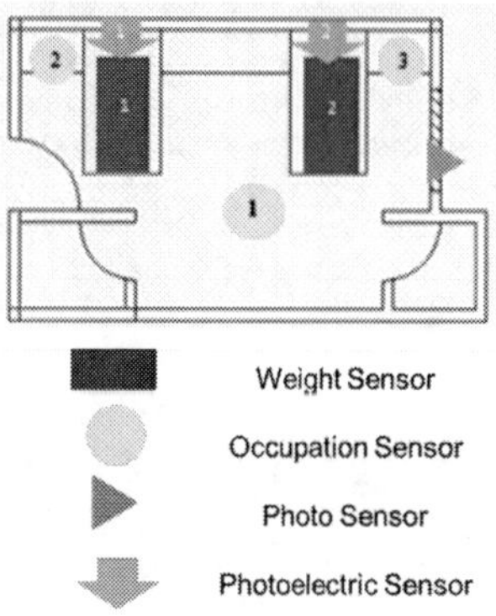

Fig.7. Sensor placement in the living area.

3.4 Control scenarios

Some activity's scenarios were represented by the combination of the arranged sensor output and are used as the data for the system control input. Inputs from the sensor are binary numbers. Using all 10 inputs discussed above, we will have a combination of 2^{10} or 1024 types of input. In this case, each input is not necessarily representing a certain occupant's activity. Activities that are given priority are the ones that require a higher illumination level. For example, in the living area, there might exist a person studying and a person sleeping, therefore, all the light fixtures will be turned on to support the studying activity. Meanwhile, if there are 2 or more simultaneous input from the remote, then none of the input is taken into consideration since the remote might be accidentally pressed.

Fig.8. Flowchart control scenario.

The controller has eight output scenarios relevant to the lighting system manipulation. Each scenario relates to a specific activity and the amount of daylighting available in the living area. Also to keep in mind that all scenarios should comply with the SNI requirements as shown in Table 4.

Table 4. Scenario of control system outputs.

Scenario	Lighting condition					Average illuminance (lux)	Comply to standard?
	L1	L2	L3	M1	M2		
Scenario 1	1	1	1	1	1	215	Yes
Scenario 2	1	1	1	0	1	157	Yes
Scenario 3	1	1	1	1	0	156	Yes
Scenario 4	1	1	1	0	0	140	Yes
Scenario 5	0	0	0	1	1	32	Yes
Scenario 6	0	0	0	0	1	15	Yes
Scenario 7	0	0	0	1	0	17	Yes
Scenario 8	0	0	0	0	0	0	Yes

3.5 Artificial neural network (ANN)

The neural network model developed is the artificial neural network using Backpropagation training method. The training function used is the trainlm, which during the training, the weighting and bias values of the ANN model are refined using the Levenberg-Marquardt optimization technique. Figure 9 shows the ANN architecture. Input and output of the model are 10 neurons each, which are data output of the sensors. Five of them are on and off condition of 4 fixtures in the dorm. The model also has two hidden layers with each n and m neurons. A number of neurons and the activation function of the hidden layers are varied to obtain a network with a high R correlation values. It is shown from the test result that the most optimal network architecture is the network with 20 neurons at the first hidden layer and 20 neurons at the second hidden layer. The first hidden layer activation function is tan-sigmoid, and the second is log-sigmoid while the output layer activation layer is pure linear. This architecture has R values for the training, validation, testing, and total of 0.98174, 0.97346, 0.99641, and 0.97814, respectively.

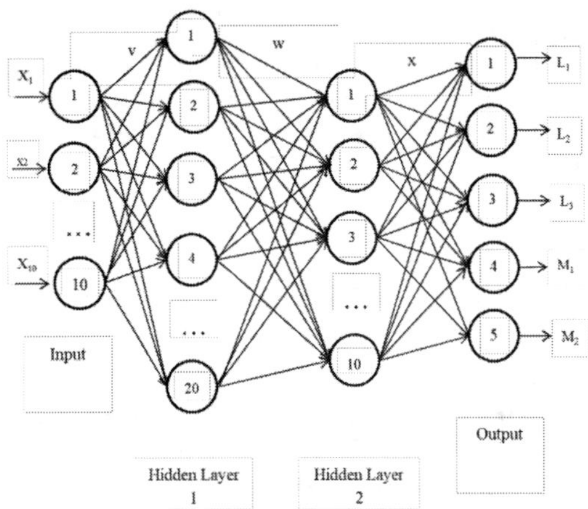

Fig.9. Architecture of the ANN Model.

The input-output pattern recognition of the network is showing that among 1024 data, there are 2 data which outputs are not suitable with the expected results during the data training process. This means that the system's performance is quite good with a system error of 0.2%. Table 5 shows the unwanted data.

Table 5. Error percentage of the designed neural network.

Scenario	Amount of Data	Amount of Error	Error Percentage (%)
Scenario 1	64	0	0
Scenario 2	64	0	0
Scenario 3	64	0	0
Scenario 4	46	0	0
Scenario 5	1	0	0
Scenario 6	2	0	0
Scenario 7	2	1	50
Scenario 8	781	1	1.28
Total	1024	2	0.2

The unrecognized input and output data are given in Table 6. After a close look, it can be seen that actually, the two unrecognized activity data are data that are hardly happening in the real world. In the first case, sensor SO1, W1, P2, and R2 are turned on, which means at the same time, there is activity in the center of the room, on bed 1, on desk 1, and R2 is being pressed. Meanwhile, for the second case, R1 and R2 are on at the same time, most likely the remotes were press at the same time. Therefore, the control scenario flowchart needs to be revised by adding a filter to select impossible scenarios. The refined flowchart is shown in Figure 10.

Table 6. Data with error output.

Input										Expected ouput					Output from iteration process				
SO1	PS	SO2	SO3	W1	W2	P1	P2	R1	R2	L1	L2	L3	M1	M2	L1	L2	L3	M1	M2
1	0	0	0	1	0	0	1	0	1	0	0	0	1	0	0	0	0	0	0
1	0	0	0	1	1	0	0	1	1	0	0	0	0	0	1	1	1	0	0

With the revised control scenario flowchart, another training of the system was done to improve the weighting values of the network. Comparison of the system's performance before and after the improvement is shown in Table 7. The R-value of the revised system is now larger, which is 0.99892 with an error of 0%.

Fig.10. Control scenario flowchart with an added filter.

E3S Web of Conferences **43**, 01017 (2018) https://doi.org/10.1051/e3sconf/20184301017
ASTECHNOVA 2017

Table 7. Comparison of the system's performances before and after improvement.

	Before improvement	After improvement
Amount of data	1024	1022
Epoch	58	30
Training performance (MSE)	$5{,}66.10^{-7}$	$2{,}72.10^{-8}$
Gradient	$7{,}88.10^{-6}$	$9{,}09.10^{-6}$
Error percentage	0,2%	0%
R train	0,99999	0,99973
R validation	0,9944	0,99681
R tes	0,98383	0,99734
R total	0,9965	0,99892

4 Conclusion

1. The existing artificial lighting condition has not yet met the standard SNI and therefore, optimization is required by replacing the luminance of the main light from 350lm to 1521 lm. Also by adding a luminaire at the desk to 500 lumens. By replacing and adding the fixtures, illumination in the Kinanthi dormitory room is already fulfilling the visual comfort standard with an illumination average of 172 lux on the workplane.
2. Based on results from the survey on the occupant's activity, the lighting control system is designed by,
   a. Utilizing sensors of occupancy sensor, weight sensors, photoelectric sensor, and photosensor. Output of the sensors represents positions and types of activities of the occupants. For activities that are not specifically detected, a manual remote is use to provide inputs for the sensors.
   b. Specific activities occurred inside the room will create system output of eight scenarios that will most likely happen. The room lighting condition and the number of fixtures switched on and off, are described by these eight scenarios.
3. The smart control algorithm is using backpropagation artificial neural network model with 10 neurons as input, first and second hidden layers each with 20 neurons, and 5 neurons in the output layer. The activation function for the first hidden layer is tan-sigmoid, log-sigmoid was for the second hidden layer, and pure-linear for the output layer. Training function used is trainlm. The total R-value for the system is 0.99892.

References

1. *Commercial building energy consumption survey*, EIA, Washington DC (2003)
2. Wang X., Tjalkens T., Linnartz J.P., *Smart Office Lighting Control Using Occupancy Sensors, Networking, Sensing, and Control* (ICNSC), IEEE 14th International Conference (2017)
3. Liu J., Zhang W., Chu X., Liu Y., *Fuzzy logic controller for energy savings in a smart LED lighting system considering lighting comfort and daylight*, Energy and Buildings **127**, 95–104 (2016)
4. Miranda A.C., Castaño V.M., *Smart frost control in greenhouses by neural networks models,* Computers and Electronics in Agriculture **137**, 102–114 (2017)
5. Labeodan L., De Bukker C., Resemann A., and Zeiler W., *On the Application of Wireless Sensor and Actuators Network in Existing Buildings for Occupancy Detection Occupancy-driven Lighting Control,* Energy and Building **127**, pp 75-83 (2016)
6. Caicedo D., Pandharipande A., *Sensor-Driven Lighting Control with Illumination and Diming Constrains*, IEEE Sensors Journal **15**(9), pp. 5169-5176 (2016)
7. Wang Z., Tan Y.K., *Illumination control of LED systems based on neural network model and energy optimization algorithm*, Energy and Buildings **62**, 514–521 (2013)

Performance Test of Solar-Powered Ice Maker: Case Study in South Lampung

Putri Wullandari[1], *Arif Rahman* Hakim[1], and *Widiarto* Sarwono[1]

[1]Indonesian Research Institute for Fisheries Post-harvest Mechanization

Abstract. The performance test of solar powered ice maker is being held at the fish auction spot of Kramat, South Lampung for 6 hours. The spesification of ice maker are: having the production capacity of 105 – 120 kg/ day, producing flakes ice with 2 x 3 x 3 mm dimension. The spesification of solar panel are: having maximum power of 200 W, 1,32 m length, 0,992 m width, and 1,3094 m^2 area. In this test, we used 9 solar panel with the total area of 11,7846 m^2. During the performance test, the intensity of sunlight, battery voltage, battery capacity, power input from solar panel, battery discharge current, maximum power consumption of ice maker, and total ice production are being measured. The sunlight-intensity value (lumen / m^2) and the power input of the solar panel (watts) was plotted into a graph. Once the graph is created then add the trendline and create the equation. In a sunny-yet-cloudy weather conditions (at 24-33°C temperature and 65-95% humidity) in South Lampung, the relationship between the sunlight intensity with the power input of the solar panel is linear with the equation y = 0,008x + 143.6. Where the maximum intensity of sunlight is 121600 lm / m² with the maximum power input of the solar panel (9 x 200Wp) is 1253 W. Total ice production for 6 hours is 28.42 kg or 4.74 kg per hour. This production resulted from a 760 W ice maker with a starting current of 4.6 A with a total power of 3.94 kWh.

1 Introduction

In recent years, increasing attention is being given to the use of waste heat and solar energy in energising refrigerating systems. Solar powered refrigeration and air conditioning have been very attractive during the last twenty years, since the availability of sunshine and the need for refrigeration both reach maximum levels in the same season. One of the most effective forms of solar refrigeration is in the production of ice, as ice can accumulate much latent heat, thus the size of the ice maker can be made small.

The cooling system has become an important part of human life, most of the refrigeration system is sourced from electrical energy [1]. However, electrical energy has not reached all areas of human residence, especially for people living in rural areas.

One of the alternatives to substitute electrical energy is the use of solar energy. The use of solar energy is one of the important contributions to reduce fossil fuel consumption and harmful emissions for the environment. Meanwhile, a solar-powered cooler for food, beverages and seafood or air-conditioning is an interesting solar energy application because both the supply of insulation and the need for cooling reach maximum levels in the same period [2].

The sun supplies energy to the earth in the form of radiation. Every year there is 3.9×10^{24} Joule or equal to 1.08×10^{18} kWh solar energy reaches the earth's surface. The intensity of solar radiation outside the atmosphere depends on the distance between the earth and the sun. Throughout the year, the distance between the earth and the sun varies between 1.47×10^8 km to 1.52×10^8 km. As a result, irradiance (E_0) fluctuates between 1,325 $W.m^2$ to 1,412 W/m^2. The average value of irradiance is called solar constant, E_0 = 1,367 W/m^2 [3].

The biggest difference between ice making and air conditioning is that the evaporation temperature for ice making is about -5²C, which is lower than that for air conditioning, leading to low efficiency and high pressure ratio for compressor used for ice maker [2].

* Corresponding author: utides@gmail.com

2 Methodology

2.1 Materials

Materials used in the process of machine design are stainless steel or plate, capillary pipe, compressor, condenser, condenser fan, accumulator, expansion / expansion valve, evaporator, and other mechanical / electrical materials such as cable, insulation, switch and fuse, R404a refrigerant, solar panel, battery, solar charge controller, cable, inverter, water pump, neodymium magnet, motor-generator type brush.

2.2 Equipment

Equipment used in the design process of machine is bending tools, welders, welding tools, grinding wheels, sealed cutting tools, saws, hand drills, miser, and other workshop equipment. The equipment used for performance testing are luxmeter, ampere pliers, laptop,

 https://doi.org/10.1051/e3sconf/20184301018

stopwatch, mercury thermometer, digital scales, , electric current meter, and electric voltage meter.

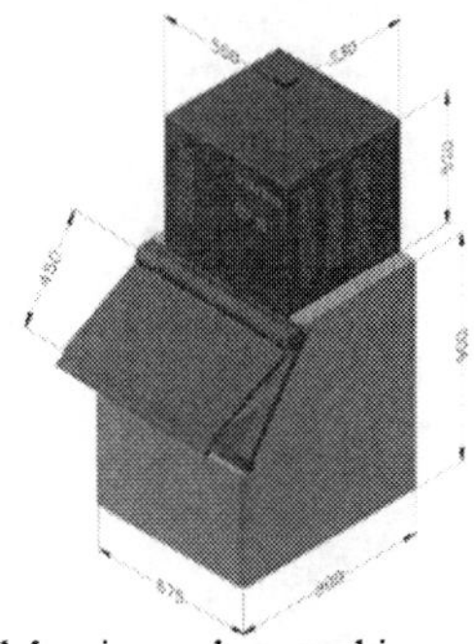

Figure 1. Flakes ice maker machine.

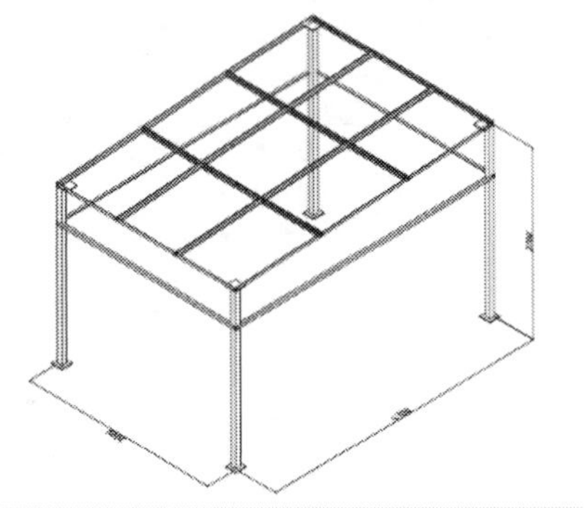

Figure 2. Solar panel's frame.

Figure 3. Box for storing inverters and batteries.

The specifications of solar panels that being used in this research are presented in Table 1.

Table 1. Solar panel's spesification.

Merk	:	ICA solar
Solar Panel Type	:	*Polycrystalline*
Maximum Power (Pmax)	:	200 W
Maximum Power Voltage (Vmp)	:	35.8 V
Maximum Power Current (Imp)	:	5.59 A
Open Circuit Voltage (Voc)	:	44.85 V
Short Circuit Current (Isc)	:	5.92 A
Nominal Operating Cell Temp. (NOCT)	:	45°C ± 2°C
Maximum System Voltage	:	1000 VDC
Maximum Series Fuse	:	10 A
For field connections, use minimum No.12AWG *copper*		
Wires insulated for a minimum 90°C		
Weight	:	15.3 kg
Dimension	:	1320 × 992 × 40 mm
Standard Test Condition	:	Temp. 25°C
	:	Air Mass = 1.5
	:	Irradiance = 1000 W/m^2

2.3 Methods

The method used in this research is presented in Figure 3.

This performance test of solar-powered ice maker is part of machine performance test by users. The details are : first we assembly the flakes ice maker machine and the solar panel, and then we operate the flakes ice maker machine and the solar panel for 6 hours while measuring the intensity of sunlight, battery voltage, battery capacity, power input from solar panel, battery discharge current, maximum power consumption of ice maker, and total ice production. After that, we process the data using Microsoft Excel. The sunlight intensity and the power input from solar panel are being plotted in the graph to determine the relationship that will be shown in the equation.

 https://doi.org/10.1051/e3sconf/20184301018

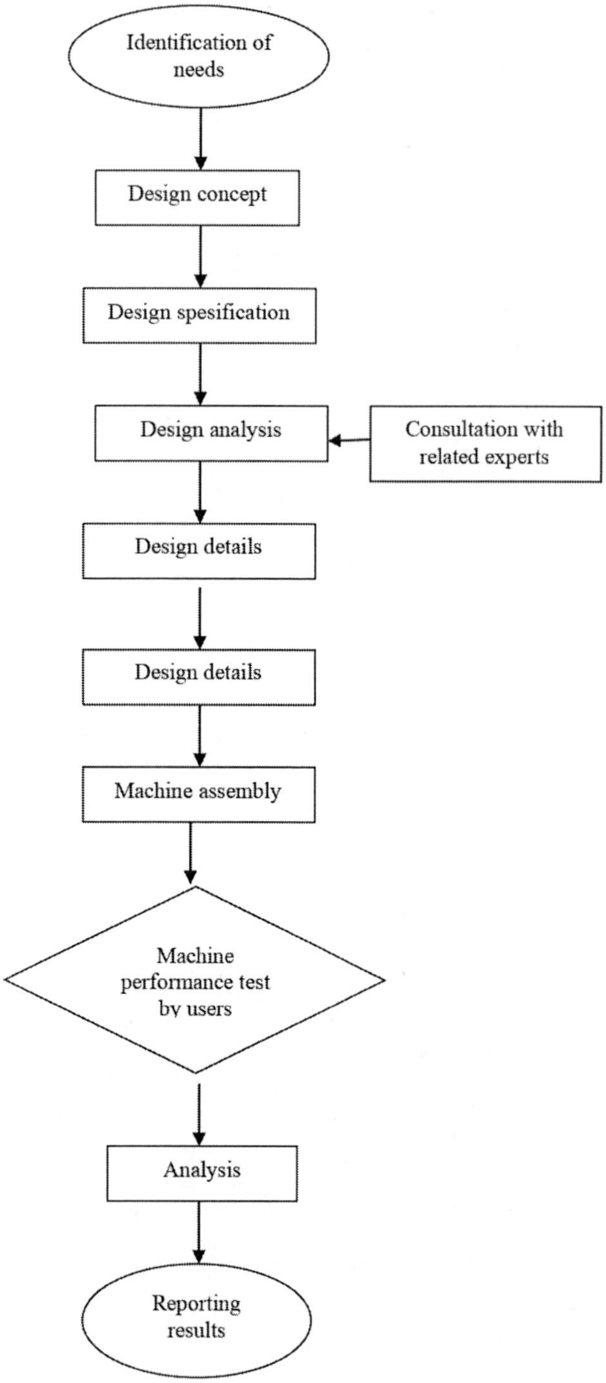

Figure 3. Flow diagram of the research.

3 Results and discussion

Tempat Pelelangan Ikan Kramat is located in Dusun Kramat, Sumur Village, Ketapang Subdistrict, South Lampung Regency. This is a landing place for local fisherman. The location of TPI Kramat is very appropriate to be a field-location test for ice maker because the supply of freshwater and electricity in this area is sufficient.

The needs of ice as a cooler in this location is not too big that only seven blocks of ice for normal conditions and at the time of fish season the required amount reaches twice as much.

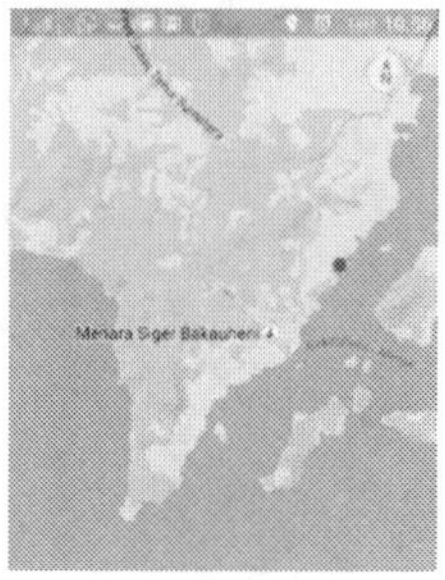

Figure 4. TPI Kramat's location.

The stages of assembly activities of the ice maker machine are :

a. Assembling 9 solar panels to get the power of 2000W.
b. Preparing the arrangement of ice maker machine and hybrid inverter / charger
c. Preparing fresh water availability for ice machine water supply. The fresh water is obtained from drilled wells that is being patched in reservoirs.

Figure 5. TPI Kramat's condition.

Figure 6. Assembling of solar panel's frame.

Solar powered-ice maker performance test is done by measuring the intensity of sunlight (lumen) per time unit, recording the power input data and voltage from solar panel, voltage and battery capacity, charging current of battery, discharge current of battery, and weather conditions at the time of testing.

Display of power input and voltage data from solar panel, voltage and battery capacity, charging current of battery, discharging current from battery is shown in Figure 7.

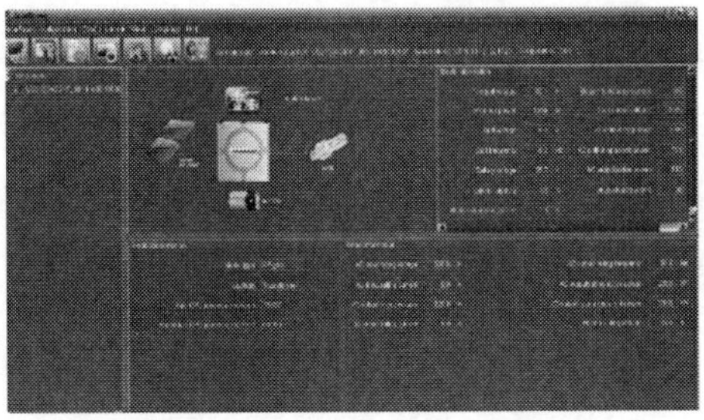

Figure 7. Display of power input and voltage data from solar panel, voltage and battery capacity, charging current of battery, discharging current from battery.

We can save this data in Microsoft Excel for further data processing. The data was recorded every minutes, but for the purposes of data presentation, data is plotted every 10 minutes. The performance test itself was being held for 6 hours, which is from 10 a.m until 4 p.m. The performance test is done at 10 o'clock because the distance is far and the team must buy some measuring instrument first.

The weather condition in Lampung, especially Kalianda, was cloudy with temperature of 27-31°C (MSN cuaca). The environmental temperature increases until 13 p.m (31°C) and then decreases until 16 p.m (27°C).

The electric voltage and current that produced from solar cell is affected by two physical variables, which are the intensity of solar radiation and environmental temperature [4]. The intensity of solar radiation. received by solar cells is proportional to the voltage and current generated by the solar cells. If the environmental temperature is higher with the fixed-intensity of solar radiation, then the solar panel voltage will decrease and the electric current generated will increase [5].

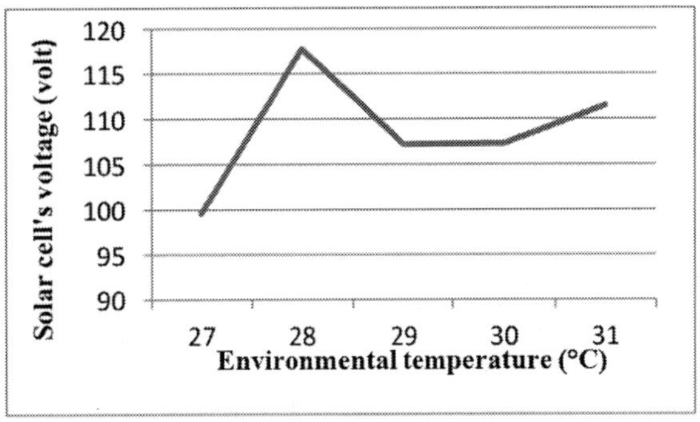

Figure 8. The effect of environmental temperature to solar cell's voltage in South Lampung.

Figure 8 shows that the solar cell's voltage increases significantly when the environmental temperatures increases from 27°C to 28°C, and the highest solar cell's voltage is 117,8 voltage at the environmental temperature of 28°C.

The intensity of solar radiation and the average environmental temperature in South Lampung is shown in Table 1.

Table 1. The intensity of solar radiation and the average environmental temperature in South Lampung on 2016 [6]

Month	The intensity of solar radiation (Joule/m^2)	Temperature (°C)
May	120	28
June	152	27.2
July	115	26.8
August	148	27.4
September	171	27.4
October	143	27.1

Table 1 shows that the highest intensity of solar radiation during May-October 2016 is on September, which is 171 Joule/m^2. Meanwhile the lowest intensity of solar radiation during May-October 2016 is on July, which is 115 Joule/m^2.

The environmental temperature also contributes to temperature's changing in solar panel. Due to the temperature rise, the electric power produced by solar panels is reduced [7].

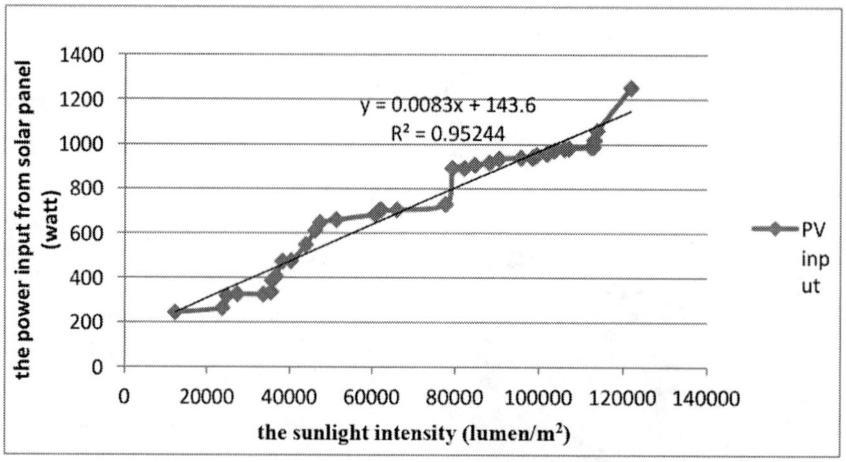

Figure 9. The relationship between the sunlight intensity and the power input from solar panel in South Lampung

From the experimental results, it can be obtained that the minimum sunlight intensity data is 12,010 lumens / m^2, while the maximum solar light intensity is 121,600 lumens / m^2.

According to [3], the earth's atmosphere will reduce the solar radiation through reflection, absorption process (by ozone, water vapor, oxygen, and carbondioxide) and scattering (by air molecules, dust particles, or pollution). For sunny weather during the day, irradiant that reaches the earth's surface is 1,000 W/m^2. Maximum insolation occurs on a clear but partially cloudy day. This is due to the reflection of solar radiation by clouds.

The minimum power input from the solar panel is 245 watts, while the maximum power input from the solar panel is 1253 watts. Minimum battery capacity is 95% at a voltage of 24.4 volt while the maximum battery capacity is 100% at 27 volt. Total ice production generated for 6 hours is 28.42 kg ice or in average per hour generated 4.74 kg.

The relationship between the sunlight intensity and the power input from solar panel in South Lampung is shown in Figure 9.

Figure 9 shows that the relationship between the sunlight intensity and the power input from solar panel in South Lampung is linear with the equation of y = 0,008x + 143,6. Then the reading of its result is as follows :

1) The power input from solar panel depends on the sunlight intensity.
2) A value of 0.008x is also called a slope that determines the direction of linear regression. The higher the intensity of sunlight then the power input from the solar panel is also higher (the relationship is positive).
3) This slope also estimates the increasing rate in power input from solar panels every day.
4) A value of 143.6 is also called an intercept. In this case the intercept means that at the value of X = 0, the power input value is 143.6 watt.
5) The determination coefficient (R^2) is 0,952, so the correlation coefficient (R) is 0.975. this means the correlation between the intensity of sunlight with the power input from solar panel is 0.975. The meaning of the correlation value can be seen in Table 2.

Table 2. The explanation of Correlation coefficient's value [8]

Correlation coefficient's value	Explanation
0-0.199	Very low
0.2-0.399	Low
0.4-0.599	Enough
0.6-0.799	Strong
0.8-1	Very strong

It means that the correlation between the intensity of sunlight with the power input from solar panel is very strong.

6) Determination coefficient of 95.2 means as much as 95.2% change in power input from solar panel is affected by the sunlight intensity. While the rest of 4.8% (100% - 952%) is another factor outside the independent variables.

4 Conclusion

1) The highest solar cell's voltage is 117,8 voltage at the environmental temperature of 28°C.
2) The power input from solar panel depends on the sunlight intensity. The higher the intensity of sunlight then the power input from the solar panel is also higher (the relationship is positive).

3) The correlation between the intensity of sunlight with the power input from solar panel is very strong.
4) Ninety five point two percent changes in power input from solar panel is affected by the sunlight intensity. While the rest of 4.8% is another factor outside the independent variables.
5) Solar-powered ice maker produces 4.74 kg per hour.

in further research, it is better to use solar power meter to measure the intensity of solar radiation.

References

1 N.A.A. Qasem, and M.A.I. El-Sharawi, Solar Energy, *Improving ice productivity and performance for an activated carbon/metanol solar adsorption ice-maker*, **98,** 523-542 (2013)
2 X.B Bu, H.S Li, and L.B. Wang, *Performance analysis and working fluids selection of solar powered rankine-vapor compression ice maker*, Solar Energy **95**, 271-278 (2013)
3 Muchammad dan E. Yohana, *Pengaruh suhu permukaan photovoltaic module 50 watt peak terhadap daya keluaran yang dihasilkan menggunakan reflektor dengan variasi sudut reflektor 0°, 50°, 60°, 70°, 80°*, Rotasi **12**, 14-18 (2010)
4 D. Suryana and M.M, Ali, *Pengaruh temperatur/ suhu terhadap tegangan yang dihasilkan panel surya jenis monokristalin (studi kasus : Baristand Industri Surabaya)*, Jurnal Teknologi Proses dan Inovasi Industri **2**, 49-52 (2016)
5 J. Jiang, T. Huang, Y. Hsiao, and C.Chen, *Maximum power tracking for photovoltaic power system*, Tamkang J. Sci. Eng. **8**, 147-153 (2005)
6 Annual report of the Meteorology and Geophysics Agency, South Lampung
7 K. H. Khwee, *Pengaruh temperatur terhadap kapasitas daya panel surya*, Jurnal ELKHA **5**, 23-25 (2013)
8 M. Anggraeni, *Kajian Penggunaan Poly Alumunium Chloride (PAC) Dalam Proses Pemurnian Nira Aren dan Lama Pemurnian Terhadap Karateristik Nira Aren (Arenga pinnata Merr)*, Skripsi Fakultas Teknologi Industri Pertanian, Universitas Padjadjaran (2008)

 https://doi.org/10.1051/e3sconf/20184301019

Performance Test of VHF-HF-based Emergency Data Communication System between the Sardjito Hospital and District Hospital in the Isolated and Disaster-prone Areas

Singgih Hawibowo[1], *Sunarno*[1], *Memory Motivanisman Waruwu*[1,*], *Yakub Fahim Luckyarno*[2], *Rony Wijaya*[3], and *Samuel Kristiyana*[4]

[1]Department of Nuclear Engineering and Engineering Physics, Universitas Gadjah Mada, Yogyakarta, Indonesia
[2]International College, King Mongkut's Institute of Technology Ladkrabang, Bangkok, Thailand
[3]Production House Unit, Faculty of Medicine, Universitas Gadjah Mada, Yogyakarta, Indonesia
[4]Institut Sains & Teknologi AKPRIND Yogyakarta, Indonesia

Abstract. The topography of Gunungkidul area in the Province of Yogyakarta, Indonesia, consisting of hills and limestone mountains. It becomes the potential factor of landslide occurrences. Especially in the rainy season, the soils in the hills become softer and it increases the possibility to cause a landslide. This condition has became the concern of many people, especially the residents who live in the landslide-prone area. This topography is also one of the factors for health centre in a certain area becomes isolated due to the inability of telecommunication signal services from the provider to reach that area, which is called a blank area. That situation makes an inability to contact the nearest district hospital to ask for support in an emergency case. This research focuses on the performance test of the emergency data communication system by using several types of data. The system uses Very High Frequency (VHF) 144 MHz and High Frequency (HF) 7.08 MHz as the data carrier and a software to interpret the text and image to be modulated onto the data carrier. That system is called a virtual hospital, means that even though the health center is located in a blank area, but specialists from the Sardjito Hospital can still deliver advices to support the health center in maintaining the medical service quality by using the VHF-HF-based emergency data communication system. Based on the results of the system's performance test between the Wonosari Gunungkidul District Hospital and the Sardjito Hospital with the distance of 40 km, the receiving quality of the data transmission for text and image are 95% and 55.33%, respectively.

1 Introduction

Gunungkidul is a district in Yogyakarta which prone to disaster. Until 2015, the disaster data records published by the province government shows that Gunungkidul is on the top three areas prone to disaster, after Kulonprogo and Bantul. The location and geographical of Gunungkidul becomes the main factor of several disasters, such as drought, hurricane, and landslide. The topography of Gunungkidul, consisting of hills and limestone mountains, becomes the factor of landslide occurrences. Especially in the rainy season, the soils in the hills become softer and it increases the possibility to cause a landslide. That condition has become the concern of many people, especially the residents who live in the landslide-prone area. The residents should have an understanding of how dangerous the landslide can be, so that the resident's disaster-resilience is increased and all people can be aware of the signs of landslide and do the evacuation immediately.

The awareness improvement on landslide disaster is not only done by the government, but also it should include the participation of the residents around the landslide-prone area. One way to increase the awareness of the residents and as a test for the disaster awareness plan made by the government of the Gunungkidul is through a prevention simulation of landslide disaster. By doing the simulation, the result can be an input for Gunungkidul government, hospitals, and Gunungkidul residents in order to prepare a plan of disaster prevention for the district (Regional Disaster Plan, RDP) and a disaster prevention in the healthcare section for the Ministry of Health and hospitals (Hospital Disaster Plan, HDP). Based on that story, a project from Swedish International Development Cooperation Agency (SIDA), in cooperation with Umea University, UGM, and the government of Gunungkidul, is about an activity for

* Corresponding author: morymw@gmail.com

landslide disaster prevention at Gunungkidul. In this case, we take a role in the training process of how to use the emergency communication system for all staffs in the Ministry of Health in Gunungkidul so that it can be a pioneer in the idea of virtual hospital. Briefly, a virtual hospital is a health consultation service between patient in the isolated area (due to the topography so the telecommunication signal cannot achieve the area) and doctors in the central hospital in downtown (Sardjito Hospital).

2 Theoretical Background

On 2005 hurricane Katrina rendered unusable a significant percentage of the voice and data communication infrastructure in coastal communities of Louisiana and Mississippi. In the days following the hurricane, amateur radio operators arrived in the devastated areas, provided local emergency communication, both voice and data, and established communication with other regions of the United States, using primarily mobile radio transceivers [1].

For decades, the high-frequency (HF) band has been recognized as the primary means of long-range wireless communications. The HF band lies within 2-30 MHz of the electromagnetic spectrum. In this part of the spectrum, propagation via *direct wave*, *surface wave*, *near-vertical incidence skywave* (NVIS), and *skywave* provides a means of communication, from line-of-sight (LOS) to beyond-line-of-sight (BLOS) and over-the-horizon (OTH) ranges [2].

All standard SSTV modes utilize a unique digital code that identifies the mode to a receiving system. The code is called the VIS, or Vertical Interval Signal code. Although the entire calibration header is often referred to as the "VIS code", the code itself is only a part of it. The seven-bit code is transmitted least-significant bit (LSB) first, and uses "even" parity [3].

There are series of experiments and attempts by different radio amateurs but most of them use proprietary transmission systems. One of the oldest and wide spread methods is slow scan television (SSTV). There are different modes of transmission. The team decided to focus in ROBOT36 because is relatively wide spread and it able to transmit colour images. In order to achieve the desired system in small, lightweight system an raspberry Pi was used for image capture, signal processing and signal generation and a custom RF transmitter was developed [4].

In recent years, small satellites have begun to draw attention due to reconstruction in low cost of manufacture and faster building time. In this project, radio transmitter system is built for nano and micro satellites. Transmitter broadcasts 1W power in 435 MHz free amateur satellite communication band, and also has software defined radio structure in all modules except up-conversion part. Design was capable of DE0-Nano evaluation board connection; it is regard of PC104 connector structure standard. FSK modulation and AX.25 data link layer protocol are designed using digital circuit in FPGA and implemented with VHDL [5].

The radio-based data communication subsystem (DCS) between trains and wayside is a key factor for safe and efficient operation of a communication-based train control (CBTC) system. The train control system is now able to provide efficient, flexible, and cost-effective operation by using cutting-edge information and communication technologies collectively referred to as CBTC. CBTC technology has become standardized, and its commercialization trend has witnessed a drastic growth in recent years [6].

Secret sharing is an important cryptographic primitive for enabling multi-party computation. A secret image sharing (SIS) scheme encrypts the secret image into few meaningless shadows or share images. These shadows are distributed among the legitimate set of participants. Only the qualified subsets of participants can co-operate to reveal the secret image and forbidden subset fails to retrieve any information about the same [7].

3 Proposed Method

This research focuses on the performance test of the emergency data communication system by using several types of data. The system uses Very High Frequency (VHF) 144 MHz and High Frequency (HF) 7.08 MHz as the data carrier and a software to interpret the text and image to be modulated onto the data carrier. That system is called a virtual hospital [8], means that even though the District Hospital or Public Health Centre is located in the blank area, but the specialists from the Sardjito Hospital can still support them to maintain the medical service quality by using the VHF-HF-based emergency data communication system. Based on the results of the system's performance test between the Wonosari Gunungkidul District Hospital and the Sardjito Hospital with the distance of 40 km.

Primary data retrieval uses repeater signal test data by sending the audio data, sending the text data and sending the image data. Software MixW 3.11g is used for data communication experiments. Software Mobile Radio is used to perform robust design of emission and coverage of repeater, base station, and remote station.

4 Result and Discussion

4.1. Propagation Analysis between RSUP Sardjito and RSUD Wonosari using 144 MHz

The two-way communication between RSUP Sardjito and RSUD Wonosari with the distance of 33.17 km can be worked but it is not continuous, with the electromagnetic field power is only 3.52 dBμV/m. In the TX power of 5 Watt, it only produces S5 on the S-meter scale due to the blockage of the southern mountain in the Patuk area.

To get a continuous system, it needs a transmit power above 60 Watt with the antenna tower height above 50 meter. The best solution in this case is to use a repeater stations at the top of the southern mountain around Patuk area or Ngelanggeran area.

 https://doi.org/10.1051/e3sconf/20184301019

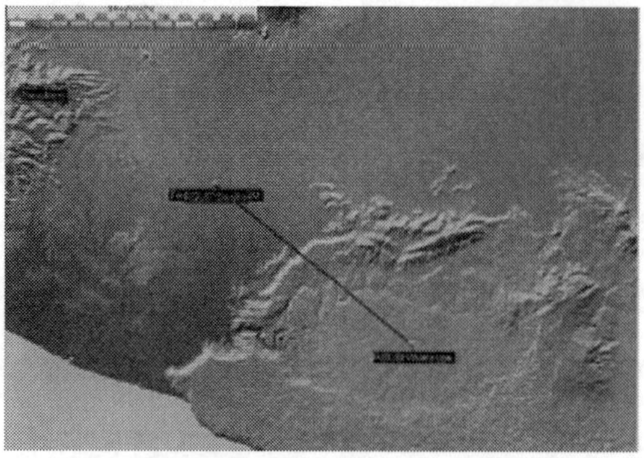

Fig. 1. Simulation using *radio mobile* to show the communication quality between Sardjito Hospital and District Hospital in Wonosari.

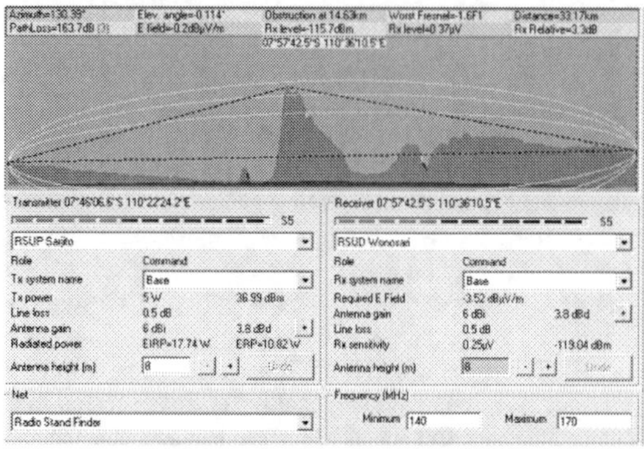

Fig. 2. The technical data of simulation using *radio mobile* to show the communication wuality between Sardjito Hospital and District Hospital in Wonosari on the frequency band of 144 MHz.

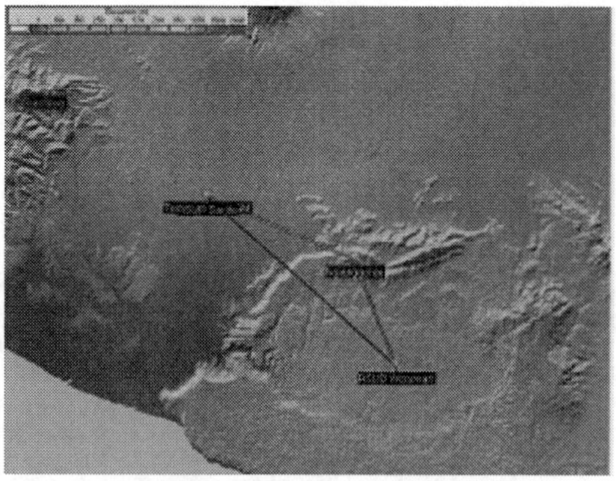

Fig. 3. Simulation using *radio mobile* to show the quality of communication between Sardjito Hospital and District Hospital in Wonosari in the frequency band of 144 MHz with a repeater located at Ngelanggeran (green line indicates the successful communications)

4.2 Propagation Analysis Sardjito Hospital and District Hospital in Wonosari using 7.08 MHz

The two-way communication between RSUP Sardjito and RSUD Wonosari with the distance of 33.17 km can be done continuously, with the electromagnetic field power reaches 30.37 dBμV/m in the TX power of 100 Watt. It produces S9 on the S-meter scale. Although it is blocked by the southern mountain in the Patuk area, this communication can be continuous by using skywave as the same as the Fressnel postulate for the HF frequency.

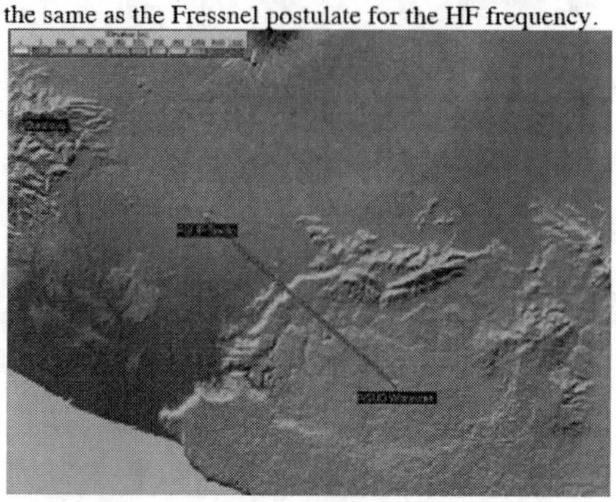

Fig. 4. Simulation using *radio mobile* to show the quality of communication between RSUP Sardjito and RSUD Wonosari in the frequency band of 7.08 MHz (green line indicates that the communication is successfully done)

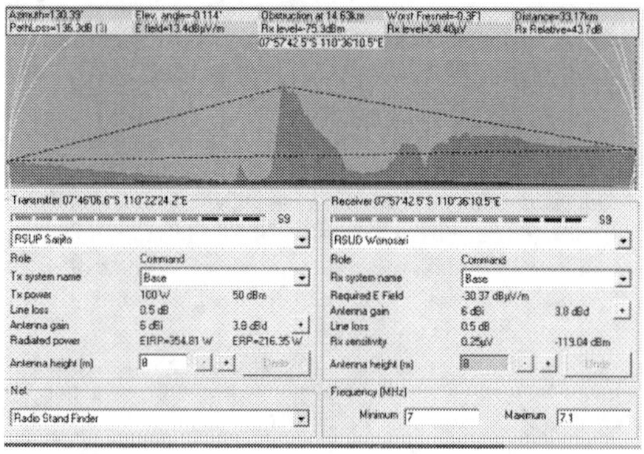

Fig. 5. The technical data of simulation using *radio mobile* to show the quality of communication between Sardjito Hospital and District Hospital in Wonosari in the frequency band of 7.08 MHz.

4.3 Text Data Transmission

The text data transmission is done using Mix.W 3.11g with the transmission mode in RTTY on the frequency of 144.148 MHz and 7.08 MHz. The data needs to be sent is a text that converted into voice (digital-to-analog). The result of text data transmission is shown in Figure 6 and 7.

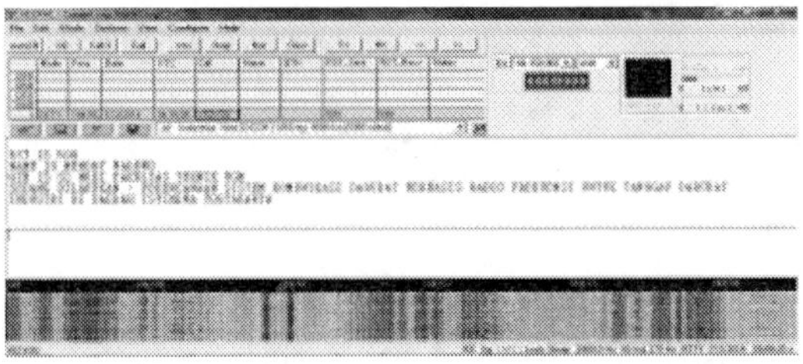

Fig. 6. The process of receiving text data using MixW 3.11g in the mode of RTTY (maximum signal power)

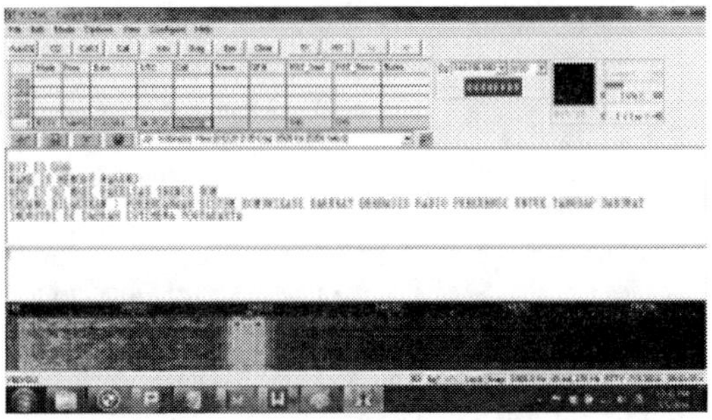

Fig. 7. The process of receiving text data using MixW 3.11g in the mode of RTTY (minimum signal power)

Table 1. Result of text data transmission

Distance (km)	Percentage of text data reception (%)	Error (%)
1	100.0	0.0
5	100.0	0.0
10	100.0	0.0
15	97.0	3.0
20	98.0	2.0
25	96.0	4.0
30	98.0	2.0
35	94.0	6.0
40	95.0	5.0
Average	97.56	2.44

Based on the Table 1, it shows that the result of data text transmission in the range of distance between 0 and 40 km reaches the average value of 97.56 %. The average error value produced by the system is 2.44 %. From those values, it can be concluded that the text data transmission is successfully conducted.

4.4 Image Data Transmission

The image data transmission is done using MixW 3.11g in the mode of SSTV. The result of image data transmission are shown in Figure 8 to Figure 10. Those images are analysed for the image quality after they are received by the receiver. The image quality can be measured by using a standard image pattern developed by EIA (*Electronic Industries Association*). This pattern is used to test several parameters in the image quality, as what explained by Grob (1984): "*This standard pattern provides a reference for checking resolution, scanning linearity, interlacing, additional characteristics of the reproduced picture*".

Here are some parameters to test the image quality:

a. Horizontal linearity
This parameter is tested by comparing the lines or horizontal shapes with the number and the size similar to the image data before and after the data transmission.

b. Vertical linearity
This parameter is tested by comparing the lines or the vertical shapes with the number and the size similar to the image before and after the data transmission.

c. Aspect ratio
This parameter is tested by comparing the ratio of the image size before and after the data transmission.

d. Contrast range
The 10-level grayscale shows a range from a maximum white to the 1/30th of scale value. Then, data is compared before and after the data transmission.

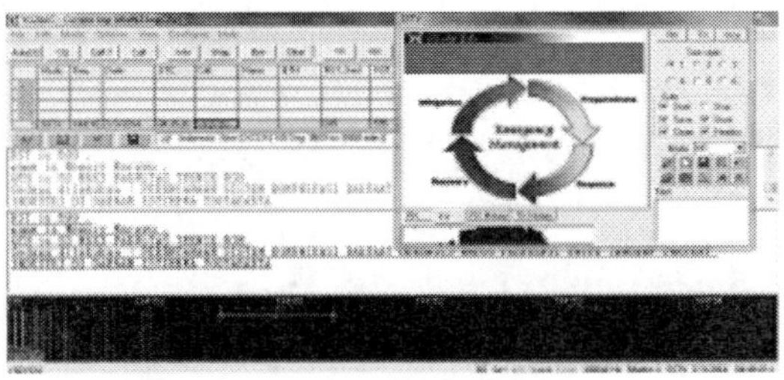

Fig. 8. The image data before it is sent using MixW 3.11g in the mode of SSTV

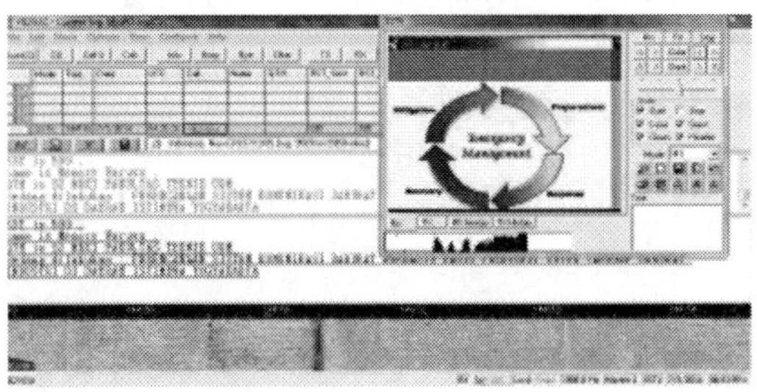

Fig. 9. The image data after it is received by the receiver (maximum signal power)

 https://doi.org/10.1051/e3sconf/20184301019

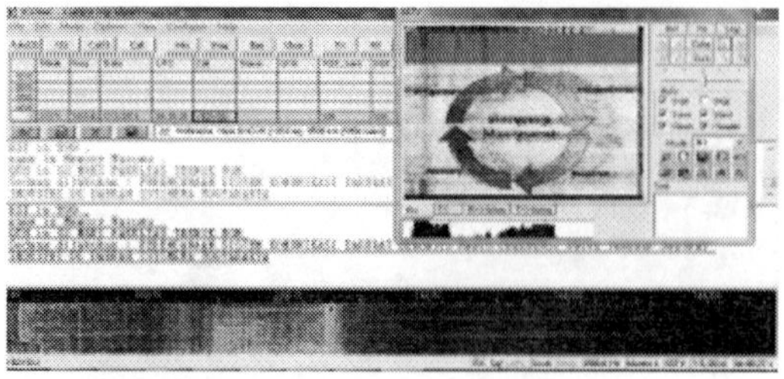

Fig. 10. The image data after it is received by the receiver (minimum signal power)

Table 2. The result of image data transmission

Quality Test Factor	Percentage of image quality score (%)				
	500 m	10 km	20 km	30 km	40 km
Horizontal linearity	90.5	80.3	78.1	75.2	57.6
Vertical linearity	95.1	80.3	78.0	72.0	58.3
Aspect ratio	99.6	94.0	84.7	80.4	75.0
Contrast range	90.2	75.4	60.2	40.5	30.4
Average	93.85	82.50	75.25	67.03	55.33

Based on the result from Figure 8 to 10 and Table 2, it is concluded that the image data can be transmitted and received in the range distance from 500 m to 40 km. But, there is a significant decrement in the number of image quality, from 93.85% to 55.33%. That decrease is caused by frequency distortion and small gain when the image is transmitted. To solve that problems, the system is added with a filter during the transmission process along with the increment of amplitude gain so that the data can be sent in the best quality it can be.

5 Conclusion

This research focuses on the performance test of emergency data communication system by using several types of data. The system uses Very High Frequency (VHF) 144 MHz and High Frequency (HF) 7.08 MHz as the data carrier and a software to interpret the text and image modulated onto the data carrier. Based on the results of the system's performance test between the Wonosari Gunungkidul District Hospital and the Sardjito Hospital with the distance of 40 km, the receiving quality of the data transmission for text and image are 95% and 55.33%, respectively. The recieved image can be recognized and identified.

We would like to thank the Laboratory of Sensor and Telecontrol System, Department Nuclear Engineering and Engineering Physics, Faculty of Engineering, Universitas Gadjah Mada, for supporting this research.

References

1. P.D. Wiedemeier, "*Transmitting Medical Imagery over 2-Meter Amateur Packet Radio Networks using TCP Reno and UDP*", IEEE-AC paper #1067 Version 5, December (2007)
2. M. Uysal, M.R. Heidarpour, "*Cooperative Communication Techniques for Future-Generation HF Radios*", IEEE Communications Magazine Topics in Military Communications, October (2012)
3. J.L. Barber (N7CXI), "*Proposal for SSTV Mode Specifications*", presented at the Dayton SSTV forum in Silicon Pixels, May (2000)
4. A. Hapenciuc, P. Svasta, "*Long Range Image Radio Transmitter*", IEEE 19th International Symposium for Design and Technology in Electronic Packaging (SIITME), 24-27 October (2013)
5. S.M. Dilek, A. Ayranci, R. Ata, O. Ceylan, H.B. Yagci, "*Radio Transmitter Design including FPGA for Micro/Nano Satellite*", IEEE 978-1-4673-5563-6/13/$31.00 (2013)
6. A. Aziminejad, A.W. Lee, G. Epelbaum, "*Radio Propagation Prediction for CBTC Data Communication Subsystem Design*", IEEE Vehicular Technology Magazine 1556-6072/15, September (2015).
7. T. Bhattachatjee, S.P. Maity, A. Roy, "*Prediction Error Based Secret Image Transmission Over Radio Mobile Channel*", IEEE WiSPNET conference 978-1-4673-9338-6/16/$31.00, June (2016)
8. S. Hawibowo, Sunarno, M.M. Waruwu, R. Wijaya, "*Development of Data Communication System for Virtual Hospital Scheme between Public Health Centers at Remote Areas and Regency Public Hospital Based on Radio Frequency Communication*", Journal of Theoretical and Applied Information Technology **Vol. 84** No.2, ISSN: 1992-8645, February (2016).

Prognostic Methods on Accelerator's Anode Voltage Regulator

Robby Christian[1,*], *Balza* Achmad [2,*], and *Hyun* Gook Kang[1]

[1]Mechanical, Aerospace, and Nuclear Engineering Dept., Rensselaer Polytechnic Institute, Troy, NY, USA
[2]Nuclear Engineering and Engineering Physics Dept., Gadjah Mada University, Yogyakarta, Indonesia

Abstract. This study investigated an adaptive control, fault diagnostics and prognostics of the anode voltage regulator system at an ion implantation accelerator. The system was modeled as a 4th order AutoRegressive with eXogenous (ARX) model, controlled by a Fuzzy Logic Controller (FLC). This model was then used as a basis for constructing and updating a fault diagnosis module and a failure prognostics module. To maintain the system's performance, the controller's response was continuously re-adjusted through an optimization scheme. A Failure Mode and Effect Analysis (FMEA) was conducted resulting on five failure modes of the regulator system. Fault data were generated in MATLAB simulation to train a random forest fault classification engine. The optimal random forest classifier was 20 decision trees with a fault diagnostics accuracy of 98.06%. A Hidden Markov Model (HMM) was constructed as the system's fault progression model based on the interaction between environmental conditions and controller actions. The particle filter and Bayesian inference methods were then employed to continuously update the HMM and predict the system's Remaining Useful Lifetime (RUL). The proposed methodology was able to integrate an adaptive fuzzy logic control, prognosis and failure diagnosis altogether allowing a continual satisfactory performance of the voltage regulator system throughout its lifetime.

1 Introduction

Ion implantation accelerator is a device commonly used to fabricate semiconductor wafers by accelerating and impacting charged-particles to a target material. The implantation process is highly affected by several critical factors, namely the ion beam's energy and intensity. These two variables are directly regulated by the electrical voltage applied to the accelerator's anode. As the implantation process typically takes a long time, it requires employing a control system which can adapt to external variations induced by voltage instabilities and component wear-out.

Various methods are available to maintain the system's performance by adjusting the control law according to changes in the system. Adaptive control [1] and Model Predictive Control (MPC) [2] are typical examples of these approaches. The adjustment flexibility is, however, typically designed to handle finite variations or nonlinearities encountered under normal operation. When the variation is due to degradations within the system, there exists a limit beyond which the controller cannot sufficiently adapt to maintain the desired performance requirements. It is, therefore, a strategic interest to characterize the degradation's evolution and predict when to perform corrective maintenance to ensure a continuous satisfactory performance of the system.

Studies of prognostics give insight into the future fault evolution characteristics. An *et al.* [3] provided a review of the existing prognostic methods. Data-driven methods such as Neural Network (NN) and Gaussian Process (GP) regression use information from collected data to identify the characteristics of damage state and predict its future state without using any particular physical model. Although being simple and fast to implement, these methods do not give any useful insight of the physical cause-effect relationship. Physics-based models, such as Paris-Erdogan fatigue crack law [4], on the contrary, depart from the well-established physics laws on the fault's causality. These models consequently provide intuitive prediction results yet may be computationally expensive especially for high fidelity models.

Byington *et al.* [5] identified failure modes of an electromechanical flight control actuator and monitored the associated key parameters. They utilized a smoothing algorithm to extrapolate the associated parameters to the point of failure. This approach, however, did not consider the effect of parameter uncertainties towards the Remaining Useful Life (RUL) estimate. Shin et al. [6] proposed a combined strategy of online monitoring and surveillance tests to predict the unavailability evolution of standby equipment. The method offers an integrated prediction approach applicable for both new and aged components. However, it is constrained specifically to devices in their standby mode, and no particular examination on a controlled system was done.

Not many research has addressed the integration of

* Corresponding author: chrisr3@rpi.edu, balzach@ugm.ac.id

fault growth prognostics with adaptive control schemes. Bole et al. [7] addressed this issue by defining the fault growth physics with a Markov chain model as a function of environmental stresses and control loads. It further demonstrated the use of a dynamic programming technique to select control actions which minimizes the risk of a continuous control load to the degrading component while minimizing deviation from the system's desired performance. While the approach was not validated experimentally on a real system, it provided some valuable understanding of how exogenous and control loads and their uncertainties thereon affects the evolution of faults.

This study aims to create an adaptive controller, fault diagnostics, and prognostics scheme for the anode voltage control regulator of the ion implantation accelerator. The result is expected to ensure a continual satisfactory performance of the system throughout its lifetime. Since the adaptive control concept is already well established, in-depth analysis is instead given to identifying failure modes and modeling the fault growth propagation driven by environmental load variations and control actions.

2 Methods

2.1. Model structure

A schematic of the anode voltage control regulator is shown in Fig 1. When an operator inputs a specific setpoint through the PC interface, a Direct Current (DC) motor drives a variable transformer to scale the Alternating Current (AC) voltage from the grid. The resulting voltage is then fed to a motor generator which generates a high voltage of up to 2.5 kV. The transformer's output is likewise measured and scaled down as a feedback signal.

The first step in developing an adaptive controller is defining the model. Preliminary results showed that the system is linear throughout its operating range. It is convenient to model linear systems in the form of a transfer function derived from physical equations, to give an intuitive insight of the dynamic variations taking place inside the system. Therefore the system can be simplified into Fig 2, with differential equations describing the input-output relationship summarized in Table 1 [8].

After performing substitutions and Laplace transformations, the system was rearranged into blocks of first order transfer functions shown in Fig 3. Hence the system's output ω_L can be expressed as a function of the input variables as Eq. (1).

$$\omega_l = v_a \frac{B(s)}{A(s)} + T_d \frac{C(s)}{A(s)} \tag{1}$$

where

$A(s) = 1 + G_1(s)G_2(s)K_mK_e + G_2(s)G_3(s) + K_mK_eG_1(s)G_2(s)G_3(s)G_4(s) + G_3(s)G_4(s)$,

$B(s) = K_mG_1(s)G_2(s)G_3(s)G_4(s)$,

$C(s) = K_mK_eG_1(s)G_2(s)G_3(s)G_4(s) + G_2(s)G_3(s)G_4(s) + G_4(s)$,

s = continuous time complex variable

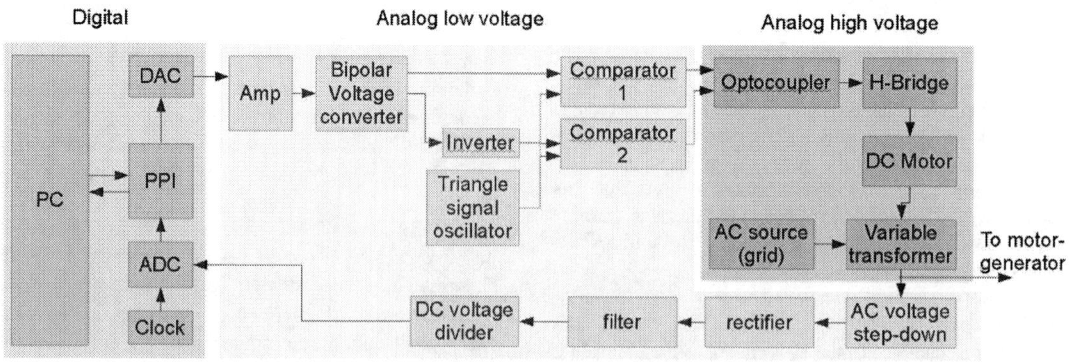

Fig. 1. System's schematics.

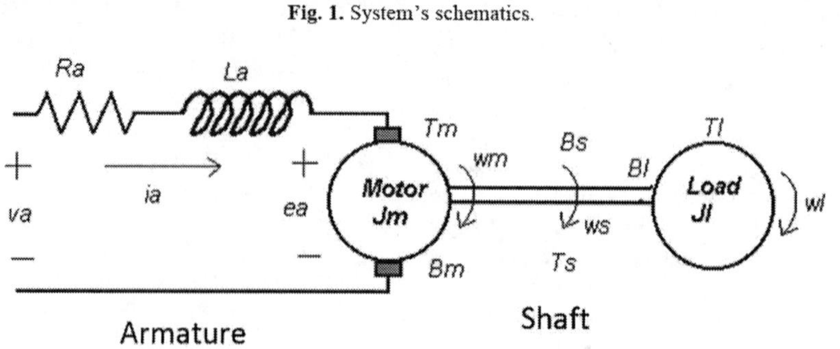

Fig. 2. Physical model of the analog high voltage domain.

Table 1. Physical input-output relationships.

Subsystem	Relations	Variables and units
Armature	$v_a = R_a i_a + L_a \frac{di_a}{dt} + e_a$	R_a = resistance (Ω) L_a = inductance (H) i_a = current (A) e_a = back Electromagnetic Force (EMF) (V) v_a = applied voltage (V)
Armature-Motor	$K_e \omega_m = e_a$	K_e = EMF constant (V/rad/sec) ω_m = angular velocity (rad/sec)
Motor	$K_m i_a = T_m$	T_m = motor torque (N.m) K_m = motor torque constant (N.m/A)
Motor-Shaft	$T_m - T_s = \left(J_m \frac{d\omega_m}{dt}\right) + (B_m \omega_m)$	T_s = transmitted torque (N.m) J_m = rotor's moment of inertia (kg.m^2) B_m = motor viscous friction (N.m.s)
Shaft	$k_s \int \omega_s \, dt + B_s \omega_s = T_s$, $\omega_s = \omega_m - \omega_l$	k_s = stiffness coefficient B_s = damping coefficient ω_s = shaft angular velocity (rad/sec)
Shaft-Load	$T_s - T_d = \left(J_l \frac{d\omega_l}{dt}\right) + (B_l \omega_l)$	J_l = load's moment of inertia (kg.m^2) ω_L = load angular velocity (rad/sec) B_l = viscous load friction (N.m.s) T_d = load torque disturbance (N.m)

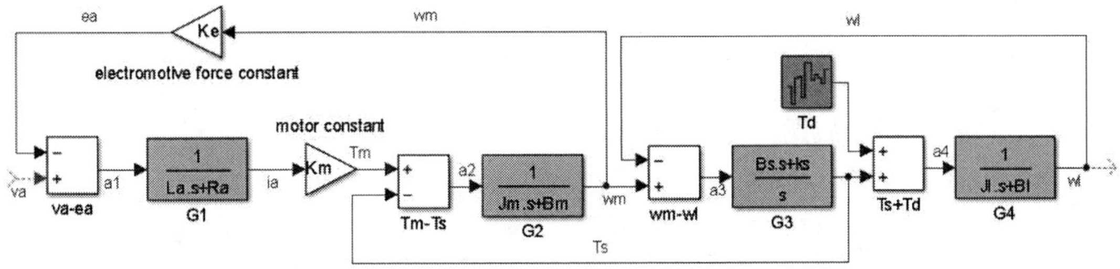

Fig. 3. Transfer function representation of the analog high voltage domain.

By applying a low-pass filtration with a transfer function *D(s)* equation (1) then becomes:

$$\omega_l = v_a \frac{B'(s)}{A(s)} + T_d' \qquad (2)$$

where

$B'(s) = K_m G_1(s) G_2(s) G_3(s) G_4(s) D(s)$

$T'_d = T_d C(s) D(s) / A(s)$, a noise spectrum with reduced correlation to the process function. It can be considered as an output or measurement noise spectrum.

The system represented by Eq. (2) can then be written in an autoregressive model with exogenous input (ARX) discrete-time model as Eq. (3), because of its digital interface.

$$A(z^{-1}) y(t) = z^{-d} B(z^{-1}) u(t) + e(t) \qquad (3)$$

where

$u(t)$ = digital input signal
$y(t)$ = digital output signal
$e(t)$ = digitalized exogenous noise signal
$A(z^{-1}) = 1 + a^1 z^{-1} + a^2 z^{-2} + \ldots + a_{na} z^{-na}$
$B(z^{-1}) = b_1 z^{-1} + b_2 z^{-2} + \ldots + b_{nb} z^{-nb}$
d = delay time
$z = \rho e^{j\psi}$, a discrete-time complex variable with magnitude ρ and phase ψ.

2.2. Model parameter estimation

System identification methodology was used to obtain the ARX model's order and parameters. in order to find the model's order, a residual analysis method was performed by using two statistical criteria namely the Akaike Information Criterion (AIC) and Final Prediction Error (FPE). Both criteria were used because the former may at times overestimate the model's order while the latter the otherwise [9].

When the model's order is correct, it will minimize the values of AIC and FPE calculated using Eq. (4) and (5).

$$AIC(p) = \log \hat{\sigma}_\varepsilon^2 + \frac{2p}{N} \qquad (4)$$

$$FPE(p) = \hat{\sigma}_\varepsilon^2 \left(1 + \frac{p}{N}\right) \qquad (5)$$

where

$\hat{\sigma}_\varepsilon^2$ = variance of the residual signal
$p = n_a + n_b$, the number of model parameters
N = number of data

Various recursive algorithms are available to estimate the ARX model's parameters once its order is found. Recursive Least Squares (RLS) is one of the widely used algorithms. In this algorithm, model parameters $\hat{\theta}(t)$ at a certain time-instant are recursively corrected from the previous estimate $\hat{\theta}(t-1)$ in proportion to the difference between model and plant output $\varepsilon(t)$ as Eq. (6)

$$\hat{\theta}(t)=\hat{\theta}(t-1)+F(t-1)\phi(t-1)\varepsilon(t) \quad (6)$$

$$F(t)=\frac{1}{\lambda_1(t)}\left[F(t-1)-\frac{F(t-1)\phi(t-1)\phi(t-1)^T F(t-1)}{\frac{\lambda_1}{\lambda_2}+\phi(t-1)^T F(t-1)\phi(t-1)}\right] \quad (7)$$

$$\varepsilon(t)=\frac{y(t)-\hat{\theta}(t-1)^T\phi(t-1)}{1+\phi(t-1)^T F(t-1)\phi(t-1)} \quad (8)$$

where
$\hat{\theta}(t)^T=[\hat{a}_1(t),\dots,\hat{a}_{na}(t),\hat{b}_1(t),\dots,\hat{b}_{nb}(t)]$, estimates of model parameters
λ_1 and λ_2 = forgetting factor variables.
$\phi(t)^T=[-y(t),\dots,-y(t-na+1),u(t-d),\dots,u(t-d-m+1)]$
$u(t)$, $y(t)$ = input and output signal respectively

Because the system's degradations are relatively slower time-varying dynamics compared to other dynamics found in normal operation condition, λ_2 was set to 1 while λ_1 was set at around 0.95 and 0.99. These parameters introduced a bias in covariance matrix F which enables a selective re-convergence of parameter estimates when the system slowly degrades while tolerating high-frequency variations such as noise. Other algorithms applicable for an ARX model are Instrumental Variable with Auxiliary Model (IVAM), Output Error with Fixed Compensator (OEFC) and Output Error with Adjustable Compensator (OEAC). These algorithms are not explained in detail in this paper due to their similarity with the RLS. Interested readers are invited to look up reference [9] instead. All of these algorithms were used to generate a pool of models from which the one that has the best validation results was selected.

Lastly, the system identified model needs to be validated. Provided that the model is correctly identified, the residual signal which is the remainder of system and model's output is expected to be purely a real-world random noise spectrum having a zero mean with low correlation to the model's output. The correlation could indicate that the experiment is poorly designed which causes divergence and random fluctuations of the model's parameters. Based on the aforementioned hypotheses, the tests were expressed as Eq. (9) and (10).

$$NCC(i)=\frac{\lim_{t\to\infty}\frac{1}{N}\sum_{t=1}^{N}\varepsilon(t)\varepsilon(t-i)}{\sqrt{\left[\left(\frac{1}{N}\sum_{t=1}^{N}\hat{y}^2(t)\right)\left(\frac{1}{N}\sum_{t=1}^{N}\hat{\varepsilon}^2(t)\right)\right]}} \quad (9)$$

$$ZC=\frac{\tau_x-N/2}{N/2} \quad (10)$$

where τ_x is the total count of events when the residual signal crosses zero value, ZC is the zero count criteria and NCC is the normalized cross-correlation spectrum. The acceptance criteria for the tests was taken to be at 3% significance level Gaussian hypothesis testing, which gives

$$\forall\ i: NCC(i),|ZC|\le\frac{2.17}{\sqrt{N}} \quad (11)$$

System identification and process control were done online through an 8-bit analog-to-digital (ADC) and digital-to-analog (DAC) converters. A preliminary test indicated that the system's cutoff frequency is 5 Hz, hence, to fulfill the Nyquist criteria, the sampling frequency was set to be 200 Hz.

2.3. Adaptive FLC

Based on the identified model, an adaptive FLC was constructed to meet the performance requirements, i.e., good corrective response to voltage instability, low maximum overshoot, low steady-state error, small rise and settling time. Because the actuator is a Single Input Single Output (SISO) type system, the fuzzy inference approach was chosen to be Mamdani. To get a responsive controller action with a fast rise time, the Middle of Maximum (MOM) defuzzification method was selected. The bell-shaped curve was chosen as the fuzzification membership function as can be seen in Fig. 4a. Due to the linearity of the system, the defuzzification membership function was also taken to be linear as shown in Fig 4b. Given that the controller's input is y(t) – setpoint, the control rules were constructed as follows:

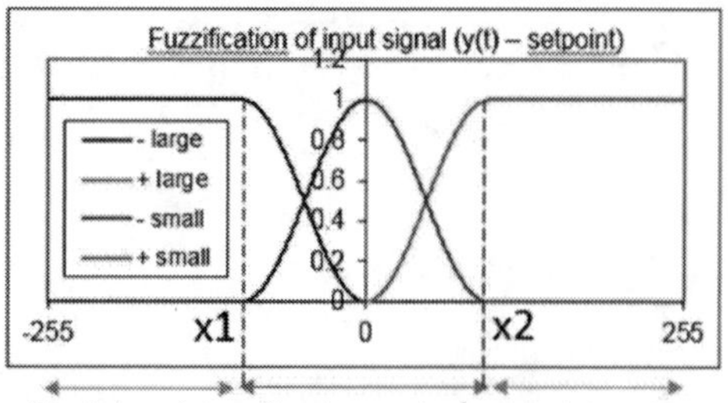

(a) bell-shaped input membership function.

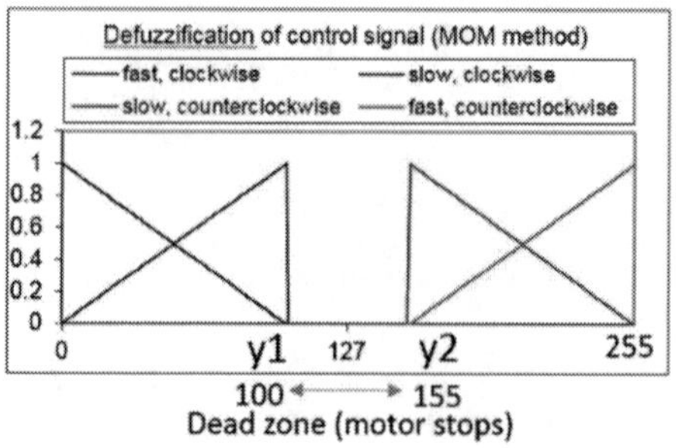

(b) control signal membership function.

Fig 4. FLC membership functions showing the precise-control zone in the input function (a) and a dead-zone in the output function. (b)

1. If e(t) is positive large, rotor is to rotate clockwise quickly
2. If e(t) is positive small, rotor is to rotate clockwise slowly
3. If e(t) is negative large, rotor is to rotate counterclockwise quickly
4. If e(t) is negative small, rotor is to rotate counterclockwise slowly
5. If e(t) is zero, rotor stops rotating

The input domain between x_1 and x_2 in Fig 4a is the range of input values to control the rotor's rotation carefully. When receiving input values in this region, the rotor rotates slower in proportion to the fuzzy input to meet the given setpoint. It is therefore adjusted to cover the range of expected exogenous disturbances during steady-state operations. This disturbance comes from the fluctuation in the grid's voltage which magnitude is between -10% and +5% out of 220 V. Therefore x_1 was set to -26 which corresponds to -22 V while x_2 was set to 13 which corresponds to 11 V. The control actions are adapted to compensate for actuator's degradations in order to maintain the system's performance. Thus the values are $1 \leq y_1 < 100$ and $155 < y_2 \leq 254$. The control value ranging from 100 and 155 were not used since our preliminary test revealed these values could not actuate the motor. As the system degrades, y_1 is reduced while y_2 is increased by a factor of y. Selection of y was made to minimize the differences between the desired system's performance and the ARX model's performance as Eq. (12).

$$\min_{0 \leq y \leq 100} \sum_{j=1}^{4} s(j) - m(j) \tag{12}$$

where $s(j)$ is the desired transient response performance while $m(j)$ is the ARX model's predicted performance. Four performance characteristics were considered in this study, namely rise time, settling time, maximum overshoot and steady-state error.

2.4. Fault modes and diagnostics

In this study, failure was defined as operation events which transient responses do not satisfy the performance requirements. Possible failure modes as listed in Table 2 were identified from references [5][10][11] and from analyzing the system's technical specification. Destructive or potentially fault-inducing tests were not done on the actuator system to preserve its condition. System failures were simulated in a MATLAB computational environment by modeling the system in Simulink. Since the actuator's performance is maintained by the FLC, failures were simulated by setting the controller's response y_1 and y_2 to their maximum responsive level and sequentially altering the values of the key variables in Table 2 until the time when the transient response characteristics exceed the performance specifications. The variables' value was then noted as the failure limit for such specific failure mode.

Root locus analysis was used to diagnose the modes of failure and measure its severity [12]. Each failure mode is expected to produce a unique pattern in the system's natural frequency and damping factor, which is reflected through the change in the model's pole-zero pairs within the central unit circle of the z complex plane. These pole-zero patterns were observed by simulating the faults with increasing severities. The results were fed to train a random forest classifier to enable a real-time pattern recognition and fault diagnosis. The trained classifier was then tested against another set of failure data to check its classification performance.

2.5. Fault growth modeling and prognostics

When a fault mode is diagnosed, a prediction of its evolution to estimate the Remaining Useful Life (RUL) of the actuator system is desireable to perform a timely maintenance schedule. The fault growth was modeled analytically to obtain a surrogate, probabilistic-based model. This approach was pursued instead of having detailed physics-based degradation models to reduce computational time. Fast computational predictions allow for a real-time prognostic application and real-time condition-based maintenance program. Furthermore, it averts the stacking of uncertainties from an over-modeling situation when multiple physics-based degradation models are used.

Table 2. Failure modes and associated physical variables.

Subsystem	Failure mode	Effect to performance	Fault indicator
Armature	Coil wears out leading to a break	Increased rise time and steady state error	R_a increases
	Coil insulation melts and windings are starting to short together	Decreased rise time, increased overshoot percentage and steady state error. Erratic rotations when coils are shorted.	R_a decreases
Motor	Bearing is damaged leading to a complete seizure	Increased rise time and steady state error	B_m increases
Shaft	Shaft cracks leading to fracture	Increased rise time, overshoot percentage and settling time	B_s & k_s increases

https://doi.org/10.1051/e3sconf/20184301020

Load	Dust and rubble fragments builds up on top of transformer's coil and increases friction between coil and rotating brush	Increased rise time and steady state error	B_l increases

The underlying principle for fault growth was adopted from the Eyring model which is based on quantum mechanics principles [12]. It recalls that environmental stresses generate the activation energy needed to cross an energy barrier at the quantum level to initiate a reaction which creates physical fault such as crack, deformation or oxidation. The environmental stress *v(k)* is characterized into deterministic and stochastic stresses as follows [7]:

$$v(k)=\rho(k)\cdot\omega(k) \tag{13}$$

where $\omega(k)$ is the stochastic environmental condition which inflict stresses upon the actuator system, while $\rho(k)$ is the deterministic multiplicative factor to $\omega(k)$ due to a more stringent control actions when faults are present. The latter is further classified into bounded and unbounded load variations which probability functions are given in equation (14) and equation (15) respectively. The variable n in equation (15) is the number of discrete states of ω. Bounded ω comes from variation of environmental condition where the actuator system is installed, i.e. temperature, pressure, humidity. Unbounded ω originates from freely-oscillating stressors with a uniform distribution over their normalized discrete states n. Several examples of stressors which fall into this category are the accelerators operation profile, demanded beam intensity, and electronic disturbances.

$$\Pr\left(\omega(k)=\omega\right)=Pr\left(\omega(k)=\omega|\omega(k-1)=v\right), v\in W \tag{14}$$

$$\Pr\left(\omega(k)=\omega\right)=\frac{1}{n} \tag{15}$$

The armature and bearing are insulated by-design thereby allowing only unbounded stress variations to affect fault growths in these subsystems. The shaft and transformer, on the contrary, are exposed to the atmosphere which causes both the bounded and unbounded stress variations to play a role in fault generations. Since the faults described in Table 2 are naturally irreversible, and because they are induced stochastically, the fault evolution satisfies a monotonically increasing function as illustrated in Fig. 5 where

$$\forall t \leq t_{maintenance} : fault_{t+1} \geq fault_t \tag{16}$$

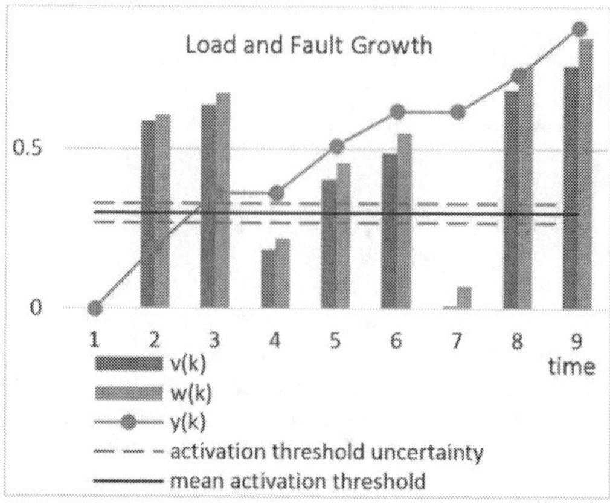

Fig 5. Monotonic fault growth $\gamma(k)$ as a function of environment and control loads

From the abovementioned analysis, the fault evolution was modeled as a second-order Markov Chain process adapted from Bole et al. [7] as Eq. (17) and (18).

$$p_{i,j}(v)=\Pr\left(\gamma(k+1)=s_j|\gamma(k)=s_i,v(k)=v\right),\quad \gamma,s_i,s_j\in S,\quad v\in V(k),\quad k\in\mathrm{N} \tag{17}$$

$$\sum_{s_j\in S}p_{i,j}=1 \tag{18}$$

where

$p_{i,j}(v)$: the probability of transitioning from one Degradation State (DS) s_i to s_j given an applied load v

γ : random variable representing system's DS

S : state space for component's DS

$V(k)$: domain of feasible loads that may be applied at a given time-instant

It is expected that there are aleatoric uncertainties in the predicted damage level should a load above its minimum energy barrier is imposed. A noise term ξ in the Markov Chain model as shown in Fig. 6, which is a similar practice to the reference studies [13,14], is then introduced to incorporated this uncertainty in the model. In contrast, the damage level is expected to remain the same when the load is below the minimum energy barrier. However in order to accommodate incipient faults and epistemic uncertainties in the fault growth physics, a relatively small probability value C was additionally introduced which may lead to an increase in the DS. For this reason, the fault evolution mechanism was then mathematically expressed as Eq. (19) and (20).

$$p_{i,j}(\rho)|\omega\geq\beta=\sum_{\omega\in W}\sum_{\xi\in\Xi}\Pr\left(\omega(k)=\omega\right).\Pr\left(\xi(k)=\xi\right).Pr\left(\beta=\beta_{true}\right) \tag{19}$$

$$p_{i,j}(\rho)|\omega<\beta=C.Pr\left(\beta=\beta_{true}\right)\simeq C\ll 1 \tag{20}$$

where W and Ξ are the state spaces for ω and ξ respectively, while β is the normalized minimum energy barrier zone.

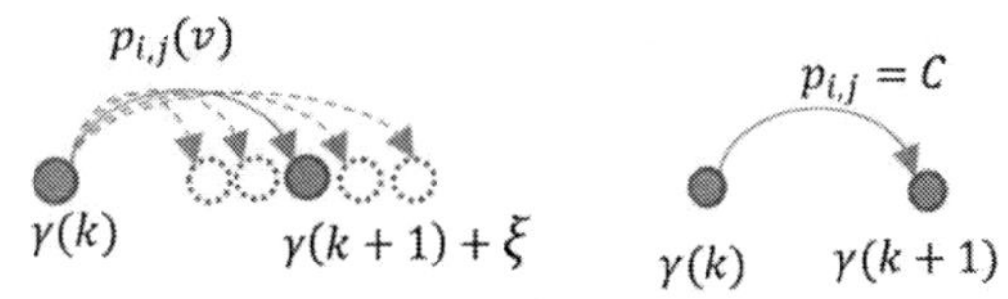

Fig 6. Markov Chain model of fault growth when the load exceeds the energy activation threshold (left) and when it does not (right)

The failure is measured in terms of the FLC's capability to maintain the system's performance. Hence, it is desirable to map the relationship between the internal degradation state and the controller's capability to compensate it. This fault compensation capability is reflected by the term Adaptive Space (AS), which is the remaining space between 0 and y_1 or 255 and y_2 in Fig 4b. Thus AS in this digital FLC setting has an integer value of 100 when the system is healthy, and 1 when the system fails. The objective of this prognostics module is, therefore, to continuously predict when AS' value drops to 1. However, it was not hypothesized as a straightforward task. Because the system identification method used as the basis of the adaptive FLC is a statistical method, it may not produce a perfect model. There was incompleteness throughout the finite observation domain deemed negligible within the boundaries of a 3% Gaussian significance level. As a consequence, the variable AS as a function of DS was expected to be noisy. This phenomenon where DS, a stochastic state satisfying the Markov property, generates AS, another stochastic variable of observational interest, can be conveniently portrayed as a Hidden Markov Model (HMM) shown in Fig. 7. The DS transition probability $p_{i,j}$ was given in equation (19) and equation **Error! Reference source not found.** where $\sum_{s_j \in S} p_{i,j} = 1$. The observed AS probability density $q_{k,l}$ was determined by experiment which fulfills equation (12) and $\sum_{\zeta(l) \in Y} q_{k,l} = 1$.

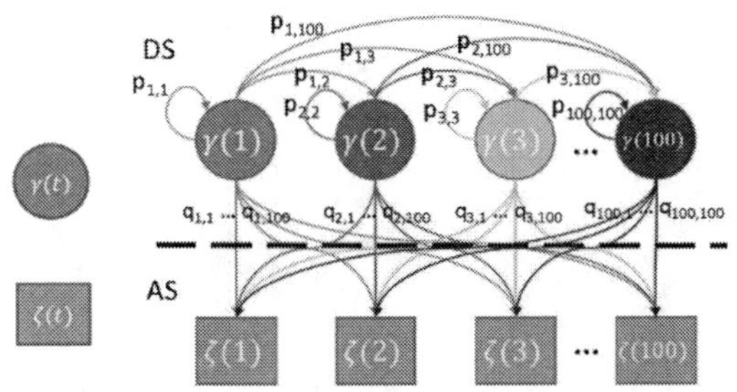

Fig 7. Hidden Markov Model (HMM) structure revealing the monotonic fault progression and its relationship with the controller's adaptiveness

Simulations with the MATLAB Simulink model were carried out to obtain $p_{i,j}$ and $q_{k,l}$. Several variables were not available due to the absence of destructive experiments and were assumed in a reasonable manner, i.e., the activation energy threshold for each of the fault modes β_{true}, β and ξ density functions. Afterwards, a normalized bounded and a normalized unbounded environmental conditions based on equation (14) and equation (15) were generated. The fault indicators in Table 2 were then affected according to equation (19) and equation **Error! Reference source not found.**. These indicators were normalized in a piecewise manner into 100 discrete DS' states. The indicators were fed to the closed-loop system model until the system failed. This run-to-failure data was then used to estimate $p_{i,j}$ and $q_{k,l}$ by using the Baum-Welch's algorithm [15].

After the HMM's parameters were identified, another run-to-fail data sequence was generated where the DS and failure mode were not revealed. The fault diagnosis module identified the active failure mode from which the prognostics module continuously updated the predictions of future AS probabilities. The prognostics module used the HMM as a prior model for DS evolution and corrected it with the observed AS. This approach was made numerically by using a particle filter algorithm. A set of 1000 particles was employed at each time step t to estimate future degradations $Pr(\gamma(t+k))$, the corresponding AS $Pr(\zeta(t+k))$ and the RUL $Pr(t|\zeta_1)$. The prediction result was then compared with the true DS to investigate its accuracy.

3 Results

The AIC and FPE values for an ARX model with increasing model order are shown in Fig. 8. AIC shows a continually decreasing trend. However since FPE shows a minimum at the 4th order model, and the further increase of model order does not reduce AIC significantly, the model's order was decided to be 4.

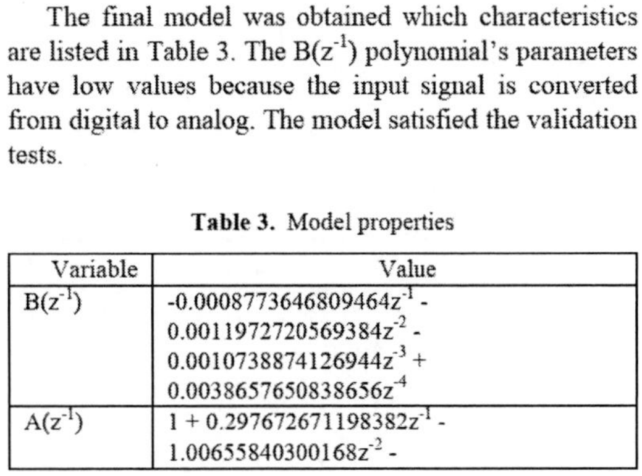

Fig 8. FPE and AIC values suggesting a 4th order ARX model

The final model was obtained which characteristics are listed in Table 3. The B(z^{-1}) polynomial's parameters have low values because the input signal is converted from digital to analog. The model satisfied the validation tests.

Table 3. Model properties

Variable	Value
$B(z^{-1})$	$-0.0008773646809464z^{-1}$ - $0.0011972720569384z^{-2}$ - $0.0010738874126944z^{-3}$ + $0.0038657650838656z^{-4}$
$A(z^{-1})$	$1 + 0.297672671198382z^{-1}$ - $1.00655840300168z^{-2}$ -

https://doi.org/10.1051/e3sconf/20184301020

	$0.308618060141436z^{-3}$ + $0.018009547975964z^{-4}$
NCC	0.00787 < 0.033653 (pass)
ZC test	| 0.186097 | < 2.17 (pass)

The multistep response characteristics of the FLC is shown in Fig. 9. The figure shows that the FLC performs satisfactorily to changes in the AC grid voltage.

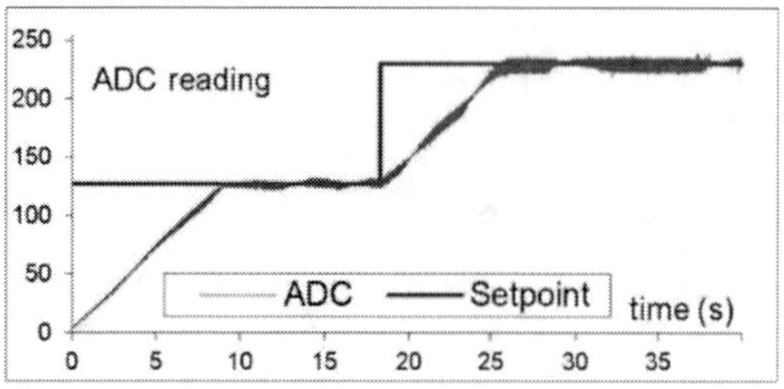

Fig 9. Multistep input response from the FLC showing a good response to setpoint changes

The performance of the random forest classifier in classifying the model's pole-zero pairs into corresponding fault modes is expressed in the form of a matrix in Table 4. Headings of the matrix are set in numbers representing the fault modes and severity classes. The physical interpretation of these classes are shown in Table 5. The performance matrix approached an ideal classifier which has diagonal matrix values of 1. It expresses the random forest's classification accuracy of 98.06%.

Results of fault generation simulation and the ensuing AS are visualized in Fig 10. The exponentially increasing DS was caused by the fault acceleration effect due to the increasing controller's action as the system degrades. Furthermore, it is observed that the AS becomes noisier along with the DS. This might be caused by decreasing Signal-to-Noise Ratio (SNR). As the system degrades, its output level decreases while the measurement noise remains the same. This phenomenon creates variations in the ARX model parameter's which were propagated to the calculated AS.

Results of prognostics are given in Fig 11. True AS and the estimated AS are plotted together for a comparison. The predictions are updated at every time-step. Fig. 11 shows the results taken at the beginning, middle and end of the system lifetime period. The significance level on predicted AS was taken to be 3%. Predictions made at t=1 describes the HMM's forecast performance given an initial calibration only. Although it gave a good RUL estimate approaching the true RUL at t=514, the uncertainty band was relatively wide. A hybrid prognosis combining both HMM's prior likelihood and online monitoring information gave a more accurate RUL estimate with reduced uncertainty as shown at t=200 and t=400. Further reduction of uncertainty was observed as more AS data became available.

Fig 12 compares the performance characteristics of the system when this methodology was applied and when it was not. The comparison was started from when degradation was first observed. Although the steady-state error of the static FLC system started to exceed its threshold at t+348, the rise time and settling time had already deviated earlier at t=1. Therefore, in this case, it failed at t=1. Increasing the controller's action inevitably increased the system's maximum overshoot. However, it is still below the prescribed threshold of 1%. The figures showed that the adaptive FLC was successful in maintaining the required system's performance despite degradations. Its lifetime was consequently extended for an additional 514 cycles.

Table 4. Performance matrix of the random forest classifier

Probability of fault class predicted by random forest

True Fault Class	1	21	22	31	32	41	42	51	52	61	62
1	1	0	0	0	0	0	0	0	0	0	0
21	0	1	0	0	0	0	0	0	0	0	0
22	0	0.05	0.95	0	0	0	0	0	0	0	0
31	0	0	0	1	0	0	0	0	0	0	0
32	0	0	0	0	1	0	0	0	0	0	0
41	0	0	0	0	0	0.91	0.09	0	0	0	0
42	0	0	0	0	0	0	1	0	0	0	0
51	0	0	0	0	0	0	0	0.98	0.02	0	0
52	0	0	0	0	0	0	0	0	1	0	0
61	0	0	0	0	0	0	0	0	0	1	0
62	0	0	0	0	0	0	0	0	0	0	1

Table 5. Fault Modes And The Corresponding Values Of Physical Variables

Subsystem	Healthy	Fault	Fail
Armature	R_a=0.8 L_a=0.01	Class ID 21 $.8 < R_a < 2$	Class ID 22 $R_a \geq 2$
		Class ID 31 $.3 < R_a < .8$	Class ID 32 $R_a \leq .3$
Motor	B_m=0.1, J_m=0.01 K_m=1, K_e=1e^{-4}	Class ID 41 $.1 < B_m < .85$	Class ID 42 $B_m \geq .85$
Shaft	B_s=0.2 k_s=1	Class ID 51 $2.5e^{-3} < B_s < .2$ $1.25e^{-2} < k_s < 1$	Class ID 52 $B_s \leq 2.5e^{-3}$ $k_s \leq 1.25e^{-2}$
Load	B_l=0.4 J_l=0.02	Class ID 61 $.4 < B_l < 1.2$	Class ID 62 $B_l \geq 1.2$

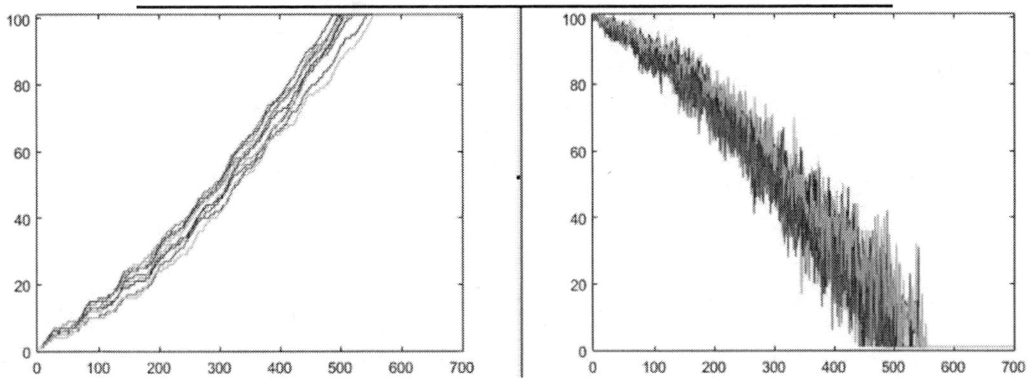

Fig 10. DS (left) and AS (right) progression from 10 run-to-fail data sequences

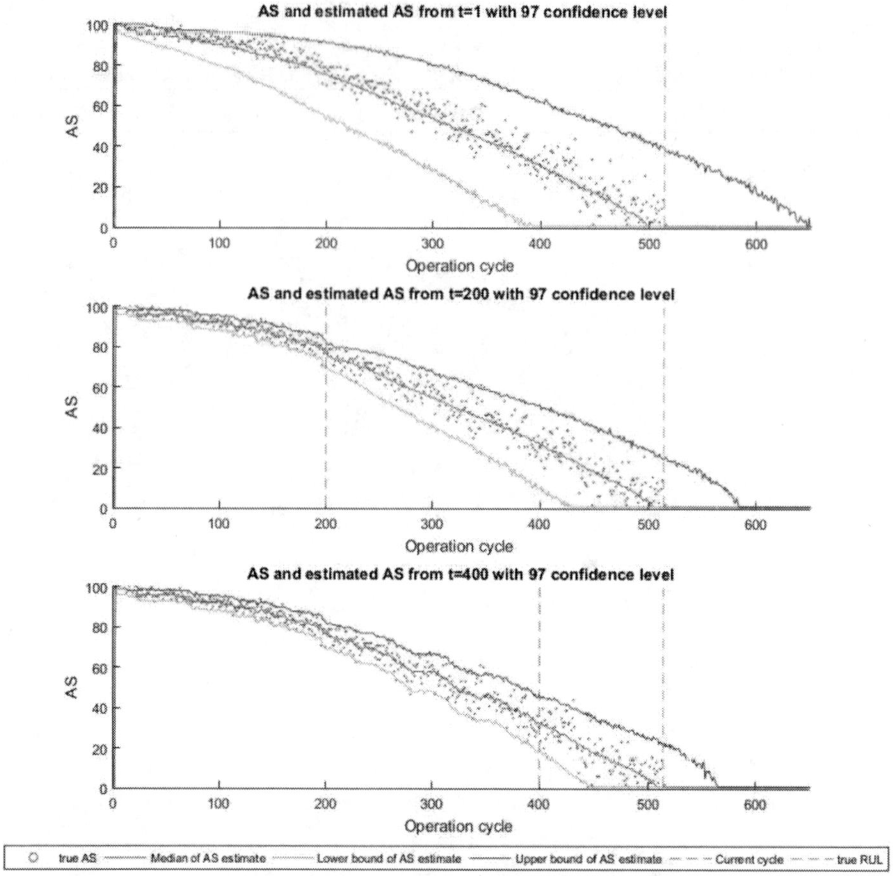

Fig 11. Prognostic results at different time steps showing the convergence of RUL estimation

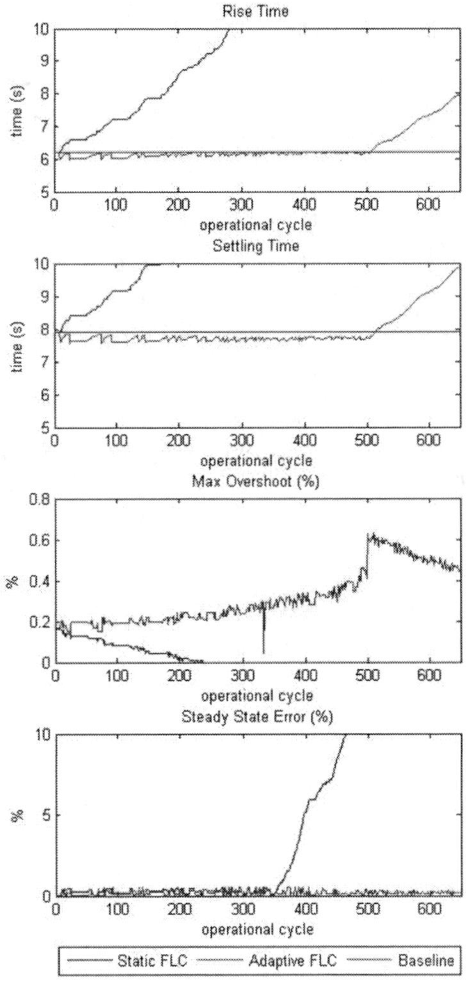

Fig. 12. System performance showing satisfactory transient characteristics during degradation period

4 Conclusion

This study developed an integrated approach to control the performance of the anode voltage regulator system at an ion implantation accelerator throughout its lifetime despite degradations. The approach relied on the modeling the system empirically through a linear system identification method. The obtained model was a fourth order ARX model validated with a 3% significance level Gaussian hypothesis testing. A fault diagnostics module based on the random forest algorithm was developed to classify the fault mode and severity. It has a classification accuracy of 98.06%. An adaptive FLC was created to compensate for faults and maintain system's performance. Results show that the performance was satisfactorily achieved for an additional 514 cycles thereby prolonging its useful life. The progression of faults and its relation to the FLC's adaptive capability was characterized by a Hidden Markov Model. The trained model provided a prior likelihood to the system's RUL on which it was updated with the Bayesian method and a particle filter method. Results implied that this approach was successful in predicting the system-specific RUL with a 97% confidence level.

References

1. K. Nouri, R. Dhaouadi, N. Benhadj Braiek, *Adaptive control of a nonlinear dc motor drive using recurrent neural networks*, Appl. Soft Comput. **8**, 371–382 (2008)
2. A. Jain, G. Schildbach, L. Fagiano, M. Morari, *On the design and tuning of linear model predictive control for wind turbines*, Renew. Energy. **80**, 664–673 (2015)
3. D. An, N.H. Kim, J.-H. Choi, *Practical options for selecting data-driven or physics-based prognostics algorithms with reviews*, Reliab. Eng. Syst. Saf. **133**, 223–236 (2015)

4. F. Cadini, E. Zio, D. Avram, *Monte Carlo-based filtering for fatigue crack growth estimation*, Probabilistic Eng. Mech. **24**, 367–373 (2009)
5. C.S. Byington, M. Watson, D. Edwards, P. Stoelting, *A model-based approach to prognostics and health management for flight control actuators*, 2004 IEEE Aerosp. Conf. Proc. (IEEE Cat. No.04TH8720), 6 (2004)
6. S.M. Shin, I.S. Jeon, H.G. Kang, *Surveillance test and monitoring strategy for the availability improvement of standby equipment using age-dependent model*, Reliab. Eng. Syst. Saf. **135**, 100–106 (2015)
7. B.M. Bole, K. Goebel, G. Vachtsevanos, *Using Markov Models of Fault Growth Physics and Environmental Stresses to Optimize Control Actions*, pp. 1–7 (2012)
8. İ. Eker, *Experimental on-line identification of an electromechanical system*, ISA Trans. **43**, 13–22 (2004)
9. I.D. Landau, Y.D. Landau, *System identification and control design: using P.I.M.+ software* 1990, (2015)
10. K. Shwetha, N.S. Narahari, C.S. Prasad, *Failure Mode Effect and Criticality Analysis Performance Test on Dc Brush Motors Used in Spacecraft Applications*, Int. J. Res. Eng. Technol. **1**, 169–176 (2013)
11. L. Tóth, T. Tóth, *Construction of a realistic signal model of transients for a ball bearing with inner race fault*, Acta Polytech. Hungarica. **10**, 63–80 (2013)
12. A. Saxena, K. Goebel, D. Simon, N. Eklund, *Damage propagation modeling for aircraft engine run-to-failure simulation*, 2008 Int. Conf. Progn. Heal. Manag. IEEE, pp. 1–9 (2008)
13. B. Saha, K. Goebel, *Uncertainty management for diagnostics and prognostics of batteries using Bayesian techniques*, IEEE Aerosp. Conf. Proc., pp. 1–8 (2008)
14. M. Orchard, G. Kacprzynski, K. Goebel, B. Saha, G. Vachtsevanos, *Advances in uncertainty representation and management for particle filtering applied to prognostics*, 2008 Int. Conf. Progn. Heal. Manag. IEEE, pp. 1–6 (2008)
15. S. Theodoridis, *Machine Learning*, Elsevier (2015)

Research and Development of a Power Monitoring System for the Sustainable Energy Management System Implementation at Green School, Bali, Indonesia

Dita Anggraini[1,2], *Nazrul* Effendy[1,*], *Muhammad* Ihsan Al Hafiz[1], and *Daniel* Ojeda Luviano[2]

[1]Department of Nuclear Engineering and Engineering Physics, Faculty of Engineering, Universitas Gadjah Mada, 55284 Yogyakarta, Indonesia
[2]Green School, 80352 Bali, Indonesia

Abstract. Green School Bali, an international school which provides education for sustainability through entrepreneurial learning and natural environment won the Zayed Future Energy Prize for developing the new renewable energy projects. Currently, Green School is supported by 2 main power sources which come from Indonesian State Electricity Company (PLN) and Photovoltaics (PV) system electricity with power reach 23.1 kW and 21.6 kWp respectively. Both of the sources are distributed to the load based on the manual balancing and daily manual switching, which create the unbalanced load and high power loss. The key to achieving sustainable energy management is the availability of accurate measurement, recording, and analysis system of the power quality in the whole buildings. To meet the requirement of the sustainable energy management system at Green School, the monitoring system of the power and energy at Green School is developed. This system contains 2 data loggers and an Internet of Things (IoT) platform. The fixed data logger provides routine measurement and recording of power quality at the PLN meter, whereas the portable data logger provides the same function for each electricity panel. The monitoring system and human interface system integrate the intelligent human-machine interface to an Ubidots platform and completed with online and offline data storage. This system is useful for supporting the sustainable energy management goal of Green School and the daily energy practitioner's work with providing a proper advance power monitoring system that has low cost and compatible with Green School system's demand. This project will also be used as a platform and teaching tool for students to develop similar projects and learn about IoT and cloud-based monitoring solutions with other measurements in the classroom.

1 Introduction

Green School is an international school which is located in Jalan Raya Sibang Kaja, Banjar Saren, Abiansemal, Kabupaten Badung, Bali, Indonesia and it provides education for sustainability through community-integration, entrepreneurial learning in a natural wall less environment [1]. Green School has specific characteristics of green building model such as no wall in the buildings, no air conditioning except for several rooms, low current system, and goals to implement sustainable energy solutions. The existing main power sources at Green School come from a mix of the public PLN grid and Photovoltaics (PV) system electricity. The available grid power is 23.1 kW and the PV system has 21.6 kWP capability. Both of the sources are switched daily with manual switching and balancing. Due to the inaccurate switching and balancing processes at Green School, it is estimated that 40% of the available PV power is not used. Although the majorities wiring systems of Green School are in 3-phase, several buildings still use separate single phase. The unbalance system will affect not only the harmonic distortion but also can cause the electrical equipment damage. To attain efficient power management, synchronization of the power is needed between the renewable energy and PLN grid based [2]. Green School won the Zayed Future Energy Prize from Ruler of Abu Dhabi and Founding Father of the United Arab Emirates, Sheikh Zayed bin Sultan Al Nahyan at the beginning of 2017. With this funding, Green School receives a 100,000 USD funding to create the sustainable renewable energy system and energy hub at Green School. This renewable project has been named as Operation Rain or Shine (OROS) Project. To achieve the project Green School also created collaboration with TH Köln University and Engineering Studio in Germany and Universitas Gadjah Mada to create the learning experience opportunity for the students in conjunction with the technical development. During this project, Green School will be expanding the renewable energy contribution to the school and environment, also build an energy hub which can be the energy and related technologies center at the school.

One of the key goals of the OROS project is to implement the sustainable energy management system at Green School. To support the sustainable energy management system, an instrument system that can

* Corresponding author: nazrul@ugm.ac.id

measure the required parameters and provide data storage, backup, visualization, and analysis is required and the basis of this paper. Smart monitoring is used to provide an efficient energy management for sustainable and energy-efficient building [3]. In this paper, the design, assembly, evaluation and implementation of a monitoring system of the PLN power and usage in Green School is discussed. The development of this monitoring system includes hardware, software and web-based monitoring and is aimed to achieve the sustainable energy management system implementation at Green School, Bali, Indonesia.

2 Methodology

This research is aimed to build the instruments that can provide the measurement of power quality at Green School, the data storage system, and the visualization platform to facilitate the sustainable energy management process. To collect the information and theoretical background, literature study about power quality in a building is conducted by the authors. The development of hardware and monitoring website will be conducted with quantitative research methods and supported by experiment and participant observation. This project is located at Green School, Jalan Raya Sibangkaja, Banjar Saren, Abiansemal, Badung Regency, Bali, Indonesia. All of the project cost is funded by OROS Project and should be inherited into Green School's students learning activities.

3.1 Literature Study

3.1.1 Power Quality in a Building

There are several international standards which define the power or voltage quality parameters. The IEEE Standard Definitions[1] defined the power measurement when the voltage and current are not sinusoidal, when the energy is dissipated when the load is unbalanced and when the voltage is asymmetric. The standard power resolution has the key concept which is the separation of the fundamental components of voltage and current from the entire of the harmonics components. It can improve the measurement quality of the instrument. The IEEE Trial-Use Standard gave a great explanation of the power quality parameters. In order to measure the 3-phase current into separated 1 phase current system, the system in each phase is assumed to be single-phase sinusoidal.

A sinusoidal voltage source can be represented as:

$$v = \sqrt{2}Vsin(\omega t) \tag{1}$$

If a linear load is supplied, will produce a sinusoidal current of:

$$i = \sqrt{2}Isin(\omega t - \theta) \tag{2}$$

With V is the Root Mean Squared (RMS) value of the voltage (V), I is the RMS value of the current (A), ω is the angular frequency or $2\pi f$ (rad/s), f is the frequency (Hz), θ is the phase angle (rad), and t is the time (s). The instantaneous power p is given by:

$$p = p_a + p_q \tag{3}$$

$$p_a = P[1 - \cos(2\omega t)] \tag{4}$$

$$p_q = -Q\sin(2\omega t) \tag{5}$$

$$P = VIcos\theta \tag{6}$$

$$Q = VIsin\theta \tag{7}$$

Power on the system consists of 3 kinds of power, the apparent power S in VA, the active power P in W, and the reactive power (Q) in var.

- *Apparent Power (S)*

The apparent power (S) is the power that we can measure it with common measurement device. The apparent power S is the product of the RMS voltage and the RMS current. The equation can be shown as follows:

$$S = VI \tag{8}$$

$$S = \sqrt{P^2 + Q^2} \tag{9}$$

- *Active Power (P)*

The active power (P) is the average value of the instantaneous power that obtained during the observation time interval from τ to τ + kT. Where:

$$P = \frac{1}{kT}\int_{\tau}^{\tau+kT} pdt \tag{10}$$

$$T = \frac{1}{f} \tag{11}$$

Where T is the cycle or period (s). The active power (P) can be described with the following equation:

$$P = VIcos\theta \tag{12}$$

- *Reactive Power (Q)*

The reactive power (Q) can be obtained by measuring the amplitude of the oscillating instantaneous power p_q.

$$Q = VIsin\theta \tag{13}$$

If the load is inductive then $Q>0$, but if the load is capacitive, then $Q<0$.

At Greenschool Bali there are very few inductive loads, nearly all loads are simple resistive loads. For this

 https://doi.org/10.1051/e3sconf/20184301021

project, it has been decided to measure only the active power based upon current measurement assuming a stable grid voltage of 220VAC.

3.2 Green School's Energy System

Green School has 2 main electricity sources, those are PLN (Indonesian government-owned corporation of the national electricity) and Photovoltaics (PV) system. The main PLN power at Green School is 35 Ampere, 220 Volt with 23,100 Watt power. The PV system contains 6 arrays with 18 panels and produces 3 phase AC power. Each panel has 200 Watt capacity, therefore the total PV system's capacity at Green School is 21,600 Watt. The PV energy is stored in a 72 kWh battery bank. Green School also had a micro hydro system (Vortex) that operated from October 21, 2016, until December 20, 2016. After a major flooding event this system was destroyed and cannot operate anymore. During 2 months operation, the vortex system produced only 761watts of average power. This system is currently under rebuild with a goal to achieve more than 8kW of power. Based on the current inventory data, Green School has 90 kW potential loads from all installed equipment, with measured daily loads reaching 44 kW while the average daily usage is 300 kWh. The Green School's energy profile can be shown in Figure 1.

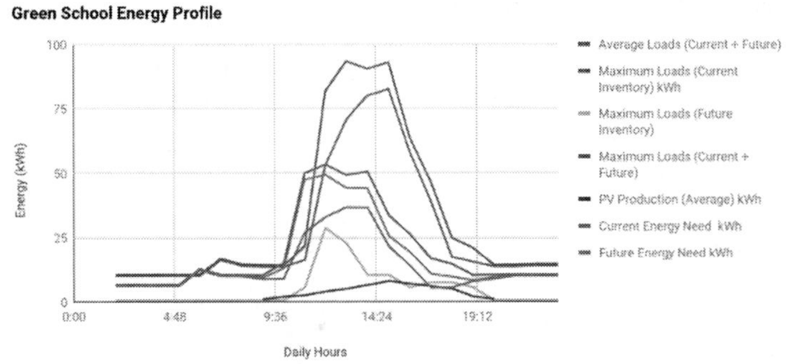

Fig. 1. Green School Daily Energy Profile [6]

Green School electricity circuit cabling is predominantly 3 phase (Phase-R, Phase-S, and Phase-T) system except for a few buildings connected as single phase. There are 30 electricity panels at Green School which distribute the PLN and solar system's electricity around Green School. The biggest problem of the 3 phase electrical system is the manual balancing. Currently, the manual balancing is only conducted in the main PLN metering by Green School's maintenance team and creates an unbalanced load resulting in inefficient use of the PV power. Therefore, the sustainable energy management system should be developed at Green School.

3.3 System Requirements

The monitoring system which contains 2 data loggers and an IoT dashboard are designed with specific requirements from Green School, Bali, Indonesia. Appropriate with the information in the introduction, Green School has 2 main power sources, the PLN and PV system. The PV system at Green School consists of 6 sets of 18 panel-arrays with 200 W capacity per panel.

The maximum PV power is 21,600 W. Green School has average of 300 kWh daily usage.

The monitoring system consists of two data loggers linked to an IoT server delivering a dashboard to the users. There is one fixed data logger and one portable data logger developed for this project. The fixed data logger is implemented on the main PLN meter, whereas the portable data logger will be able to do remote measurement at each distribution panel. Currently, there are 30 building and distribution panels at Green School. The data loggers utilize clamp type current sensors and are designed to support 24/7 continuous monitoring with 15 minutes default time interval for each-recording. Although the default is 15 minutes, the recording interval is changeable if required.

For the General User Interface (GUI) the data loggers use intelligent Nextion HMI TFT Human Machine Interfaces. The measurement result is displayed on the Nextion screen, and transferred to the cloud in the Ubidots platform. Ubidots is Internet of Things (IoT) Host specializing in hardware and software development. Ubidots application enablement platform effortlessly integrates device production, cloud computing, and deployment[2]. Both the portable and fixed data loggers have two current measurement ranges and sensors. For the fixed data logger, the low current range is 0-30 A and the high current range is 0-100 A. For the portable data logger, the low current range is 0-10 A and the high current range is 0-30 A. The lower current range will have higher sensitivity, the utilization will depend on the user's requirements.

3.4 Overall Monitoring System Design

The power monitoring system at Green School is proposed to provide the more flexible, reliable, and

 https://doi.org/10.1051/e3sconf/20184301021

economical power monitoring system. This system is the part of the evolution from metering to monitoring on the energy management system. The classification of the power measurement system according to their objectives can be shown as follows:

Table 1. Classification of the Power Measurement System[3]

Group	Objective	Measurement
Meter	Instantaneous	V,I,P
Analyzer	Periodic, Reactive	V,I,P, harmonics, waveform, event
Monitor	Continuous, Proactive	Above contents+statistis+indices

The proposed system is developed to continuously measures the current and power at from PLN Grid at Green School and the each building consumption and provide the statistic and indices. Appropriate for the previous discussion about system requirement, the monitoring system of power quality at Green School consists of 2 data loggers and 1 monitoring website. The simple diagram of the system design can be shown as follows:

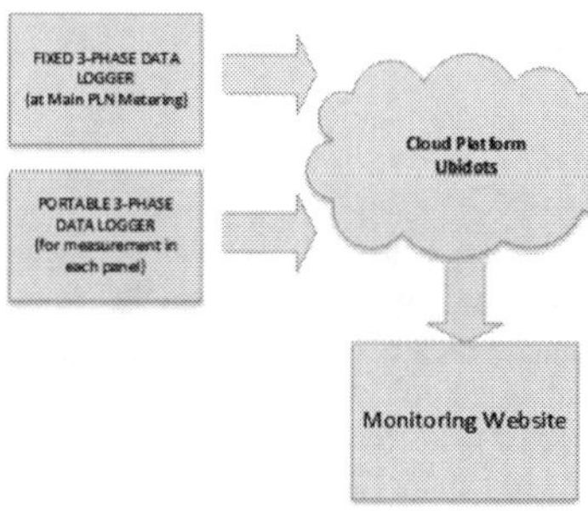

Fig. 2. Green School's power quality monitoring system design

The hardware development using Arduino Mega 2560 and interactive Human Machine Interface display Nextion 3.2" . The hardware systems were selected to achieve complex connections and codes, Arduino Mega 2560 is used for the complex projects and larger space for the sketch. Arduino Mega 2560 is a microcontroller board based on the ATmega2560. It has 54 digital input/output pins (of which 15 can be used as PWM outputs), 16 analog inputs, 4 UARTs (hardware serial ports), a 16 MHz crystal oscillator, a USB connection, a power jack, an ICSP header, and a reset button[4]. Both of the data loggers use the Nextion NX4024T032-3.2" HMI TFT Intelligent LCD Touch Display Module. The nextion HMI TFT facilitates the development of an analog touch screen user interface with programmable function buttons. It facilitates string instructions, assigns values on run time and provides the automatic updates for the editor and firmware[5]. The IoT platform that used for this project is Ubidots. Ubidots is an IoT Hosting Service with expertise specializing in hardware and software development which enables the platform to effortlessly integrate device production, cloud computing, and deployment[2]. The clear explanation about each system will be discussed in the following section:

3.4.1 Fixed 3-Phase Data Logger

The fixed 3-phase data logger is located in the main grid meter at Green School which comes from PLN source. This fixed data logger provides automatic recording of the current and power measurement of each phase with the adjustable time interval. The default of the time interval is 15 minutes and 5 seconds for the testing mode. This hardware consists of 3 Current Transformers (CTs) which connected into microprocessor through signal conditioning. As the Greenschool loads are mainly resistive loads with few rotating machines it was decided to measure only the current and infer no power factor loss. Two types pf CT have been selected: 3 SCT-013-030 with current range 0-30 A and 3 SCT-013-000 with current range 0-100, one sensor for each phase. A Nextion NX4024T032-3.2" LCD is used for the interactive human interface and coded with the Nextion editor. The recorded data will appear in real-time on the Nextion LCD display and be automatically recorded on the IoT platform. This device will calculate 1 second RMS power values, average the power over the recording interval and record average and peak power for each time interval. The values than will appear on the LCD every time interval loop and will be automatically clouded into the IoT platform. The data is also stored into an SD card for backup. The electrical circuit diagram of the fixed 3-phase data logger can be shown in the following figure 3:

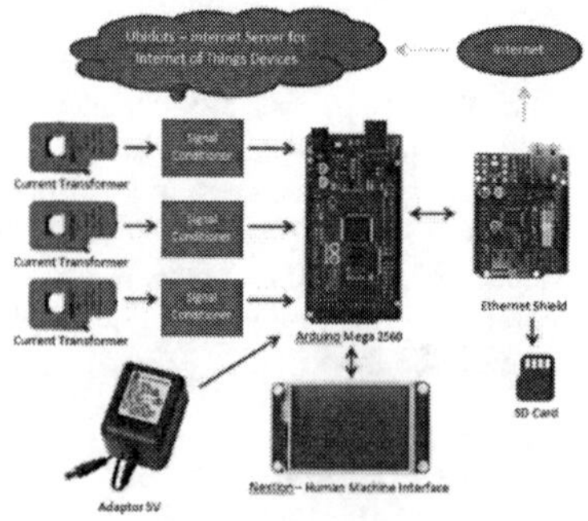

Fig. 3. System diagram of the fixed 3-phase data logger

The fixed 3-phase data logger is completed with 2 different ranges current transformers which are 0-30 A and 0-100 A. The default implemented CTs are the 0-100 A as the current at the PLN meter has been observed to fluctuate up to 35 A. The internet connection for this device is supported by the wired LAN cable connection from Green School's IT team to support a robust internet connection for the fixed system.

3.4.2 Portable 3-Phase Data Logger

The portable 3-phase data logger has the same concept with the fixed 3-phase data logger but has more features.

This data logger enables the remote measurement of RMS current and RMS power of each phase at each electricity panel. The portable 3-phase data logger uses the Arduino Mega 2560 as the microprocessor and Nextion NX4024T032-3.2" HMI TFT Intelligent LCD Touch Display Module as the interactive UI. Through the Nextion HMI TFT module, the users will be able to set the current range (low: 0-10 A or high: 0-30 A), recording duration, recording timer, and panel number. At Green School, there are 30 electricity panels which have the different storage system identification on this device, this device will provide real-time measurement and data on every electricity panel. Similar with the fixed 3-phase data logger, the portable 3-phase data logger will measure average and peak power for each time interval. The values than will appear on the LCD every time interval loop and will be automatically clouded into the IoT platform. The data also stored into an SD card for backup. This device is powered by a LiPo rechargeable battery of 1200 mAh for the power source and Arduino Wi-Fi Shield for the internet connection. This device uses the Wi-Fi connection at Green School for communication. The electrical circuit diagram of the portable 3-phase data logger can be shown as follows:

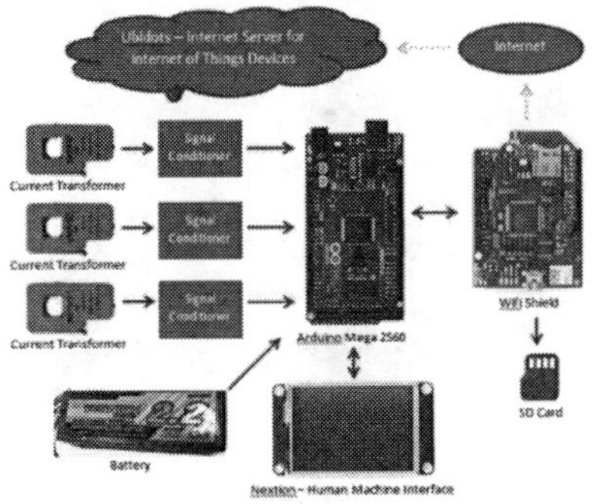

Fig. 4. System diagram of the portable 3-phase data logger

3.4.3 Monitoring Website

The Green School monitoring website enables the real time monitoring of current and power values on each phase and can be accessed by all levels of Green School community. The monitoring website actually is an IoT platform hosted by Ubidots. Ubidots is not only a ready use IoT platform that can accelerate the hardware and software development, but also has several useful features such as the live dashboard, proven device libraries, time series storage and playback, trigger events, math and statistical engine, multi protocols and comprehensive API. Ubidots supports the most general protocols for IoT such as HTTP and MQTT, it can securely clouding the data with the utilization of a token-based authentication and an optional HTTPS/SSL encryption layer, where it can remain in renowned cloud providers[2]. This website enable the user to see a live dashboard of recorded data, arrange and analyze the raw recorded data to see the average, sum, and other arithmetic calculation. Also the platform provides the capability to download the csv file source of the raw data. This monitoring system also provides different statistical graphs that can help the energy practitioners at Green School to analyze the energy condition.

4 Results and Discussion

4.1 Green School's Existing Energy Performance Analysis

The total energy yield of the PV system from 2012-2017 can be shown as follows:

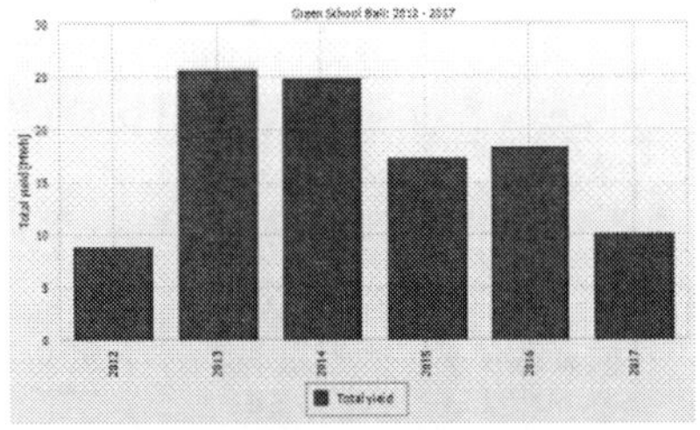

Fig. 5. Total yield of the PVs farms[6]

From the following picture, it can be shown that the PV system's performance is decreasing from 2013 till 2017. This condition can happen because of several possible conditions such as climate, shading, or improper energy management system. The big problem on the energy management at Green School is the manual switching between PLN electricity and the PV system electricity production. Both sources are connected into a Master Distribution Panel (MDP) and divided into seven 3-phase switching relays. The 7 switches allow the distribution of electricity into all Green School's buildings. Every day, the maintenance team will review the PV production and monitor the battery percentage. Then the maintenance team will do select switching between PLN and PV electricity manually. On the weekend commonly there is no body that will do the switching so the PV power on the weekend is unused [6]. This creates a big power loss and has the potential to damage the batteries of the PV system if switches are let connected. In 2013 at the beginning of the PV system implementation PV provided for 35% of Greenschool power needs as follows:

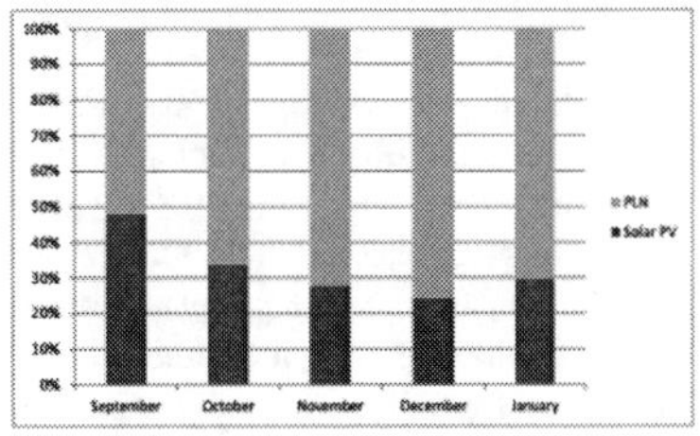

Fig. 6. Green School's Energy Consumption 2013[7]

Whereas the data from the last energy audit shows that the contribution of the renewable energy system is decreased to less than 20%. It is clearly represented by the following data:

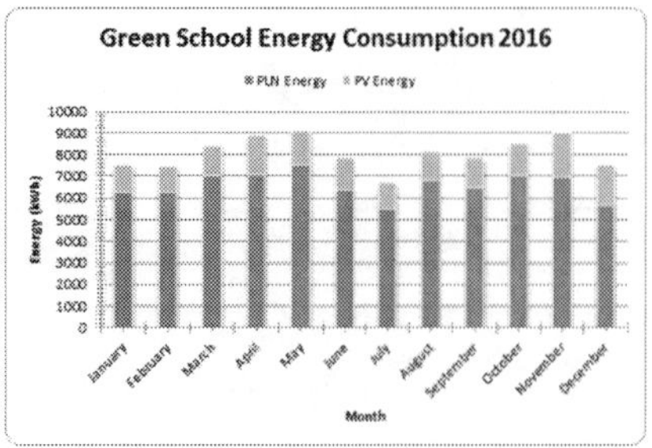

Fig. 7. Green School's Energy Consumption 2016[6]

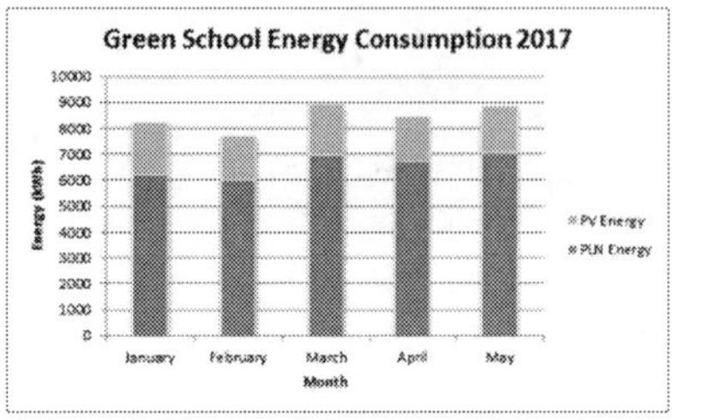

Fig. 8. Green School's Energy Consumption 2017[6]

Based on this data it is important for Greenschool to focus on improving the amount and use of renewable energy. A better energy management system is extremely important to improve the renewable energy system performance, reduce the energy cost and carbon footprint, and increase the positive impact of energy efficiency on the society and environment.

4.2 The Sustainable Energy Management System Goal at Green School

Currently, the sustainable energy management topics are concerned to the rising costs, energy efficiency and environmental impact in energy generation, distribution, management and usage. The conditions make the innovation on the energy management system is needed. On this project, Green School developing a team of dedicated scientists, engineers, practitioners, researchers, and students to make the energy cleaner, more reliable, more efficient, more secure and gives the positive impacts around Green School.

Energy consumption is a dominant component of the organization's sustainability cost structure. In case of the increasing energy cost, the energy efficiency currently becomes an important thing that should be realigned to accommodate the volatile energy prices in the future. Information technology, advanced automation, and the trained practices for energy management are the key factor for the success of sustainable energy management[8].Obtaining the visibility into energy reduction goals and empowerment of the workers at every level in the organization will bring positive impacts on the energy consumption, therefore the education of energy efficiency should be given to all level of students and staffs of Green School.

The first step of creating the sustainable energy management is the energy audit. Energy audit provides a tool that enables an institution to track the progress of the energy management program. It can identify which is the most inefficient equipment or process, and how they affect the energy performance in an institution [8]. The routine energy audit is one of the goals that must be achieved by Green School. By performing the routine energy audit, it will provide a good assessment of the effectiveness of energy management programs compare to past performance.

Siemens create the strategies for sustainable energy management which can be summarized in Table 1. A monitoring website is important on the energy management system to enable the workers at an organization to monitor their energy consumption in a real time manner, therefore they can evaluate their usage and improve their management system. At Green School, a monitoring system of current and power in each phase also important for the balancing and switching process. On the previous discussion, the energy system at Green School is already discussed and it is very clear that the balancing and switching process of Green School's electricity sources are conducted manually.

Table 2. Strategies for Sustainable Energy Management [8]

Domain	Key Attributes
People	- Management commitment and Resolve - Corporate energy coordinator which directs and coordinates the energy management activities across the sources - Energy management considered a core competence
Processes	- Energy management initiative in place - Routine energy audit to measure the progress against a benchmark - The capital projects need to be justified based on energy saving purpose
Technologies	- Investment of Research & Development (R&D) to change the energy generation process and reduce the energy consumption - Electrical automation integrated with process automation systems - Advanced simulation and process control or homegrown application for the energy management -
Information	- Monitoring the processes and equipment to track energy usage and problem areas identification - More granular accounting methods for energy use - Energy prices and energy

	consumption consistently monitored

At Green School, the sustainable energy management goal is to connect the people, processes, technologies, and information to create a system that can cover the followings steps:

- Has the standard method of energy efficiency which economically feasible for Green School
- Developed the energy efficient buildings both for the existing buildings and the future buildings.
- Perform the routine energy audit to find the equilibrium between the present and future energy necessity and demand.
- Maximize the usage of clean energy and high-efficiency energy conversion which give positive impacts to the environment.
- Develop the short and long term high-efficiency energy storage.

To start all the goals, the developed power monitoring system is proposed to provide the real-time measurement devices that can help the realization of the routine energy audit and support the decision making for the energy practitioners at Green School. This proposed system will provide the continuous data availability, statistical analysis, and real-time monitoring dashboard.

4.2 Power Monitoring System Implementation at Green School

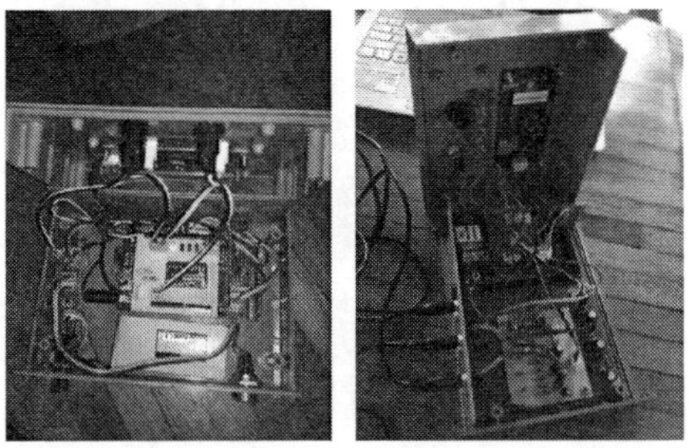

(a) (b)

Fig. 9. a. Portable data logger electrical circuit b. Fixed data logger electrical circuit

On September 3rd, 2017 this Power Quality Monitoring System was installed at Green School, Bali. The fixed 3-phase data logger is located beside the main grid meter (PLN meter) and the system is successfully recording data and clouds them into the IoT platform. The electrical circuit physical forms of both fixed and portable 3-phase data loggers are shown on the figure 9. The fixed 3-phase data logger is connected beside the PLN meter and the clamp sensors are calibrated against a commercial clamp meter as shown in the figure 10. The Nextion design examples are shown on the figure 11 and the real display snapshot of the Nextion at the device is shown in the figure 12.

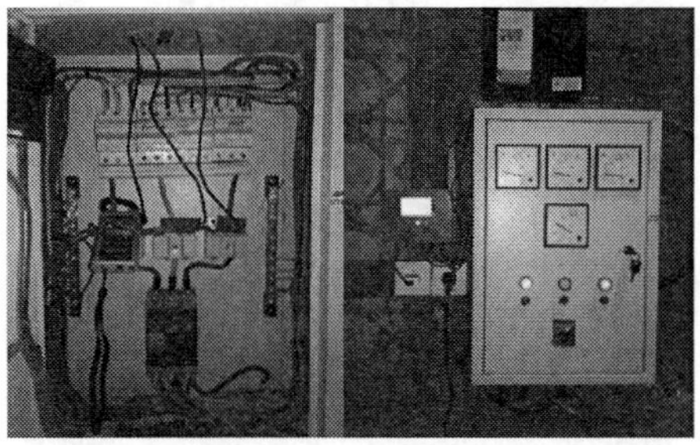

(a) (b)

Fig. 10. a. Portable data logger electrical circuit b. Fixed data logger electrical circuit

The HMI provides the interactive way for the users to monitor the system and give control and feedbacks to the energy management system. One of the excellences of Nextion display is its ability to create a user interface without coding but more based on design. With importing the image files into the Nextion Editor software, a developer can easily design their interface display. In figure 11, the design example of user interface is shown and in figure 12 the real display of the Nextion is shown. From those pictures, it can be seen that what we get on the display is similar with what we design on the editor.

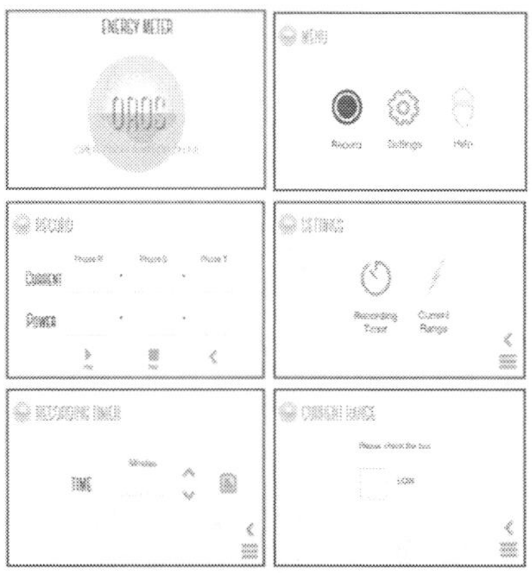

Fig. 11. Nextion display design examples

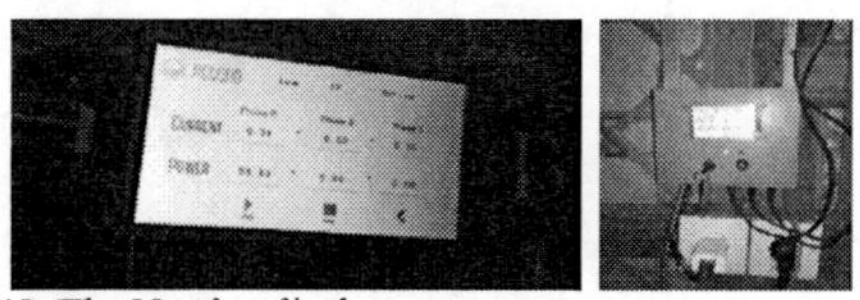

Fig. 12. The Nextion display

After the data is recorded, the data is automatically uploaded into the IoT platform. The Green School's IoT server login can be shown as follows figure 13:

https://doi.org/10.1051/e3sconf/20184301021

Fig. 13. Green School's IoT system server login

The Green School's IoT server dashboard will be the real-time monitoring system that can be accessed by the society at Green School, the device administration will be managed by the IT team and energy audit team. The examples of the recorded data dashboard are shown on the following figure 14 and figure 15:

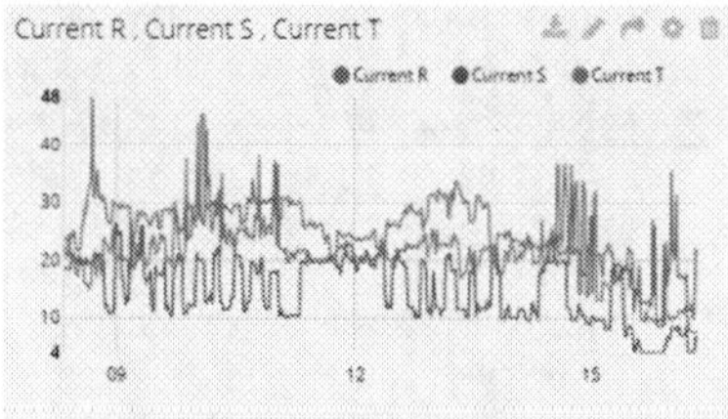

Fig. 14. Current Ubidots Dashboard-Every Minutes Recording

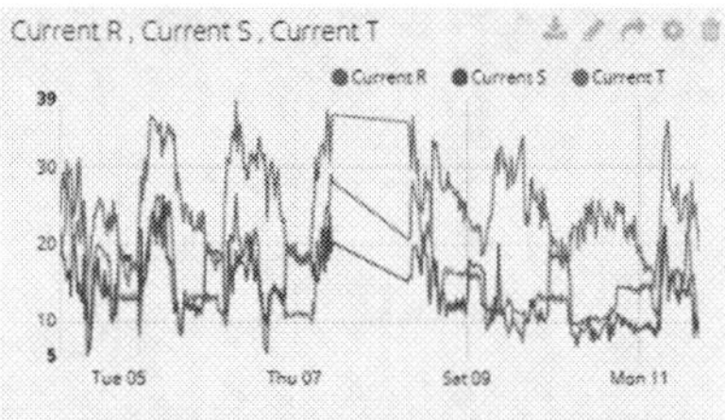

Fig. 15. Current Ubidots Dashboard-Hourly Recording

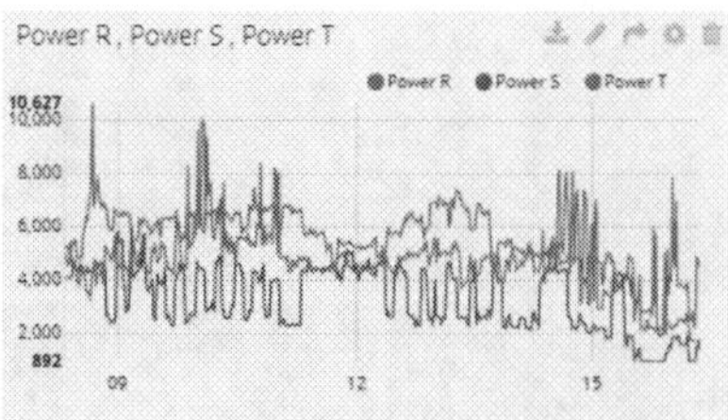

Fig. 16. Power Ubidots Dashboard-Every Minutes Recording

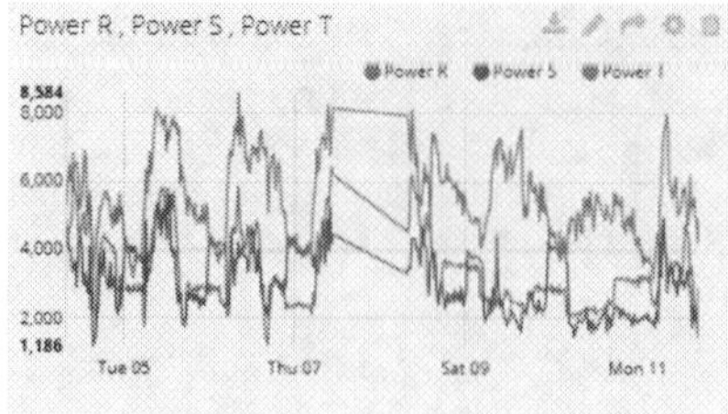

Fig. 17. Power Ubidots Dashboard-Hourly Recording

From the power value per phase, the total power that consumed from PLN grid can be calculated by adding all the power from phase-R, phase-S, and phase-T. The total power real-time data visualization can be shown as follows:

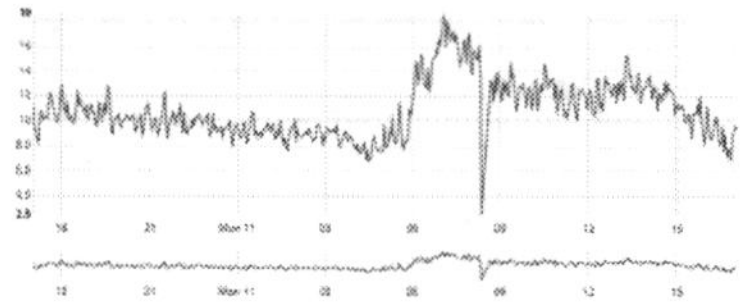

Fig. 18. Hourly Total Power Data (kW)

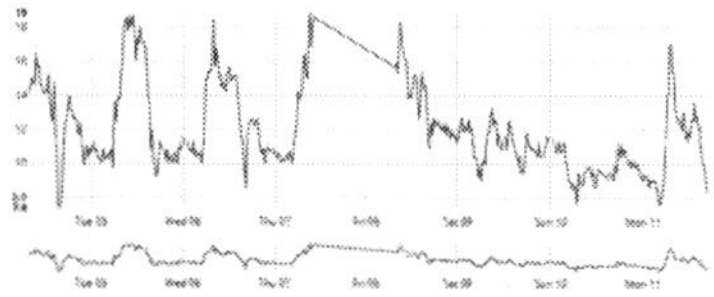

Fig. 19. Daily Total Power Data (kW)

The energy data every hour can be shown in the following picture:

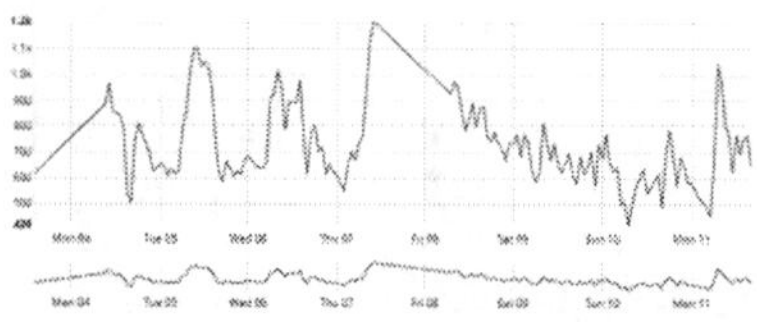

Fig. 20. Hourly Energy from PLM Grid Data Visualization (kWh)

To support the proper balancing method, the peak current measurement is also important. The data of the peak current from the electricity grid can be shown as follows:

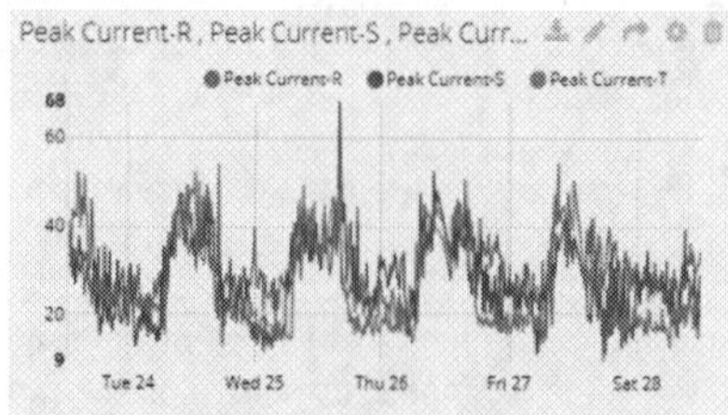

Fig. 21. Peak Current Data (Ampere)

Based on the pictures, the real-time current and power on each phase can be obtained and the energy practitioners at Green School can get the better understanding of the energy performance at Green School. From the data, it can be concluded that the electricity load at Green School is a relatively unbalanced load system. The data from the monitoring system shows that the average current on the phase-R, S, and T are 24.46 A, 14.80 A and 21.74 A respectively. The average power on phase R, S, and T are 5,374 watt, 3,253 watt and 4,784 watt respectively. The data shows that the fluctuation of the current in each phase is not balance each other. From figure 21, it can be concluded that the difference between 1 phase to the other phases is quite big. This information should be one consideration for the future expansion of existing electrical system performance. For the future building and inventory, the load should be distributed equally. Likewise if there are any projects that need the additional equipment, with this real time data the workers can analyze the energy condition and have the good understanding to choose where they can connect the equipment to approach more balance load system. The full picture of the Green School's IoT system dashboard is shown in the figure 21.

Having a monitoring platform through a website based system gain several benefits compare than using software. The following benefits can be got through Ubidots application:

- Open system architecture support: Ubidots Support the most general protocols for IoT (such as HTTP and MQTT). It supports several programming language such as Python to accelerate the application development time. From the API authentication, to data transport and storage, it adheres to web security best practices. Provides the user-friendly and consistent HMI: with internet based platform especially Ubidots service, the users do not need to learn the difficult commands for any developed application.
- Good scale ability:[9] it can be developed gradually. Ubidots also provides the system expansion and the addition of the organizations and devices.
- Time series storage and playback: the backend is a time series that can ingest or retrieve the millions of data-points in short time, helping the developer to scale the application easily.
- Math and Statistical Engine: It allows statistical operation, will give benefit to the data processing.
- Provide the ready trigger events: Create SMS, Email, Telegram or Web hook alerts based on your sensor data that can trigger thousands of events, even if they happen at the same time[2].

5 Conclusion

5.1 Summary of Findings

On this paper, the power monitoring system as the part or OROS project is described. This monitoring system will be useful for supporting the sustainable energy management goal of Green School and the daily energy practitioner's work at Green School. Since Green School is an environmentally conscious school which consumes relatively low energy, this project provides the proper advance power quality monitoring system which is suitable with Green School system's demand and has low cost. The monitoring system is developed by the reliable microprocessors and sensor devices which are connected into advance usable human interface and IoT platform. More than that, this project also gives place for the students learning experience by provide new

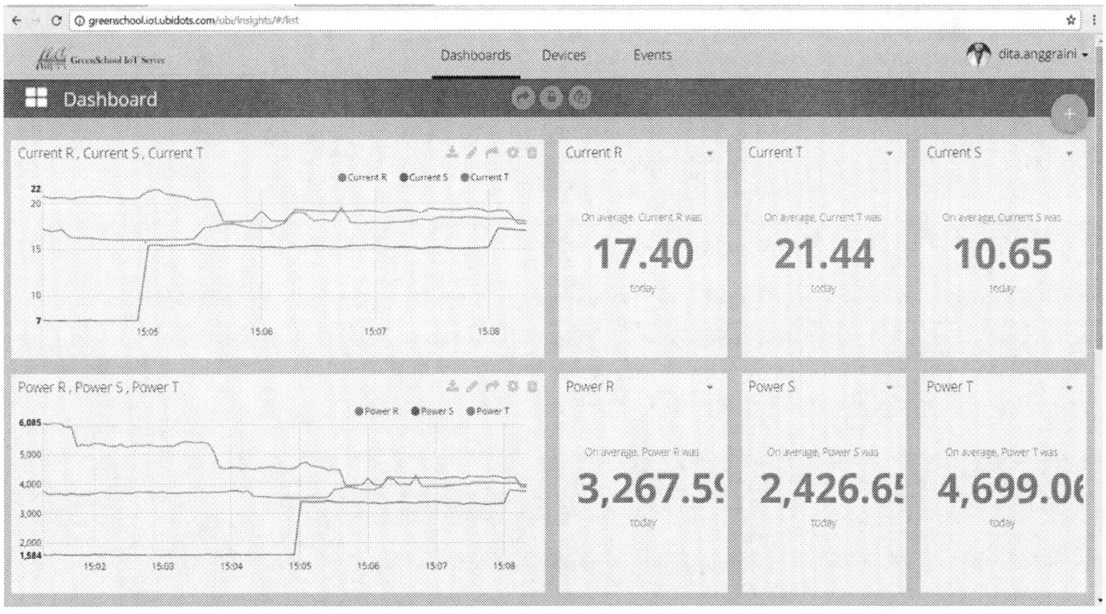

Fig. 22. Green School's Monitoring System Dashboard

https://doi.org/10.1051/e3sconf/20184301021

technologies that can be easily adopted and developed. This system has been used to integrate the technology and human sources to support and improve a regular energy audit and sustainable energy management system. With the implementation, the proper data storage, visualization, and analysis are available for further development.

5.2 Future Development Plan

The power quality monitoring system should be improved by better calibration and testing process and improved analysis of the sampling and statistical calculations. The improved selection of RMS, average, or peak sampling that will give better results. Management of the data set and handling of internet outages.

This research was fully funded by Operation Rain or Shine (OROS) Project, Green School, Bali, Indonesia. The authors are thankful to the OROS team, Green School's society, Engineering Studio Berlin, and Green School's Maintenance Team who provided expertise, materials, learning opportunity, knowledge and support for this research. The authors also like to show the gratitude to Tony Vovers for the guidance, evaluation, and advices that greatly improved the project and manuscript.

References

1. M. Committee, *IEEE Trial-Use Standard Definitions for the Measurement of Electric Power Quantities Under Sinusoidal , Nonsinusoidal , Balanced,* The Institute of Electrical and Electronics Engineers, Inc., **vol. 40**, no.3. (2000)
2. "Ubidots Official Website." [Online]. Available: https://ubidots.com/.
3. D.-J. Won, I.-Y. Chung, J.-M. Kim, S.-I. Moon, J.-C. Seo, and J.-W. Choe, *Development of power quality monitoring system with central processing scheme,* IEEE Power Eng. Soc. Summer Meet. 2002, **vol. 2**, pp. 915–919 (2002)
4. "Arduino Website." [Online]. Available: https://www.arduino.cc/en/Main/arduinoBoardMega2560/.
5. "Nextion Intelligent Display."
6. D. Anggraini, *Green School's Energy Audit Report,* (2017)
7. Akuo Energy Indonesia, "*Annual review - PV power station*, (2014)
8. A. A. Group, *A Strategic Roadmap for Sustainable Energy Management and Energy Efficiency for Industrial, Commercial, Municipal and Manufacturing Operations*, (2009)
9. R.-C. Leou, Y.-C. Chang, and J.-H. Teng, *A Web-based Power Quality Monitoring System*, **vol. 0**, no. C, pp. 1504–1508 (2001)

Response to Pontoon and Pendulum Motion at Wave Energy Converter Based on Pendulum System

I. S. Arief [1], *I. K. A. P.* Utama[2], *R.* Hantoro[3], *J.* Prananda[3], *Y.* Safitri[2], *T.* Rachmattra A.[3], and *Rindu* F. K.[3]

[1] Department of Naval Architecture, Institute of Technology Sepuluh Nopember, Surabaya, Indonesia
[2] Department of Physics Engineering, Institute of Technology Sepuluh Nopember, Surabaya, Indonesia
[3] Department of Marine Engineering, Institute of Technology Sepuluh Nopember, Surabaya, Indonesia

Abstract. Energy conversion technology derived from ocean wave energy has been developed, one of them is the Power Wave sea- Pendulum System (PLTGL-SB). PLTGL composed of a pontoon which is subjected to the excitation force of ocean waves and will move the pendulum on the top of the pontoon. This study aims to analyze the best shape for PLTGL- SB's pontoon and determine the largest deviation generated due to the movement of the pendulum vertical pontoon. Pontoon shapes studied are pontoon consisting of a large cylinder in the middle and there are two boats on the right and left, like a trimaran ship. Variations were made in this study consisted of variations of length and height boats, the draft height, wave period, the mass and the length of the pendulum arm. Best pontoon shape is determined by simulating the shape of a pontoon with Computational Fluid Dynamic (CFD). Pendulum deviation obtained by mathematical modeling pontoon two degrees of freedom (roll) and pendulum. Based on the chart Response Amplitude Operator (RAO) pontoon shape is best for PLTGL-SB is a pontoon with a 2/3 full large cylinder diameter, 1.5 cm height boats, catamarans length of 41.5 cm. Based on the results of a mathematical model of the largest deviation of the pendulum is generated when the pontoon is placed in the period of 0.8 s, with a mass and pendulum arm lengths are 19.9 g and 10.6 cm.

1 Introduction

According to General Directorate (Ditjen) of Renewable Energy and Conservation Energy (EBTKE) in 2013, hydro energy (ocean) is the one of potential renewable energy in Indonesia. Indonesia has large sea three times of its land, in the amount of 3,544,743.9 km2. Ocean energy source consists of three types, i.e. ocean heat energy, tidal energy, and ocean wave energy.

Ocean wave energy started to study more than 30 years ago (Hongda Shi, 2014) [1]. Nowadays, conversion energy technology from ocean wave energy is already expanded, i.e. oscillating water column installed pendulum system in the shoreline, shelter duck system, and Pelamis water converter (Rodrigues, 2010) [2].

Ocean wave energy device such as Oscillating Water Columns (OWC) changes the movement of wave in air cylinder column to produce pressured air and furthermore it used to move a turbine and electric generator. OWC has an advantage because it has a little-submersed part in the water. OWC technology also has a disadvantage because it limited by water depth so it is not allowed to be installed in the deep sea. Appropriateness study of OWC has done by Shi et al (Hongda Shi, 2014) by modeling air to optimize room scale and turbine pressure so that produce a large efficiency of electrical power (Hongda Shi, 2014). Another technology created to resolve the limitations of OWC is floating dish. Floating dish first time installed by the Ocean University of China. Floating dish consists of underwater part i.e. piston to get ocean wave and upper part i.e. generator to convert the movement of piston become electrical energy (KS. Choi, 2012). Floating dish use Heave movement from an ocean wave. The technology which uses piston movement to get ocean wave i.e. Pelamis, Duck and Eagle converter developed Guangdong Institution of Energy Conversion, China (Hongda Shi, 2014).

In Indonesia, OWC technology started in 2004 built by BPPT especially BPDP (Balai Pengkajian Dinamika Pantai). OWC prototype built at Parang Racung Beach, Gunung Kidul. Before OWC, the forerunner technology is Ocean Wave Energy Power Plant – Pendulum System (PLTGL-SB). Study of PLTGL-SB start in 2002 by Zamrisyaf, an employee of the research center and development PT. Perusahaan Listrik Negara (PLN).

Ocean wave energy power plant using pendulum system from Zamrisyaf consist of pontoon and pendulum system with pontoon dimension i.e. L: 6m, B: 3 m, Tp: 1.5 s and D: 2 m with a slope angle of the hull is 450. There are two types of pendulum based on position i.e. pendulum placed vertical and horizontal. The first test of PLTGL-SB was done in Ulank Karang beach, Padang in 2002 and got a patent in 2009 with horizontal and vertical pendulum placed on drums.

The concept of PLTGL- SB continues to enhance by some studies. In 2012, PLTGL-SB was developing by the varied shape of octagonal pontoon arranged become

* Corresponding author: kutama@na.its.ac.id, teuku14@mhs.ne.its.ac.id

https://doi.org/10.1051/e3sconf/20184301022

triangle (tripod) in the height of wave 0.5 m and 1.5 m (Yasin Bestari, 2013) [3]. This research based on simulation using software Numeca to get the optimum motion of pontoon for PLTGL-SB. In 2013, research of PLTGL-SB developed by Saadilah Akbar by varied the slope of pontoon hull in height of ocean wave i.e. 1m and 1.5m. In 2014, research is continued by doing an experimental study in pontoon shape. Pontoon shape i.e. oval, ball, and tube. The biggest force production obtained in a ball shape, because in a ball shape has largest water surface area than another shape. Known that by the increase of length in water exposed section then it will affect power that received pontoon (Cahyadi, 2014).

The electricity produces from pendulum system lack to fulfill society energy needs. So, the development of this wave energy converter requires an acceleration process so that the scale-up process is required through the increase of PLTGL-SB power capacity[4].

Research i.e. analyze PLTGL-SB with pontoon shape consist of three cylinders; one big cylinder and two small cylinders (jukung) as a balancer. Variation in this research is length variation of jukung arm, the height of jukung and height of water draft. Then the influences of that variation will analyze by software Ansys Aqwa until found the best pontoon model for PLTGL-SB. Next, make the formulation of the mathematic model to know the slope of the pendulum in one degree of freedom pontoon motion one-degree pendulum movement. Validation of simulation using experiment study.

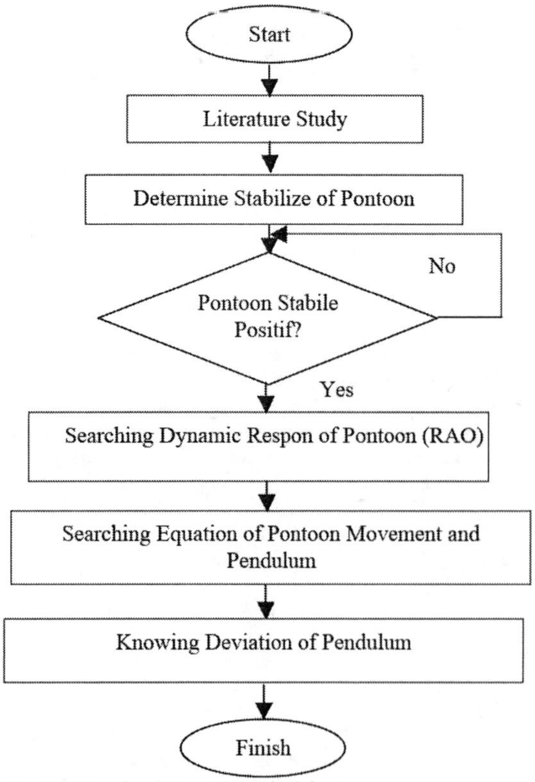

Fig 2.1 *Flowchart* Research Methodology.

2 Methodology

2.1 Research Flow Diagram

Generally, the steps are a draw in flow diagram i.e. Picture 2.1 below. The process of this research begins with a study of literature on the research of Marine Wave Power Generation Pendulum System (PLTGL-SB) which already existed before. The basic things that the author studied at the stage of literature study are the science of ocean waves, the movement of marine buildings, the response of marine buildings to the wave excitation style. After that, the authors determine the stability of the pontoon. Determination of stability is determined by three points namely, Center of Gravity (COG), Center of Buoyancy (COB), and point metacenter. Pontoons that can already be said to be positively stable if the position of the metacenter point above the point of COG and COB. Stable positive means that the pontoon will return to its original position if the pontoon is exposed to the excitation force of the ocean waves. After that, the author also studied the theory of CFDs for floating marine buildings. After that, the authors retrieve the data into the towing tank where the instrument test, the wave period, and the wave amplitudes through the experimental circuit. After performing the data retrieval, the writer analyses the pontoon response to the wave excitation style to achieve the first goal of this final project, which is to know the best pontoon shape for PLTGL-SB with various variations. Pontoons motion response is obtained by simulating the pontoon using Ansys Hydrodynamic Diffraction or Ansys Aqwa. The simulation results on the hydrodynamic diffraction will be known response value amplitude operator (RAO) or pontoon response due to the coming ocean waves. The results of the simulation of RAO pontoons analysis conducted as many as 16 variations. Variables that are varied include water laden height. Different water will affect the magnitude of the movement of ships (marine buildings) floating (Mulyawan, 2010). The high water-laden difference will affect the COB point value and the Metacentre point. Jukung variation is done because the length of jukung will affect the stability of the ship (Barras, 2006).

The number of variations shown in table 3.1. The variations used to analyze the variations of pontoon 1, 2, 9, 11, and 15. While the pendulum mass variation and the height of the pendulum arm will affect the COG pontoon.

The second objective of this study, namely to know the size of the pendulum, is achieved by performing mathematical models until the equations of motion and the pendulum response is obtained. Mathematical modeling is performed on one degree of free pontoon (roll) and one-degree pendulum free. Table 3.2 is a detail of pendulum variations. Variations of pendulum mass and pendulum sleeves were performed to find out the deviations to be generated pendulum and sought the largest deviation to be converted into electrical energy. Variations used to analyze data are variations to 5, 6, 9,

10, 21 and 23.

Table 1. Variation of Pontoon.

Variation	Draft	Height of Outrigger (cm)	Length of Outrigger (cm)	Pendulum Arm (cm)	Weight of Pendulum (gr)
1	1/2 D	4	49.5	16.5	40.6
2	1/2 D	4	49.5	16.5	19.9
3	1/2 D	4	49.5	10.67	40.6
4	1/2 D	4	49.5	10.67	19.9
5	1/2 D	4	41.5	16.5	40.6
6	1/2 D	4	41.5	16.5	19.9
7	1/2 D	4	41.5	10.67	40.6
8	1/2 D	4	41.5	10.67	19.9
9	2/3 D	1.5	49.5	16.5	40.6
10	2/3 D	1.5	49.5	16.5	19.9
11	2/3 D	1.5	49.5	10.67	40.6
12	2/3 D	1.5	49.5	10.67	19.9
13	2/3 D	1.5	41.5	16.5	40.6
14	2/3 D	1.5	41.5	16.5	19.9
15	2/3 D	1.5	41.5	10.67	40.6
16	2/3 D	1.5	41.5	10.67	19.9

Table 2. Variation of Pendulum.

Variation	Draft	Wave Period (s)	Mass (gr)	Height of Arm (cm)
1	1/2 D	0.4	40.6	16.5
2	1/2 D	0.4	40.6	10.67
3	1/2 D	0.4	19.9	16.5
4	1/2 D	0.4	19.9	10.67
5	1/2 D	0.6	40.6	16.5
6	1/2 D	0.6	40.6	10.67
7	1/2 D	0.6	19.9	16.5
8	1/2 D	0.6	19.9	10.67
9	1/2 D	0.8	40.6	16.5
10	1/2 D	0.8	40.6	10.67
11	1/2 D	0.8	19.9	16.5
12	1/2 D	0.8	19.9	10.67
13	2/3 D	0.4	40.6	16.5
14	2/3 D	0.4	40.6	10.67
15	2/3 D	0.4	19.9	16.5
16	2/3 D	0.4	19.9	10.67
17	2/3 D	0.6	40.6	16.5
18	2/3 D	0.6	40.6	10.67
19	2/3 D	0.6	19.9	16.5
20	2/3 D	0.6	19.9	10.67
21	2/3 D	0.8	40.6	16.5
22	2/3 D	0.8	40.6	10.67
23	2/3 D	0.8	19.9	16.5
24	2/3 D	0.8	19.9	10.67

2.2 Design of PLTGL

The design consists of a pontoon with three cylinders. One large cylinder and two small cylinders (jukung) located on the right and left as a counterweight. The pendulum form studied is a vertical pendulum. Pontoons with such shapes are the object of CFD simulation and observed the motion response due to the excitation force of the incoming waves. The following is a detailed measure of the pontoon and pendulum used in this study:

Table 3. Geometry of PLTGL.

Geometry of PLTGL	
Diameter of Big Cylinder (D)	16.5 cm
Thickness of Big Cylinder	0.7 cm
Diameter of Small Cylinder (Jukung)	4 cm
Thickness of Small Cylinder	0.2 cm
Length of Big Cylinder	32 cm
Length of small Cylinder	32 cm
Cylinder Density	1.39 gr/cm^3
Water Draft	1/2 D and 2/3 D
Height of Jukung	1.5 cm and 4 cm
Length of Plat	32 cm
Breadth of Plat	17 cm
Plat Density	1.2 gr/cm^3
Pendulum Pole Height	2.75 cm (from plate)
Pendulum Pole Arm	23 cm
Pendulum Mass	40.6 gr and 19.9 gr

The next stage is to calculate the stability of the pontoon, by determining the point of the centre of gravity, the centre of buoyancy, and the point of metacentre. The meta centrum point is searched using the following formula,

$$M = \frac{I_O}{V} - A_O B_O \quad (2.1)$$

By looking for a meta centre point, we can see whether the pontoon is stable or not. Based on meta Centrum values it can be determined that, if M is above the point of weight then the object is said to be stable, M = centre of gravity than the pontoon is said to be neutral stability, and if M is below the point of weight it is said to be unstable. While the buoyancy point sought by using software Maxsurf. COB is likened to COG from the submerged parts, so the search will be the same as COG with a record of just the submerged pontoon that is used as a review. The COG (Centre of Gravity) point is searched by using equation (2.2).

$$x_0 = \frac{\sum_{i=1}^{n} w_i x_i}{\sum_{i=1}^{n} w_i} \quad y_0 = \frac{\sum_{i=1}^{n} w_i y_i}{\sum_{i=1}^{n} w_i} \quad z_0 = \frac{\sum_{i=1}^{n} w_i z_i}{\sum_{i=1}^{n} w_i} \quad (2.2)$$

The results of the calculation of the three points on pontoon ½ D and 2/3 D are as follows:

Table 3. Calculation of Pontoon.

Position from Keel	Draft 1/2 D	Draft 2/3 D
Draft	8.25 cm	11 cm
COB	4.752 cm	6.345 cm
Metacentre	12.17 cm	11.78 cm
Dyed Volume	0.00342 m^3	0.00485 m^3
Dyed Mass	3.42 kg	4.85 kg

https://doi.org/10.1051/e3sconf/20184301022

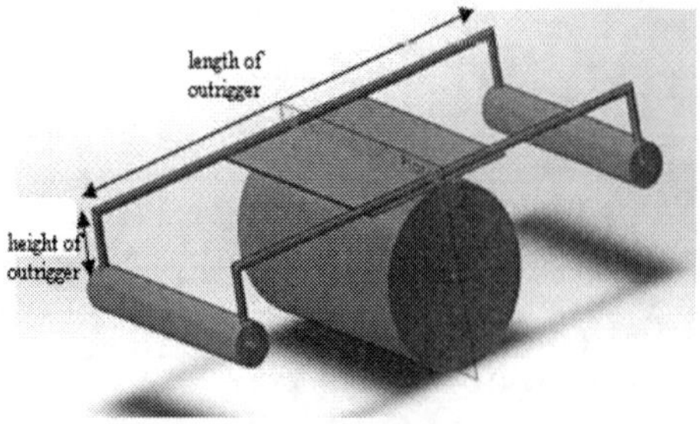

Fig 2.2 Picture of PLTGL-SB 3D

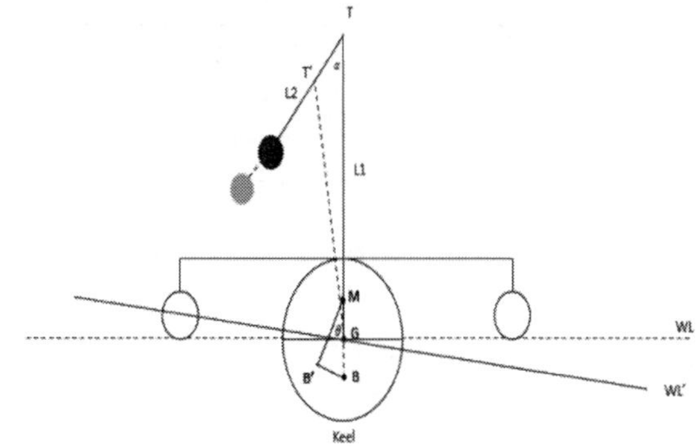

Fig 2.3 PLTGL-SB Front Side

2.3 CFD Simulation of Pontoon

2.3.1 Pontoon Geometry

Large cylinders and two jukung are made of one unit in the workbench. The geometry of the pontoon is made up of several models based on the variations to be performed. The geometry of the pontoon is made up of 4 models. The geometry of the pontoons is made with a ½ D load with a jukung length of 49.5 cm and 41.5 cm. Then made also a pontoon geometry with laden 2/3 D with jukung length 49.5 cm and 41.5 cm. Figure 2.3 is an example of a pontoon geometry image in a workbench.

2.3.2 Pontoon Model

After going through the stage of making geometry the next stage is to model the pontoon. Stage modeling pontoons things that must input, including:

- Water depth, on the geometry
- Mass fluid type (water), on the geometry
- Water size on the x and y-axes, in the geometry section
- COG point (Centre of Gravity), at the point mass
- Grating radius, at the point mass.

The depth of the waters input according to the depth of the tank where the experiments tool is done, i.e. as deep as 2 meters. Water size on the x axis is 50 m and y-axis of 3 m, according to the tank where the experiment is conducted. Determination of COG point and girth radius is done by manual calculation using Microsoft excel for each variation of the pontoon.

2.3.3 Meshing

Meshing is the process of splitting the domain into small parts or become elements. Meshing on Ansys Hydrodynamic Diffraction can be automatically generated by inputting defeaturing tolerance and maximum element size. In the Generated Mesh information section will display the number of nodes and elements meshing results. The maximum number of elements in Ansys Hydrodynamic Diffraction is as many as 18,000 elements and 12,000 elements of diffracted elements. A number of nodes and element is the number of nodes and elements that meshing results. The number of nodes (diffracting body) and a number of an element (diffracting body) are the numbers of nodes and elements on a water-immersed pontoon.

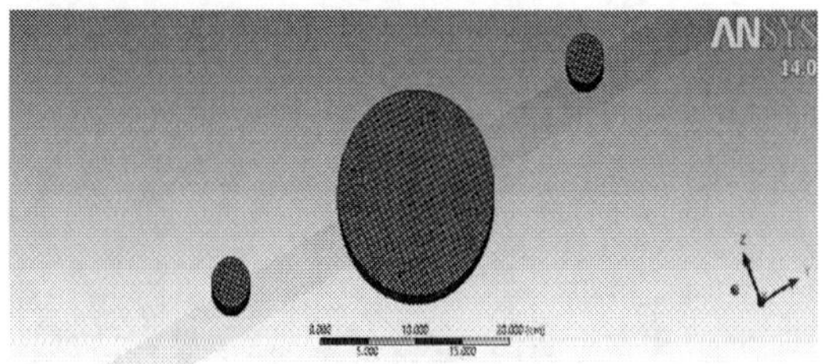

Fig 2. 4 Result of Meshing Pontoon

Table 4. Detail of Mesh.

Detail of Mesh	
Max Element Size	1 cm
Defeaturing Tolerance	0.5 cm
Number of Nodes	8910
Number of Elements	8952
Number of Nodes (Diffracting Bodies)	3127

2.3.4 Pre-Processing

After doing meshing the next stage is the setup stage. At this stage, the input is coming to the wave direction. The direction of the wave coming is adjusted by experiment, i.e. the direction of 900 or the direction of the wave coming from the body of the pontoon. After inputting the wave direction the next stage is to input the wave frequency range. The default wave frequency conditions can be defined in range and single. At this stage, the wave frequency is input manually within a range. The frequency range used is between 1 Hz and 5 Hz.

 https://doi.org/10.1051/e3sconf/20184301022

2.3.4 Post-Processing

The final stage of simulation on hydrodynamic diffraction is solution or result. In this section, the author can determine what results want to appear. To know the response of the pontoon due to ocean waves, the RAO graph (Response Amplitude Operator) is displayed. This study will only analyze the results of RAO graphs contained in the solution section.

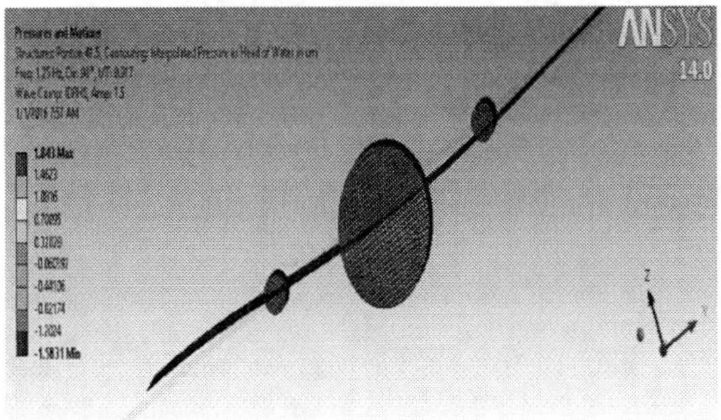

Fig 2. 5 Simulation result with Incoming Wave

2.4 Mathematics Formulation of Pontoon Movement and Pendulum

Mathematical modelling of pontoon and pendulum motion is obtained from the calculation of the length of point B (Buoyancy), Metacentre (M), and point G (Gravity). The third distance of this point is calculated from the point of the pontoon keel. The equation of motion is obtained by simulating the motion of one degree of freedom pontoon (roll) and one degree of freedom pendulum with Lagrange law. At ½-diameter draft large cylinders, the following values are known:

B (Centre *of Buoyancy*): 4.752 m from keel
M (*Metacentre*): 12.17 cm from keel
G (Centre *of Gravity*): 8.25 cm from keel
GT (Length from G point to Pendulum Pole): 42.05 cm

At draft 2/3 large diameter cylinder, known value as follows:

B (Centre *of Buoyancy*): 6.345 m from keel
M (*Metacentre*): 11.78 cm from keel
G (Centre *of Gravity*): 11 cm from keel
GT (Length from G point to Pendulum Pole): 31.05 cm

The method of searching the equation of motion using Lagrange method, that is from the energy side of kinetic and the potential energy of the system, the equation of system motion is derived from the decrease as follows.

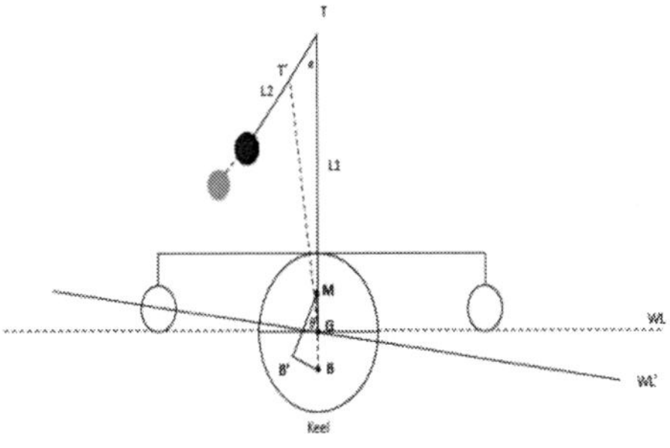

Fig 2. 6 Cylinder Pontoon and Pendulum from Front View

$$BB' = BM \sin\theta \tag{2.3}$$
$$x_1 = -BB' \cos\theta \tag{2.4}$$
$$x_1 = -BM \sin\theta \cos\theta \tag{2.5}$$
$$z_1 = BB' \sin\theta \tag{2.6}$$
$$z_1 = BM \sin\theta \sin\theta \tag{2.7}$$
$$z_2 = -\frac{GT}{BM}(-BM \sin\theta \cos\theta) - l_2 \sin\theta \tag{2.8}$$
$$z = l_2(1 - \cos\alpha) + \left(-\frac{GT}{BM}\right) BM \sin\theta \sin\theta \tag{2.9}$$

Then the values of x1 and y1 are lowered once to get the speed function, after which it is fed into kinetic and potential equations. EK1 is the kinetic energy of the pontoon, EK2 is the kinetic energy of the pendulum, EP1 is the potential energy of the pontoon and EP2 is the potential energy of the pendulum.

$$EK_1 = \frac{1}{2} m_1 (\dot{x}_1{}^2 + \dot{z}_1{}^2) \tag{2.10}$$
$$EK_2 = \frac{1}{2} m_2 (\dot{x}_2{}^2 + \dot{z}_2{}^2) \tag{2.11}$$
$$EP_1 = m_1 g y_1 \tag{2.12}$$
$$EP_2 = m_2 g y_1 \tag{2.13}$$

Then EK and EP become,

$$EK_1 = \frac{1}{2} m_1 (BM^2) \dot{\theta}^2 \tag{2.14}$$
$$EK_2 = \frac{1}{2} m_2 (GT^2 \dot{\theta}^2 + L_2{}^2 \dot{\alpha}^2 - 2\, GT L_2 \dot{\theta} \dot{\alpha}) \tag{2.15}$$
$$EP_1 = m_1 g\, BM \sin\theta \sin\theta \tag{2.16}$$
$$EP_2 = m_2 g\,(-GT\, BM \sin\theta \sin\theta + L_2(1 - \cos\alpha)) \tag{2.17}$$

The Lagrange equation is searched by the following formula Langrange for θ is

$$\frac{\partial}{\partial t}\left(\frac{\partial EK}{\partial \theta}\right) + \frac{\delta EP}{\delta \theta} - \frac{\delta EK}{\delta \theta} = Q \tag{2.18}$$

Q is the excitation force of the ocean waves

$$Q = A \sin \omega t \tag{2.19}$$

Langrange for α is

$$\frac{\partial}{\partial t}\left(\frac{\partial EK}{\partial \alpha}\right) + \frac{\delta EP}{\delta \alpha} - \frac{\delta EK}{\delta \alpha} = 0 \tag{2.20}$$

The equation of motion obtained after using Lagrange and after replace it for pontoon and pendulum movements are as follows

The first equation

$$(m_1 BM^2) s^2 \theta(s) + 2\, BM\, g\, \theta(s) = Q \tag{2.21}$$

https://doi.org/10.1051/e3sconf/20184301022

Value of Q after being Laplace is

$$(A\cdot\sin(w\cdot t))\ \text{laplace}\ \rightarrow \frac{A\cdot w}{s^2 + w^2} \tag{2.22}$$

The second equation of the pendulum after being Laplace as follows,

$$(m_2GS^2 - m_2\,GT\,l_2)s^2\theta(s) + m_2(GS\,l_2 - l_2)s^2\alpha(s) - (2m_2g\,GT)\theta(s) - (m_2gl_2)\alpha(s) = 0 \tag{2.23}$$

Then, first and second equation become matrix, such as follow,

$$\begin{pmatrix}\theta(s)\\ \alpha(s)\end{pmatrix} = \begin{pmatrix} m_1BM^2s^2 & 0 \\ -m_2GT\,l_2s^2 & m_2l_2{}^2s^2 + m_2gl_2\end{pmatrix}^{-1}\begin{pmatrix}\frac{A\omega}{s^2+\omega^2}\\ 0\end{pmatrix} \tag{2.24}$$

$$\begin{pmatrix}\theta(s)\\ \alpha(s)\end{pmatrix} = \begin{pmatrix}\frac{A\omega}{(m_1BM^2s^2+2\,g\,BM)(s^2+\omega^2)}\\ \frac{A\,GT\,l_2\,m^2s^2\omega}{(m_2l_2{}^2s^2+g\,m_2\,l_2)(m_1BM^2s^2+2\,g\,BM)(s^2+\omega^2)}\end{pmatrix} \tag{2.25}$$

Inverse Laplace for both equation, then the equation of pontoon movement and pendulum is

$$\theta(t) = -\frac{2\,A\,\omega\sin t\,\omega\sqrt{BM}\,g - \sqrt{2}A\,\omega\sin\left(\frac{\sqrt{2}t\sqrt{BM\,g}}{BM\sqrt{m_1}}\right)BM\,\omega\sqrt{m_1}}{2\,BM^2m_1\omega^3\sqrt{BMg} - 4\,BM\,g\,\omega\sqrt{BM\,g}}.$$

$$\alpha(t) = \frac{A\,GT\,\omega\left((2\,g\,l_2)\sin\left(t\sqrt{\frac{g}{l_2}}\right)\sqrt{\frac{g}{l_2}} - 2gl_2\sin(t\,\omega)\omega + BM\,g\,m_1\sin(t\omega)\omega - BM\,l_2m_1\omega^3\sin\left(t\sqrt{\frac{g}{l_2}}\right)\sqrt{\frac{g}{l_2}} - \sqrt{2}\,BM\,g\,m_1\sin\left(\sqrt{2}t\sqrt{\frac{g}{BMm_1}}\right)\left(\sqrt{\frac{g}{BMm_1}}\right)\right)}{BM\,g\,(g - l_2\omega^2)(2g - BM\,m_1\omega^2)(2l_2 - BM\,m_1)}$$

3 Result and Discussion

This chapter contains simulation validation, data analysis, and discussion. Analysis of data performed, the first analysis of pontoon motion against variations laden (high jukung), length jukung, and pendulum mass variation. The second data analysis, is the influence of pendulum deviation on the pontoons movement with variation laden, variations in the frequency of incoming waves, the mass and length of the pendulum arm.

3.1 Validation of Simulation Result

In this study, validation is done by calculating the pioneering deviation error (rolling motion) when the pontoons are exposed to waves in the actual state and simulation. Validation is also done based on the number of meshing in the simulation. Roll movement value is validated in this final project with water-laden variation 1/2 D and 2/3 D (large diameter of pontoon) at wave period condition 0.8 seconds.

The average error obtained is 8.098% explains the difference of experimental data value and simulation less than 10% then it is considered the mesh value and the simulation boundary condition is close to the experimental code so that it can be used to simulate the variation of water, pontoon support, and wave period.

Validation of experimental pendulum deviation and simulation using mathematical modelling was performed at a period of incident wave 0.6 s with mass and pendulum arm length is 20 g and 10.67 cm. The maximum deviation of the experimental pendulum was 170 and the width of the pendulum of mathematical model was 14,290 deg.

3.2 Response Amplitude Operator (RAO) of Pontoon

The computational results of the pontoon movement response over the regular waves are arranged in the form of RAO curves as shown in Figure 4.1. Movement modes to be analyzed, roll, heave, and pitch with wave direction 900 and amplitude of 1.5 cm waves. Three Heave, roll, and pitch movement modes are three movements that have a dominant influence over the other three motions for floating free floating seas (Djatmiko, 2012). In relation to wave frequency variation, computation has been performed from incident wave frequency 1 Hz to 5 Hz, when viewed from wave period, as opposed to from 0.2 seconds to 1 second. Next will be analyzed pontoons response in the frequency range with variations of pontoon 1/2 and 2/3 large diameter cylinder, a variation of jukung 1.5 cm and 4 cm, a variation of jukung length 49.5 cm and 41.5 cm in each motion mode.

3.2.1 Respons of Pontoon with Variation of Water Draft

In this study, the authors will analyze the influence of variation on the pontoons response based on the roll, heave, and pitch graph of Response Amplitude Operator (RAO) which is done from frequency incident wave 1 Hz to 5 Hz at regular wave with an amplitude of 1.5 cm wave. The analysis of the laden influence was performed on the 1st and 9th pontoon variations, ie at 49.5 cm jukung length, 40.6-gram pendulum load, and 16.5 cm pendulum arm length. Table variations of pontoons can be seen in more detail in Table I

The pattern of roll and heave motion response shows a similar pattern for laden 1/2 or 2/3 diameter. RAO motion roll and heave on laden 2/3 diameter larger than the roll and heave on laden ½ diameter. This corresponds to the height of the hull from the position of the keel that is exposed to the water, the greater the rolling movement will increase (Yoshiho Ikeda, 2012). In addition, the higher the water the ship the resulting heave movement is also increase (Andrian, 2010). Pitch movement response is similar to roll and heave movement response, but the pitch movement has a small value compared to roll and heaves this happens because the wave comes from the 900 directions so that the pitch movement becomes non-dominant. The pitch motion on the 2/3 D load is very small, but when viewed from a smaller scale we will get the same line trend with roll, heave, and pitch motions.

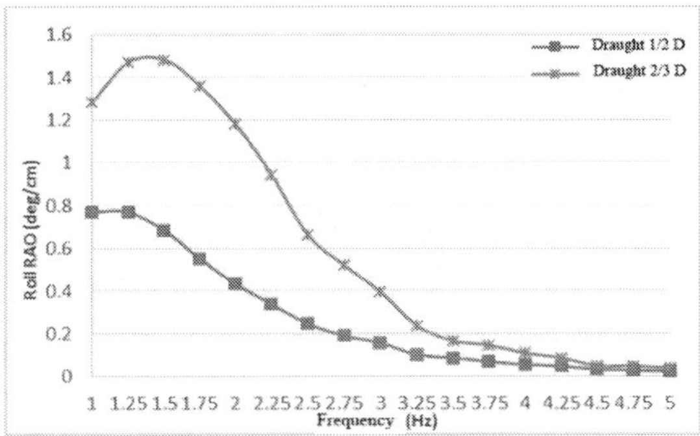

Fig 3.1 Roll RAO with Variation of Water Draft

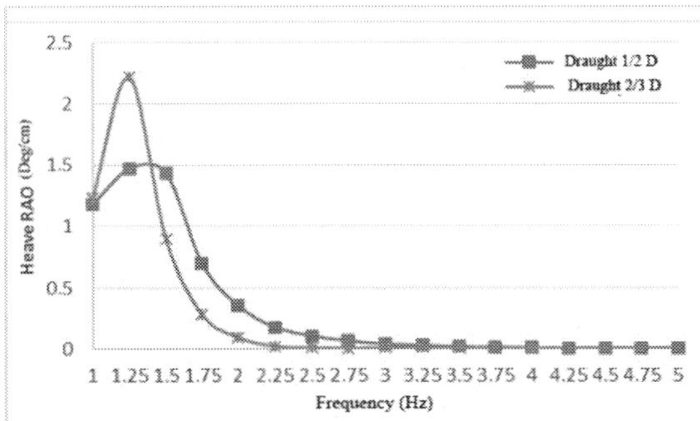

Fig 3.2 Heave RAO with Variation of Water Draft

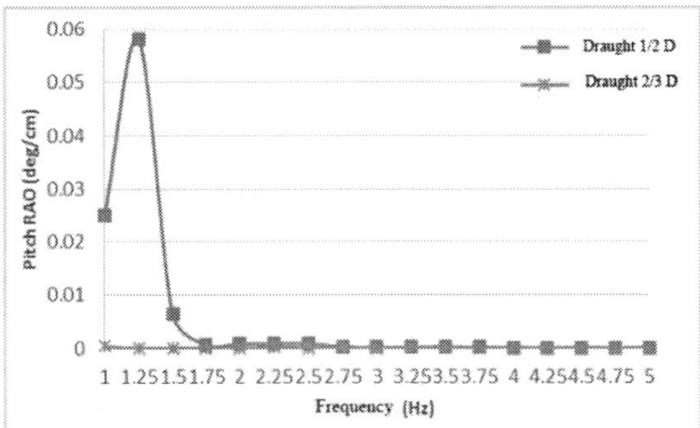

Fig 3.3 Pitch RAO with Variation of Water Draft

The graph above shows the movement response (RAO) roll, heave, and pitch on laden 1/2 or 2/3 after passing the frequency around 1.25 Hz - 1.5 Hz decreased, meaning that if the pontoon is placed on the frequency increase or small period then the movement response will decrease. Thus, with a large or denser frequency between successive wave peaks, it will give the effect of a moving pontoon over a relatively flat water (Djatmiko, 2012). In Figure 3.2 the heave movement has decreased sharply after passing the 1.25 Hz wave frequency as well as the pitch motion of Figure 4.3, while in Figure 4.1 the roll value has a gentle slope, it can be concluded if the pontoon is put on wave frequency above 1.25 Hz roll movement will be more dominant compared to heave motion, while the pitch motion remains smaller than the roll and heave movement.

As mentioned earlier, the vertical pendulum in this study moves in the y-axis direction with the direction of the incident wave 900. This causes the pontoon roll movement mode to become more dominant to move the pendulum. Therefore, referring to the RAO chart of the above roll motion, the pontoon is more appropriate if it is placed on the frequency of the coming wave of 1 Hz - 1.75 Hz.

3.2.2 Response of Pontoon with Variation Length of Jukung

Analysis of the effect of jukung length with a variation of 49.5 cm and 41.5 cm (variation 2 and 6 pontoons) on the motion of roll, heave, and pitch of pontoons performed on laden 1/2 diameter can be seen in Figure 4.4. The effect of variation of jukung on Roll, heave, and pitch motions is very small, seen from the trend of the coinciding line.

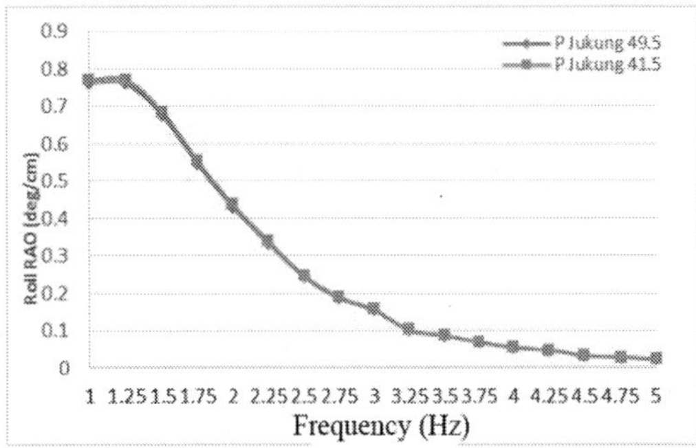

(a)

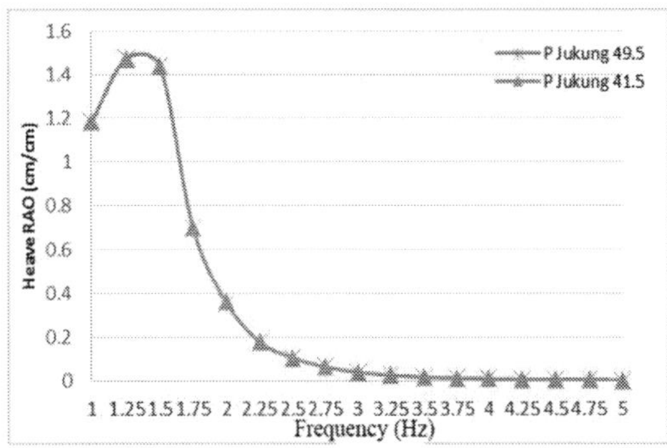

(b)

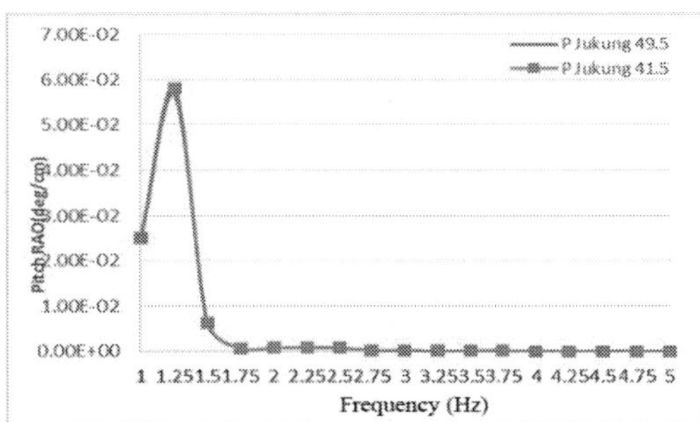

(c)

Fig 3.4 Pontoon Movement with Variation Length of Jukung at ½ D. (a) Roll (b) Heave (c) Pitch

Figure 3.5 is the motion response of roll pontoon, heave, and pitch variation of jukung length at 2/3 diameter (variations to 11 and 15). The influence of variation of jukung length on Roll motion is very small, seen from the trend of the coinciding line. If the pontoon with a length of 49.5 cm jukung is at the frequency of sea wave 1.25 Hz large rolling angle is 1.46 deg/cm and heaving 2.1978 cm/cm, pontoon with long jukung 41.5 cm large rolling angle is 1.47 deg / cm and heaving is 2,292 cm/cm . It can be concluded a shorter jukung, 41.5 cm, resulting in greater rolling and heaving motions although this movement difference is not significantly significant with 49.5 cm jukung length. In the pitching motion for 49.5 cm and 41.5 jukung length it also has a small difference, but the RAO pitch value at a different frequency of 1 Hz, meaning that at a jukung 41.5 at 1 Hz frequency the pontoon moves larger, but the pitching is small compared to the other two motions.

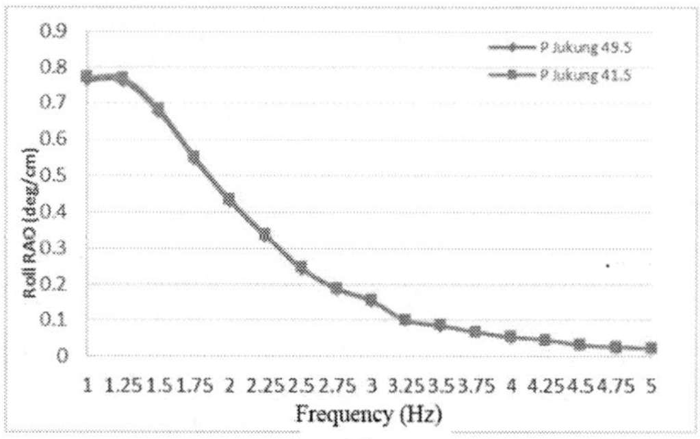

(a)

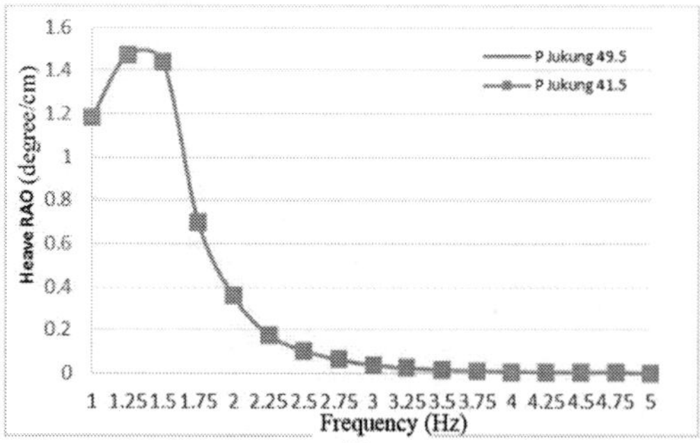

(b)

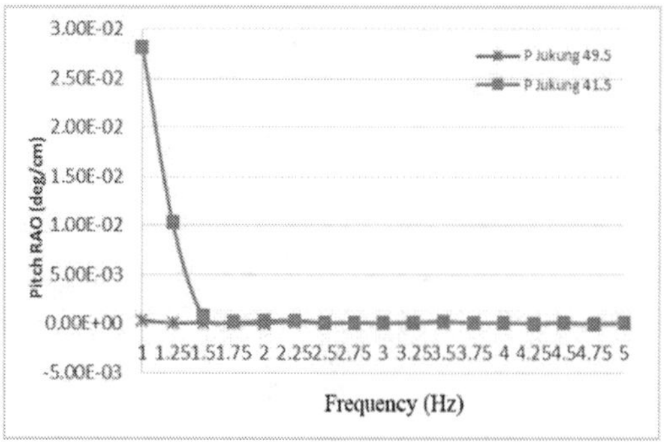

(c)

Fig 3.5 Pontoon Movement with Variation Length of Jukung at 2/3 D. (a) Roll (b) Heave (c) Pitch

It can be concluded that with the longer jukung the value of RAO roll will be smaller because with increasing length of jukung the pontoon will become more stable. This is in accordance with the principle of a trimaran vessel that has a separate hull configuration resulting in good stability with a small rolling acceleration angle with longer jukung length (Sari, 2012) and according to DR Berrett and CB Barrass (1999) by increasing the width of the ship (the length of the ship is considered fixed), then the ship will be more stable.

3.3 Analyzation of Pendulum Movement

The pendulum motion response is derived from the mathematical model of one degree of freedom of the pontoon and one degree of freedom of the pendulum. By finding the equation of motion for pendulum then plotted a graph of pendulum response in the form of pendulum deviation. The pendulum motion analysis will be based on pendulum mass variation, 40.6 gram and 19.9 gram, pendulum arm variation, 16.5 cm and 10.67 cm, placed on the pontoon with variations of water ½ and 2/3 large diameter of the pontoon.

3.3.1 Mass Variation of Pendulum Movement

In this analysis, we will explain the results of pendulum motion with a mass variation of 40.6 grams and 19.9 grams placed on the pontoon with water-laden 2/3 diameter (Figure 3.7) with a wave period of 0.8 s

Referring to the variations of 21 and 23 pendulums (seen in Table 3.2) we see a graph depicting the relationship of time and deviation with two mass variations. The larger the pendulum mass, the smaller the pendulum's pendulum. The pendulum mass of 19.9 grams has a maximum deviation of 14.3750, while the mass of 40.6 grams has a maximum deviation of 13,930.

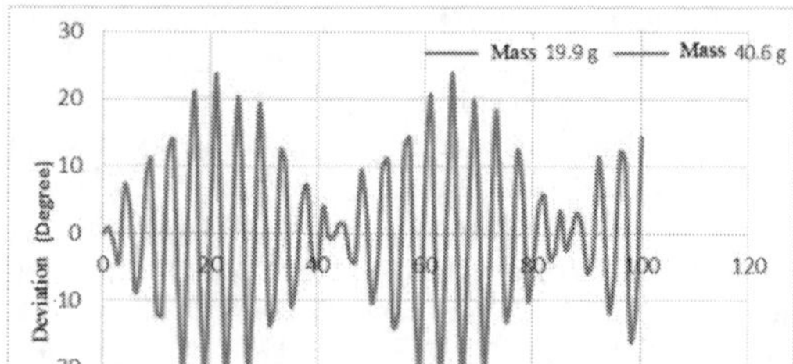

Fig 3.6 Respond of Pendulum with Mass Variation

3.3.2 Arm Length Variation of Pendulum Movement

In this analysis we will describe the pendulum motion with variation of pendulum arm length 16.5 cm and 10.67 cm placed on pontoon with water load 1/2 diameter (Figure 4.9) and water is laden 2/3 large diameter of pontoon cylinder (Figure 4.10) with variation of period waves 0.4 s, 0.6 s, and 0.8 s and 40.6 g of pendulum mass are fixed.

The figure below is a pendulum deviation graph with arm length variations at 0.4 s wave frequencies (1st and 2nd pendulum variations), 0.6s (5th and 6th pendulum variations), and 0.8s (variations of the 9th and 10th pendulums) on laden pontoons ½ large diameter cylinder. Table pendulum variation can be seen in table 3.2.

Based on the response of the mathematical model formula, the pendulum movement on the pontoon with ½ diameter of the large cylinder, when placed on the wave period 0.4 s, 0.6 s, and 0.8 s has a similar movement. The 16.5 cm pendulum arm produces a small drift compared to the sleeve length of 10.67 cm. According to the equation (2.3) where the length of the pendulum is proportional to the pendulum period, the pendulum arm is longer the pendulum period will be longer and the pendulum angle from the fixed position will be smaller. The maximum pendulum wavelength with the length of the arm 10.67 cm in the wave period 0.4 s is 2.70, whereas if the coming wave period is 0.6 s the maximum pendulum pulse is 13,130. The maximum pendulum width with the length of the arm 10.67 cm in the period of 0.8 s is 12,630.

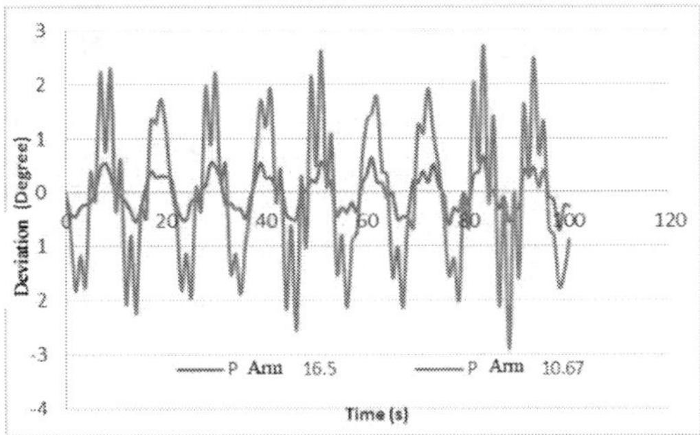

(a)

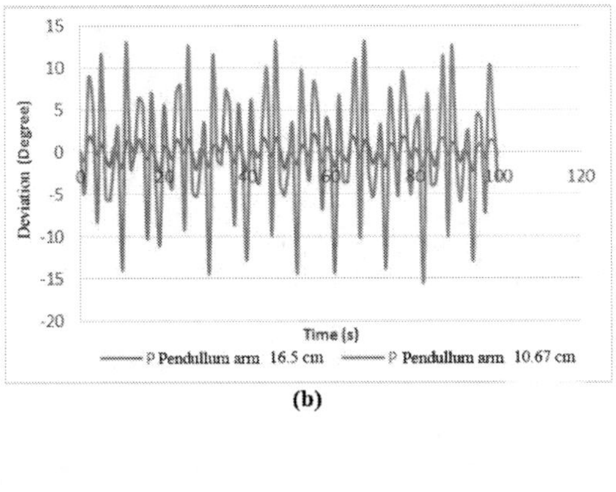

(b)

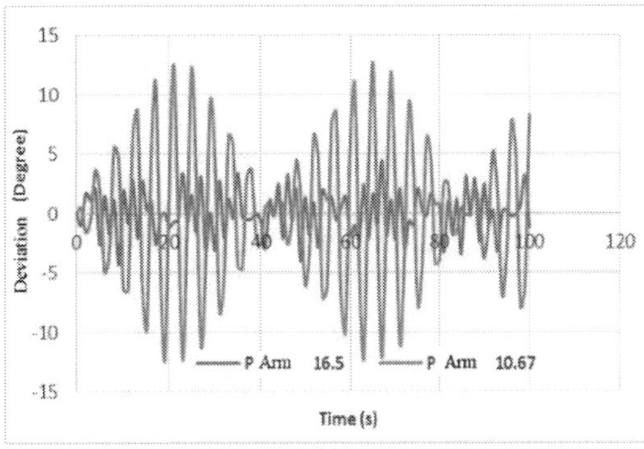

(c)

Fig 3.7 Respond of Pendulum with Length Variation with Wave Period (a) 0.4s (b) 0.6s (c) 0.8s.

The pendulum's intersection for pontoon with water-laden 2/3 large cylinder diameter and wave period 0.4 s is 2,330. If the wave period is 0.6 s, the maximum pendulum wavelength is14.290. The maximum pendulum in the wave period of 0.8 s is 23.90. According to the equation (2.3) where the length of the pendulum is proportional to the pendulum period, the pendulum arm is longer the wave period will be longer and the pendulum angle from the fixed position will be smaller.

Based on the analysis of the influence of variation of load, variation of arm length of the pendulum, and the variation in the frequency of incoming waves, it can be concluded that pendulum laid on laden 2/3 will result in larger deviation, in accordance with analysis of 4.1 on the influence of the pontoon movement against variation Laden. Period of the waves are getting bigger it will cause the pendulum shrinkage is getting bigger too. Also, the shorter length of the pendulum arm will cause larger deviations.

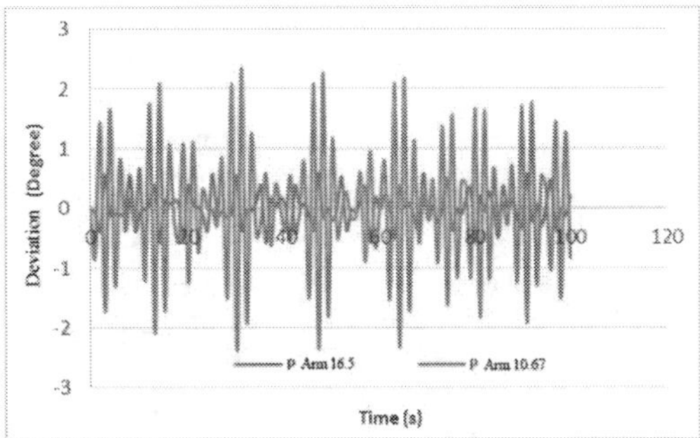

(a)

 https://doi.org/10.1051/e3sconf/20184301022

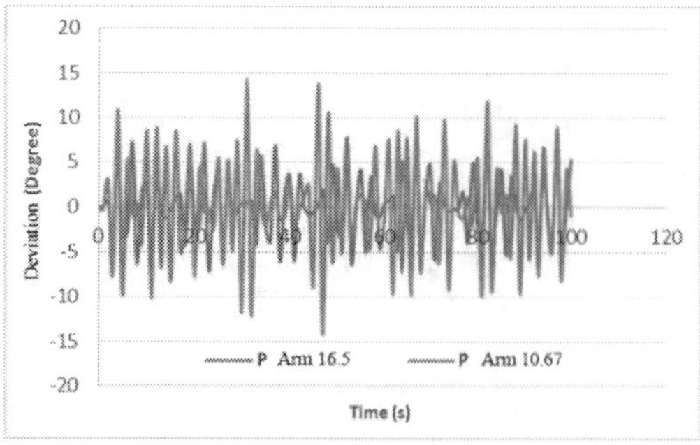

(b)

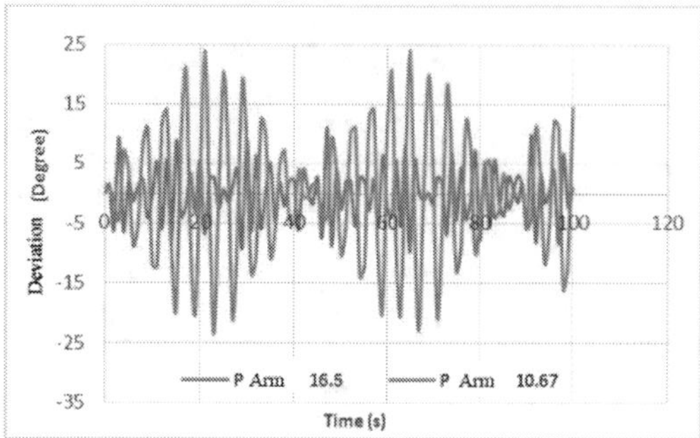

(c)

Fig 3.8 Respond of Pendulum with Length Variation in 2/3 D with Wave Period (a) 0.4s (b) 0.6s (c) 0.8s.

4 Discussion

Marine Wave Power Generation - The Pendulum System (PLTGL-SB) is one of the renewable energy conversion technologies of ocean wave motion that will become the excitement style of the pendulum, wherein the pendulum will produce a drift which will be converted into electrical energy. This final project study uses a pontoon consisting of a large cylinder and two small cylinders as an outrigger, like a trimaran ship. To answer the purpose of this final project, simulated the pontoon movement on Ansys AQWA to determine the best pontoon motion for PLTGL-SB, then to know pendulum deviation done mathematical modeling one DOF pontoons and one DOF pendulum. The variations performed on this final project are variations of pontoons, ie 1/2 and 2/3 diameter of pontoon cylinder, outrigger length 49.5 cm and 41.5 cm, pendulum mass 40.6 gram and 19.9 gram, and pendulum length 49.5 cm and 41.5 cm with amplitude of wave used 1.5 cm, wave direction 900, and frequency range of wave 1 Hz- 5 Hz.

The simulated variation of jukung length, with longer length (49.9 cm) of rolling motion obtained is smaller in accordance with the principle principle of trimaran ship (due to pontoon pontoons such as trimaran vessels) where having a separate hull configuration produces good stability with angle small rolling acceleration with longer jukung length (Sari, 2012).

The simulation results with variations of ½ and 2/3 large diameter of pontoon are obtained to get the best pontoon motion based on RAO graph, then the best move on laden 2/3 diameter. This corresponds to the height of the hull from the position of the keel that is exposed to the water, the greater the rolling movement will increase (Yoshiho Ikeda, 2012). In addition, the higher the water the ship the resulting heave movement is also increased (Andrian, 2010).

To obtain pendulum divergence, mathematical modeling was performed and a pendulum motion equation was obtained with one DOF pontoon (roll) and one DOF pendulum. The result of the plot of pendulum response of mass variation hence got bigger mass hence bigger resulted deviation small. The plot of the pendulum response variation in arms length obtained by the longer arm will result in smaller pendulum drift, according to the theory of equation 2.3.

Therefore, to get the best pontoon motion can be concluded pontoon with jukung length 41.5 cm, laden water 2/3 diameter, and placed in the period of 0.8 s wave will produce good pontoon movement so suitable for PLTGL-SB with mass 1919 gram pendulum, and a short pendulum arm, which is 10.67 cm to produce a larger deviation.

5 Conclusion

In accordance with the initial purpose of this research, which is to know the best pontoon form for PLTGL-SB and to know the largest deviation produced pendulum due to excitation style pontoon in laboratory scale, then from the analysis that has been done in the previous chapter, can be drawn conclusion as follows:

1. The best pontoon shape for PLTGL-SB based on the result, that is on the pontoon with high laden 2/3 diameter big cylinder pontoon or with height jukung 1.5 cm and jukung length 41.5 cm. Based on the response graph (RAO), the laboratory-scale pontoon in this study will result in maximum movement to move the pendulum if placed at a frequency of 1 Hz-1.75 Hz.
2. The largest pendulum deviation is generated when the pontoon is placed in a 0.8 s wave period or a frequency of 1.25 Hz, with a mass and a pendulum arm of 19.9 g and 10.67 cm with the largest deviation value being 23.50.

The authors gratefully acknowledge funding support from Ministry of Research and Higher Education in Indonesia for a research grant.

References

1 F. C. Hongda Shi, *Theoretical Study on the Power Take-Off Estimation of Heaving Buoy*, Renewable Energy, 441- 448 (2014)

2 L. Rodrigues, *Wave Power Conversion Systems for Electrical Energy Production*, Renewable Energy, 1-7 (2012)

3 Besari Y. Mustofa, *Analysis of Pontoon Model Motion on Tripod for Wave Energy Converter-Pendulum System*, Jurnal Teknik ITS.1:G-204 (2012)

4 Sony Junianto, *Analysis Of Pendulum Movements Due To The Pontoon Movements In Wave Energy Converter Pendulum Systems*, Jurnal Teknik ITS.2(1):1 (2013)

Study of Supervisory Control Implementation in A Small Scale Variable Speed Wind Turbine

Katherin Indriawati[1,*], *Ali* Musyafa[1], *Bambang* L. Widjiantoro[1], and *Anna* Milatul Ummah[1]

[1,2,3,4]Institute of Technology Sepuluh Nopember

Abstract. In relation to improve wind energy production, efforts to increase the extraction of wind energy should be done when there is a decrease in wind power. The decline occurs when there is a change in wind speed. At low wind speed operating range, such as in Indonesia, the controller optimizes power extraction through wind turbine rotor regulation following optimal rotor speed. This study proposed the use of the PI control system as an intelligent control system to solve nonlinearity problem and the setpoint adjustment mechanism to get at the problem of the uncontrolled stochastic driving force input. The PI control is called as a regulatory control while set-point adjustment is known as one mechanism in supervisory level. Thus, that control system is called as the supervisory control. This control had a task to maximize output power of a wind turbine. The technique was applied to a small scale horizontal axis wind turbine operating in wind speed range of 3-11 m/s. The applied optimization algorithm generated an optimum set-point simultaneously when there was a change of wind speed.

1 Introduction

Economic activities has made energy into a key infrastructure requirement in the modern industrial economy. A large energy deficit is expected to occur in the coming 2030s as stated by the energy outlook, the Agency for Assessment and Application of Technology. On the other while, global environmental pollution and climate change issues have encouraged various parties, both government and private, to participate in efforts to reduce the causes of pollution and carbon dioxide emissions. The situation has encouraged Indonesia to face a new era, which called the era of increasing the use of renewable energy sources.

One form of renewable energy source that is environmentally friendly is wind energy. Indonesia has made a fundamental policy specifically for wind energy by establishing wind energy as a source of clean energy and future energy sources. In 2009-2010, the world's installed renewable energy capacity has grown from 250 Giga Watt (GW) to 312 GW, with the highest EBT growth occupied by the Bayu Energy Power Plant (BEPT) of 159 GW in 2009 to 198 GW in 2010. To increase the role and contribution of BEPT to have a big role, so the effort to utilize efficient technology becomes important. Improved technological capabilities to extract wind energy into electricity need to be developed more efficiently, more reliability, more environmentally, and more affordably.

The development of wind energy in Indonesia has become a national energy program handled and developed by LAPAN. Procurement of self-made wind turbines and purchased from European countries, China, Taiwan, and others. The wind turbine has a traditional type of fixed-speed turbine or stall regulation blade fix. This type of wind turbine is more suitable for zones that have stable wind speed and wide range. While tropical wind speeds like Indonesia have a narrow range and fluctuate in the area of 4-6 m/s at night and 3 m/s in the morning until 10.00. Therefore, greater effort is needed to design wind turbines in low wind speed areas (cut-in small) where the chances of a decrease in wind turbine power production are very large. Low wind speed work area will influence cost optimization criteria in wind turbine besides will affect the cost of turbine component [1].

Currently the development of fixed-speed generator technology has shifted to variable-speed generator. The application of controls on wind turbines has been carried out with PI conventional control strategies. In this control strategy requires a detailed knowledge of system dynamics and requires the decline of complex mathematical models [2]. The perfection of conventional control is strongly influenced by system nonlinearity. In addition, in conventional control systems, control areas are divided into several regions that make the system complex and slow [3]. Therefore, nowadays intelligent control system has been developed which is able to produce optimum power as in [4] and [5].

The development of a small-scale wind turbine prototype with variable pitch angle control based on Fuzzy Logic Control (FLC) for low average wind speed has been done by one of the research team even though it has just reached the wind power production stage [6]. The integration of wind turbine and electric generator needs to be developed so that the exploitation of wind energy can be expanded. In this case, the wind energy

* Corresponding author: katherin@ep.its.ac.id

conversion system built is expected to extract the maximum possible energy in the condition of Indonesian wind farms that have a low average annual speed.

2 Problem Formulation

The characteristics of wind speed and the ability of wind turbines holds the key to successful installation of wind turbine systems. The interaction is a very complex system, but the event can be simplified through aerodynamic torque modelling or aerodynamic power placement. In relation to efforts in increasing wind energy production, efforts to increase the extraction of wind energy should be done when there is a decrease in wind power. The decline occurs when there is a change in wind speed. A suitable control strategy is to keep the value of Tip Speed Ratio (TSR) to stay within the working range of the wind turbine. The optimum TSR value will cause the wind turbine to extract maximum wind power at available wind conditions. Thus, the faced problem is choosing the optimum pitch angle position (to obtain the optimum TSR value) capable of generating maximum power, taking into account the low cut-in value so that the turbine is still capable of generating electrical energy when the wind speed is low. Furthermore, the optimum pitch angle becomes a reference value for the control system that adjusts the pitch angle position.

Power is the energy derivative of time. The wind power that can be captured and extracted by the horizontal axis wind turbine system is further defined as follows [7]:

$$P = \tfrac{1}{2}\, \rho_{air}\, C_p\, A_r\, V_w^3 \quad (1)$$

$$P = \tfrac{1}{2}\, \rho_{air}\, C_p(\lambda,\beta)\, A_r\, V_w^3 \quad (2)$$

While λ is formulated as follows:

$$P = T_g\, \omega \quad (3)$$

The relationship between power output of wind turbines and rotor speed of shaft is defined as follows:

$$\lambda = \frac{\omega \cdot R}{V_w} \quad (4)$$

If ω is constant according to the desired setpoint and R blade is constant, then C, will only depend on V. and ß. Since V. can not be controlled and is random, angle ß is used as a manipulated variable to extracted the maximum power output when there is a change in wind speed.

The relationship between the output power generated by the wind turbine, the generator rotor speed and the wind speed is indicated by the curve in Fig. 1.

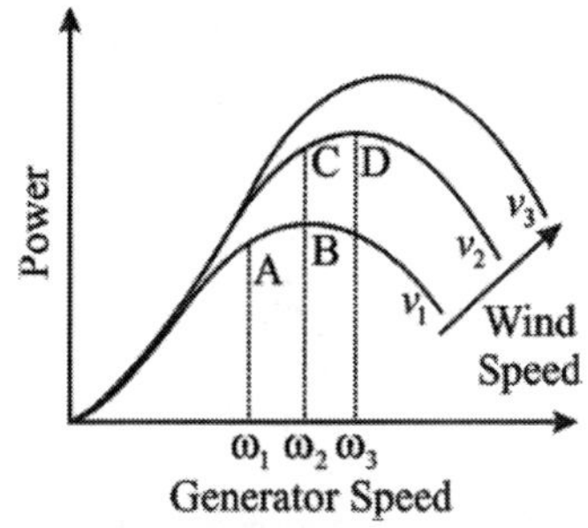

Fig. 1. The relationship between wind speed, generator speed, and the output power of wind turbines [8].

Fig. 1. Shows the presence of a rotor speed (ω) in each work area for a certain wind speed in order to maximize output power of wind turbine. If the wind speed is v_1 and the wind turbines is operated in point A (for the generator speed is ω_1), then for maximizing the output power, the generator speed should be changed into ω_2 in order to the output power is in point B. It also applies to the wind speed v_2 and v_3. When the wind speed is v_2, the wind turbines will be operating in point C, but actually the maximum output power that can be extracted in that wind speed is point D, then the generator speed should be changed from ω_2 into ω_3.

In power optimization, generators also have an important role beside the numerical value of pitch angle in the process of converting mechanical power into electrical power. Maximization of this process is influenced by the output power of turbine and the numerical value of torque generator. In this study the numerical value of torque generator and the pitch angle value are generated by PID control to maximize the electrical power output of wind turbine system.

Based on the above description, then the problem raised in this research is how to develop an intelligent control system based on supervisory control in order to obtain maximum power with minimum maintenance cost through the pitch angle variable as a manipulated variable.

Supervisory control determines the conditions for ordering lower level automation functions or referred to as regulatory control. The hierarchical structure of the power generation system looks in line with the optimization of the supervisory through the determination of the setpoint. Here, the supervisory controller continuously monitors system operating conditions and keeps the system within the specified target domain. Control operation ensures that the goal of system operation is achieved and maintained despite the constraints of uncertainty and resources. It can also act to prevent the system from being in a critical condition or to reduce the consequences of system component failure. Some of the parameters used in this study are as follows:

Table 1. Definition of parameters and their numerical value.

Parameters	Definition	Value	Unit
β_0	Blade pitch angle	0	deg
β_{max}	Maximum blade pitch angle	27	deg
β/s	Pitch rate	10	deg/s
Cp_{max}	Maximum coefficient of power	0.5	-
λ_{max}	Maximum tip speed ratio	9.95	-
V_w	Wind speed	5-11	m/s
K_{sh}	Stiffness coefficient	80.27	Pu/rad
T_s	The early torque of wind turbine	0.83	pu
F	Friction factor of generator	0.01	pu
ω_{me}	Rotor electrical frequency	1.2	pu
ω_{max}	Maximum rotor speed	1.2	Pu
ω_{min}	Minimum rotor speed	0.7	pu
P_{mec}	Mechanical power output	400	W

3 Supervisory Control

Supervisory Control (SC) is a wind speed control system. SC has the supervisory level for doing optimization at the generator speed as a setpoint in the simulation of this study in order to be found a fixed blade pitch angle. This angle will be controlled by PI control so that it can produce optimum power output of wind turbine without exceeding the speed limit of the rotor.

The PI control is used to overcome the characteristics of nonlinearity generated by the wind turbine systems. This control uses the proportional parameter and integral parameter (PI control) in the process.

The concept of supervisory control system in this study can be explained through fig. 2.

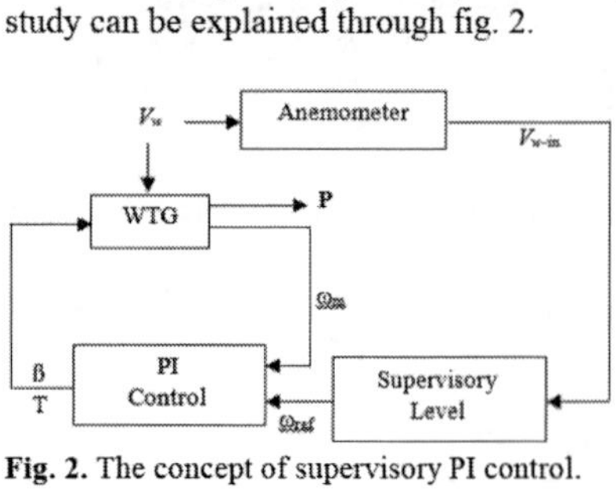

Fig. 2. The concept of supervisory PI control.

Fig. 2. Describes the use of wind speed data to be used as the reference rotor speed at setpoint. The wind speed will be measured by the anemometer and will be used as input at the supervisory level to get the certain generator speed. At this supervisory level there is an optimization of rotor speed to produce maximum power. Then the reference generator speed will be used as input on the PI control system to do the linearization and produce a certain pitch angle and generator torque. This blade pitch angle will change the geometry of the aerodynamic system and become the input of the wind turbine system so that the resulting power becomes optimal. And the torque will change numerical value of generator rotational speed.

3.1. Supervisory Control

The proposed setpoint adjustment is based on extremum seeking (ES) algorithm. This is an dynamic optimization algorithm that does not need mathematical model of electrical power as objective function. Here the power is wanted to be maximized by regulating turbine speed. Furthermore, using this method, any wind speed measurement is not necessary but only the electrical power measurement is required.

ES algorithm finds an optimizing turbine speed as a setpoint value ω_{ref} for the unknown time varying cost function $J(\omega_{ref},t)$ which is defined as:

$$J = -\frac{1}{P_0}(\hat{P} - P_0)^2 + P_0 \tag{8}$$

where P_0 is rated electrical power and $\hat{P}$ is estimated electrical power.

The ES algorithm consists of a highpass filter, a sinus signal, two lowpass filters, an integrator, and a dynamic compensator which are arranged as shown in Fig. 5. A highpass filter, a sinus signal, and one of two lowpass filters are used to extract a signal that is proportional to the gradient of the cost function J with respect to the optimizing variable ω_{ref}. Then the integrator and the compensator drive the gradient to zero and thus achieve the optimal value. The remain lowpass filter has a function to improve steady state tracking of the optimal parameter ω_{ref}.

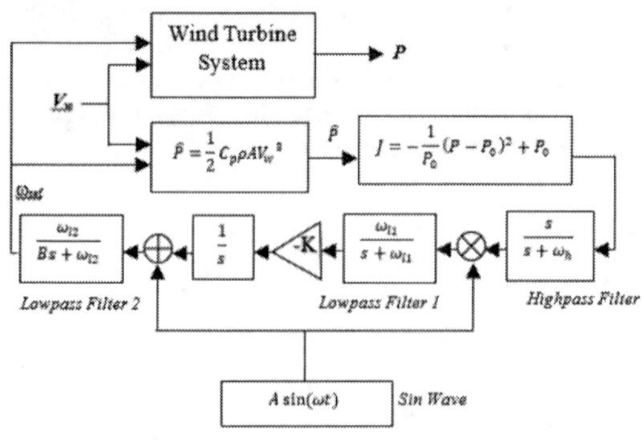

Fig. 6. The structure of setpoint adjustment in supervisory control.

From fig. 6. we know that for designing the supervisory control, there are parameters that must be required. They are the determination of ω_h value on the highpass filter, ω_{l1} on lowpass filter 1, ω_{l2} and value of B on lowpass filter 2, amplitude (A) and angular frequency (ω) in the function of signal generating (sinusoidal signal), and numerical value of negative gain.

The value of parameters are ω_h=1, ω_{l1}=0.01, ω_{l2}=1, B=1, A=0.1, ω=1, and K=3.7.

4 Simulation Results

This paper present design of supervisory control to maximize output power of wind turbine system by modelling system using MATLAB-SIMULINK software. The used rated wind speed in the system is 11 m/s. The results of simulation are as follow fig. 7., fig. 8,., and fig. 9.

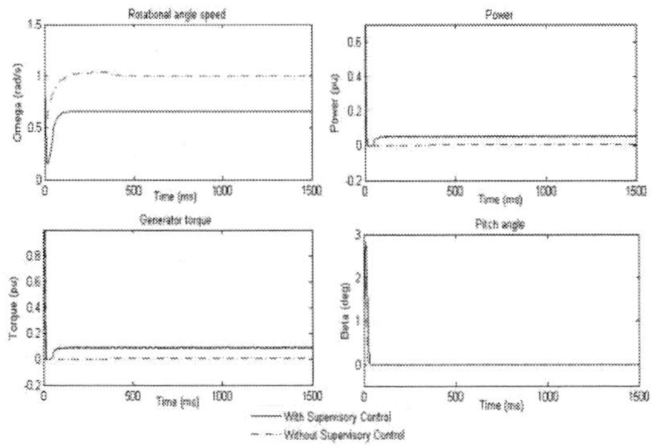

Fig. 7. The results of simulation at below rated wind speed.

The figure shows result of rotational angle speed, output power, generator torque and pitch angle at the time with supervisory control and without supervisory control. The fixed line shows the presence of supervisory control in the wind turbine system. The dots line shows result of wind turbine system without controlling method.

At the below rated wind speed, wind turbine speed will has small rotational angle speed. It makes wind turbine extract small power. Then in the certain wind speed, wind turbine system will not be suitable for generating the power. By applicating supervisory control method, it will make wind turbine system generate output power optimally at above and below rated wind speed.

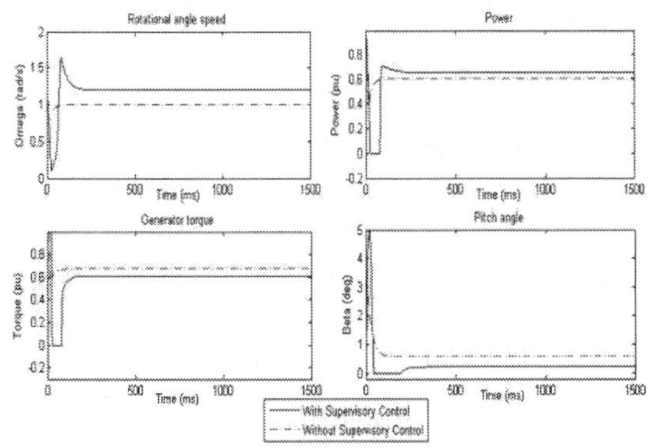

Fig. 8. The results of simulation at rated wind speed.

Fig. 8. shows the presence of increasing output power in wind turbine system by supervisory control. At the same wind speed, wind turbine system without supervisory method just generate smaller power. This process is followed by presence of decreasing torque value and pitch angle.

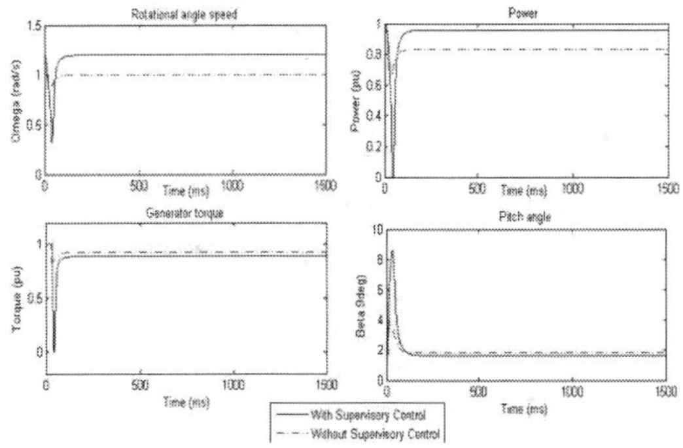

Fig. 9. The results of simulation at above rated wind speed.

Fig. 9. shows the performance of supervisory control in generating output power of wind turbine system optimally at above rated wind speed. The increasing of wind speed will affect in generating power. The rotational angle value will be increasing to generate power maximally.

If we view from the generator torque graph. This value is influenced by PID control. Parameter of Kp and Ki must be required. This numerical value is generated by tunning. At the same time, the value of pitch angle is controlled by PI controller. The PI control can minimize error in the system.

4 Conclusion

The proposed supervisory control system was proven to be successful in increasing the electrical power output of the small scale wind turbine. The supervisory level choose the optimum pitch angle position (to obtain the optimum tip speed ratio) that was capable of generating maximum power by using extremum seeking method. And the PI controller was able to handle nonlinearity behaviour of wind turbine.

References

1. H. Markou and T. J. Larsen, *Control Stategies for operation of pitch regulated turbines above cut-out wind speeds*, PSO-project (2009)
2. J. Zhang, M. Cheng, Z. Chen and X. Fu, *Pitch Angle Control for Variable Speed Wind Turbine*, DRPT2008 (2008)
3. B. Neamanee, S. Sirisumranukul and S. Chatratana, *Control performance analysis of feedforward and maximum peak power tracking for small-and medium-sized fixed pitch wind turbines*, ICARCV, IEEE (2006)

4. A. Z. Mohamed, M. N. Eskander and F. A. Ghali, *Fuzzy logic control maximum power tracking of a wind energy system*, Renewable Energy, **vol. 23**, pp. 235-245 (2001)
5. Q. Wang, *An Intelegent maximum power extraction algoritm for inverter-based variable speed wind turbine system*, IEEE (2004)
6. A. Musyafa, A. Harika, I. Negara and I. Robandi, *Pitch Angle Control of Variable Low Rated Speed Wind Turbine Using Fuzzy Logic Control*, International Journal Of Engineering & Technology, **vol. 10**, no. 05 (2010)
7. A. Pintea, D. Pepescu and P. Borne, *Modeling and control of wind turbines*, 12th Sysmposium Large Scale systems Theory and App. (2010)
8. Mullane A., Lightbody G., Yacamini R.., *Adaptive Control of Variable Speed Wind Turbine*, University College Cork (2001)
9. F. M. McNeil and E. Thro, *Fuzzy logic a practical approach*, Academic Press Ltd. (1994)

Synthesis of Fatty Acid Methyl Ester from Soybean Oil Using Electrolysis Enhanced by Treated Kaolinite as Catalyst

Rudy Syah Putra[1,2*], *Wahyu* Sriyono [1], *Mutiara* Ayu[1], and *Reina* Intan Kusumawati[1]

[1]Department of Chemistry and New and Renewable Energy Research Group, Universitas Islam Indonesia, Jl. Kaliurang km. 14, Yogyakarta 55584, Indonesia

Abstract. The performance of treated kaolinite obtained from three types treatment that was calcination at 600 °C (K600), 800 °C (K800) and acid treated of K800 (KA800). All catalysts were tested for their catalytic activity via electrolysis process through (trans)esterification of soy bean oil at room temperature. X-ray diffraction (XRD) and Fourier transform infrared (FTIR) was performed to characterize the microstructures and chemical properties of as-prepared treated kaolinite material. The effect of catalyst amount (2, 3 and 5 wt.%) was evaluated for the conversion of FAME from soybean oil and followed by GC-MS, determining of the purity of FAME at different operation variables. The result showed that for KA800 as catalyst, the FAME conversion was the highest as well as 18,85% at 3 wt.% catalyst of catalyst loading in the mixture, respectively.

1 Introduction

The depletion of fossil fuel production be accompanied with climate change in the world has encourage world researchers to find renewable energy sources namely biodiesel as an alternative energy source developed to substitute the fossil fuels. Based on data between 2005 and 2015 presented by Naylor and Higgins (2017) showed that the global biodiesel production continues to increase by ~ 23% per year despite a sharp decline in crude oil prices. This coincides with a sharp rise in the price of crude oil and diesel fuel that makes some countries support the use of biodiesel [1].

Biodiesel synthesis processes typically use renewable sources such as vegetable oils, animal fats and recycled cooking oil [2] are carried out by (trans)esterification of triacylglyceride with methanol and assisted with homogeneous alkaline catalysts, such as NaOH and KOH which produce Methyl Ester Fatty Acid (FAME) as a product main and glycerol as by-products [3]. Homogeneous alkaline catalysts are extremely sensitive to Free Fatty Acid (FFA) contained in low quality cooking oils that lead to saponification, decreased catalyst quality, decreased synthesis efficiency and increased production costs. Guan and Kusakabe [4] reported that the synthesis of biodiesel from used-cooking oil using electrolysis process have a yield percentage of > 97%. This process runs quickly at relatively low temperatures, but this method is less efficient because once the process is complete, it has to separate between the biodiesel and the homogeneous catalyst that used. In addition, the homogeneous catalyst could not be regenerated after (trans)esterification process. Furthermore, a large amount of Indonesian society on the consumption of palm oil as a vegetable oils thus increasing the availability of used-cooking oil as feed-stock in the producing of biodiesel. The conversion of used-cooking oil into biodiesel is a smart move, which gives three advantages at once namely (i) its economic value, (ii) becomes a solution for handling used-cooking oil and (iii) producing an environmentally friendly biodiesel products [3].

Jorge et al. (2011) [5] has reported the production of biodiesel from used-cooking oil by transesterification reaction using methanol and kaolinite as a solid acid catalyst. This study has become a positive alternative to utilizing used-cooking oil as feed-stock combined with a cheap kaolinite catalyst. Kaolinite is a raw material that has a selling value, cheap, available, non-corrosive and environmentally friendly materials. In this study, the enhancement of FAME synthesis from soybean oil using electrolysis process by treated kaolinite thermal and acid as catalyst in room temperature was demonstrated.

2 Methodology

2.1 Chemical and equipment

Soybean oil was obtained from supermarket. High grade of methanol, HCl, NaCl and co-solvent THF were purchased from Merck, Germany and the kaolinite was purchased from Wako Chemical, Japan. Fig. 1 shows an electro-catalytic cell (70 ml) contains two graphite electrodes (Ø 0.8 mm x 130 mm) that were separated by a distance of 15 mm [3-4]. The qualities of soybean oil were determined by physicochemical analysis and the results were shown in the Tab. 1.

* Corresponding author: rudy.syahputra@uii..ac.id

Table 1. Physicochemical properties of soybean oil

Oil	Acid Number	Refractive index	ρ, Kg/L	Moisture, %
Soybean	0.138	1.469	0.920	0.217

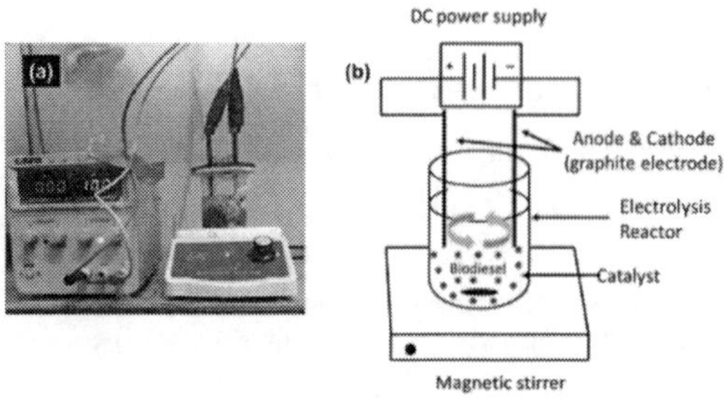

Fig. 1 Electrolysis reactor set up (a) and illustrated functional apparatus in the reactor (b).

2.2 Preparation and characterization of catalyst

Treatted kaolinite was prepared by calcination of a natural kaolinite at 600 °C and 800 °C. This kaolinite, more reactive than the parent kaolinite, was submitted to acid activations. Acid activation was carried out with 6 M HCl at room temperature with times of treatment of 24 h [6]. X-ray diffraction (XRD), Fourier transform infrared (FTIR) was performed to characterize the microstructures and chemical properties of as-prepared acid-calcination of kaolinite material. The performance of FAME synthesis using electrolysis was compared between acid activation and treated kaolinite thermal. The FAME phase was separated from glycerol and then washed with deionized water to remove the residual inorganic components. Concentration of produced FAME and unreacted oils remaining in the product were measured using GC-MS.

2.3 Catalytic testing

Soybean oil was used as a feedstock to obtain optimal condition before implemented to used-cooking oil for FAME synthesis. Effect of catalyst amount (2, 3 and 5 wt.%) were evaluated for the conversion of FAME from soybean oil. The reaction mixture was agitated using magnetic stirrer. The electrolytic cell was filled with 70 ml of reaction mixture containing soybean oil, methanol, THF as a co-solvent. The methanol-to-oil molar ratio was 6:1 and co-solvent-to-methanol molar ratio was 1:1 were added on the basis of oil weight of the entire reaction mixture, respectively. The electro-catalytic was carried out at room temperature (25 °C) using a constant voltage of 18.2 V for 60 mins [3-4]. Fig. 2 shows a step of optimization process for biodiesel production in this study. This method is free from hazardous and expensive reagents, contamination and the product can be separated by simple filtration.

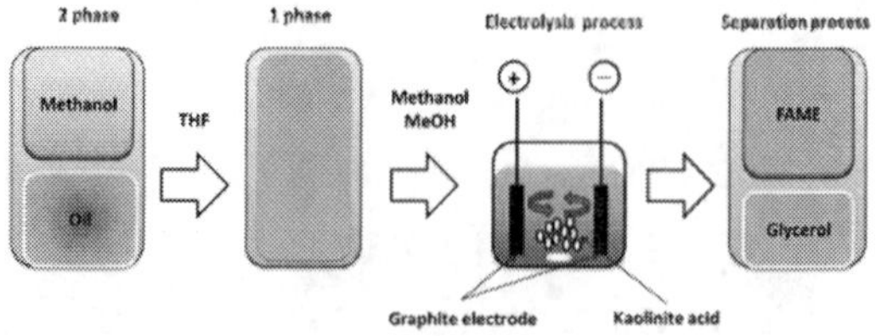

Fig. 2 Steps on FAME synthesis using electrolysis process and kaolinite acid as catalyst.

3 Results and discussion

3.1 Catalyst characterization

Generally, the structural of clay material changed after the thermal and acid treatment. Figure 3 shows the XRD profiles of the untreated and treated kaolin samples. The parent clay shows well-defined reflections at 2θ value of 12°, 25° (corresponding to the d values of 7.154 Å). These peaks correspond to the reflections from [001], which are typical characteristic peaks of kaolinite [7]. After acid treatment, the peak intensity of the clay was found to decrease, since leaching was quite severe with strong acids. This was attributed to the structural disorder that occurred owing to the acid leaching, which affects the crystalline character of the clay. The increase in intensity and/or narrowing of the peak may be related to the increase of crystallite size and/or the decrease of the mean lattice strain [8].

The FTIR spectrum of the clay peaks at 3673, 1119, 1044, 946, and 798 cm^{-1} [6] as shown in Fig. 4. There was not much variation in the peak pattern for acid treated kaolinite. However for thermal treated kaolinite, the peak intensity was found to decrease progressively indicating the loss of structural water with a reorganization of the structure; only a small part of AlO_6 octahedra is maintained, while the rest are transformed into much more reactive tetra- and pentacoordinated units [6].

3.2 Effect of treated kaolinite on the FAME synthesis

Table 2 shows the FAME conversion from soybean oil using treated kaolinite as catalyst in the electrolysis process. Generally, kaolinite which had been calcinated at 600 and 800 °C give better result in electrolysis process responded high FAME yield compared with acid kaolinite, and it was affected by 3% of catalyst loading. In K800 catalyst, all of the OH band dissapears due to dehydroxylation via by calcination, thereby producing thermally activated kaolinite. These transformation occcurs with the loss of structural water (see Fig. 4) in the kaolinite framework. The rest of the kaolinite was transformed into much more reactive tetra and pentacoordinated units [6]. Thus, a mild and efficient method has been developed for the (trans)esterification of fatty acid to their corresponding fatty acid methyl ester by using activated kaolinite clay in good yield. This method is free from hazardous and expensive reagents, contamination and the product can be separated by simple filtration.

It was showed that excellent conversion rates (as high as 18.85%) were obtained in 3 wt.% of acid kaolinite (KA800) as a catalyst loading since total FAME yield was the highest (see Fig. 3). Total FAME on the K800 catalyst was as much as 6.47% in 3% catalyst loading, while on the K600 catalyst was as much as 2.2% in 5% catalyst loading. This fact shows that acid kaolinite was a high reactive catalyst in the (trans)esterification reaction. While, thermal kaolinite (i.e K600 and K800) showed low active catalyst, resulting a low FAME conversion (1.63% for K800 and 0% for K600).

Figs. 5 show the FAME profiles from biodiesel using kaolinite thermal and kaolinite acid as catalyst in the electrolysis process. There is no sharp different of FAME yield when using thermal (K800) and acid kaolinite (KA800) at 800 °C as catalyst. However, a sharp different FAME conversion showed when using thermal kaolinite at 600 °C. High to low FAME conversion (%) from used-cooking oil was as follow methyl linolein > methyl dhypalmitine > methyl hexadecanoate using kaolinite acid, respectively. These results were different for each catalyst used in the electrolysis process. Therefore, the amount of methyl ester produced was not affected by a certain methyl ester type.

Table 2. FAME conversion from soybean oil using treated kaolinite as catalyst

Treated kaolinite	Catalyst loading (%)	FAME yield (g) [b]	Purity (%) [c]	Yield (%) [d]
Without catalyst [a]	0	2.12	42.17	3.58
Thermal (800 °C), K800	2	32.93	5.76	3.79
	3	27.53	11.75	6.47
	5	22.62	3.61	1.63
Acid (800 °C), KA800	2	26.57	9.77	5.19
	3	19.44	48.47	18.85
	5	25.57	9.35	4.78
Thermal (600 °C), K600	2	-	-	-
	3	31.38	1.57	0.99
	5	30.74	3.58	2.20

[a] The calculation was based on 25 g soybean oil as feed-stock, [b] FAME yield after washing and solvent evaporation, [c] percentage area (%) from GC-MC, [d] Yield (%) = [FAME yield (g)/Oil mass (g)] × Purity (%).

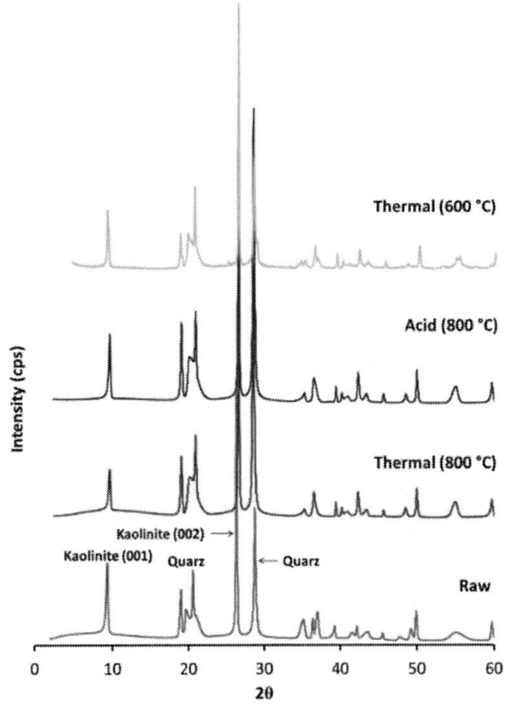

Fig. 3 XRD diffractogram of different treated kaolinite clay

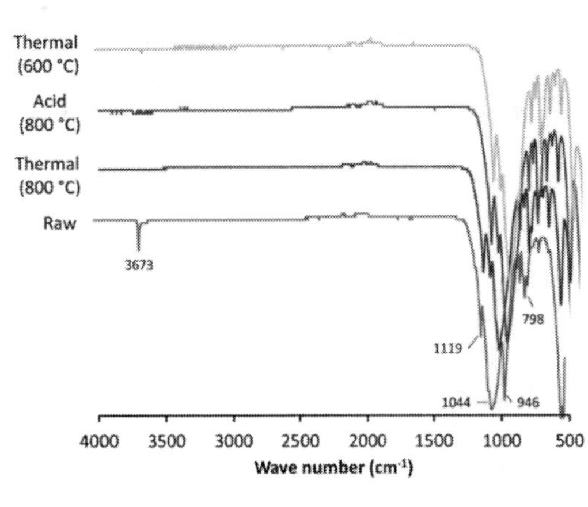

Fig. 4 FTIR spectra of different treated kaolinite clay

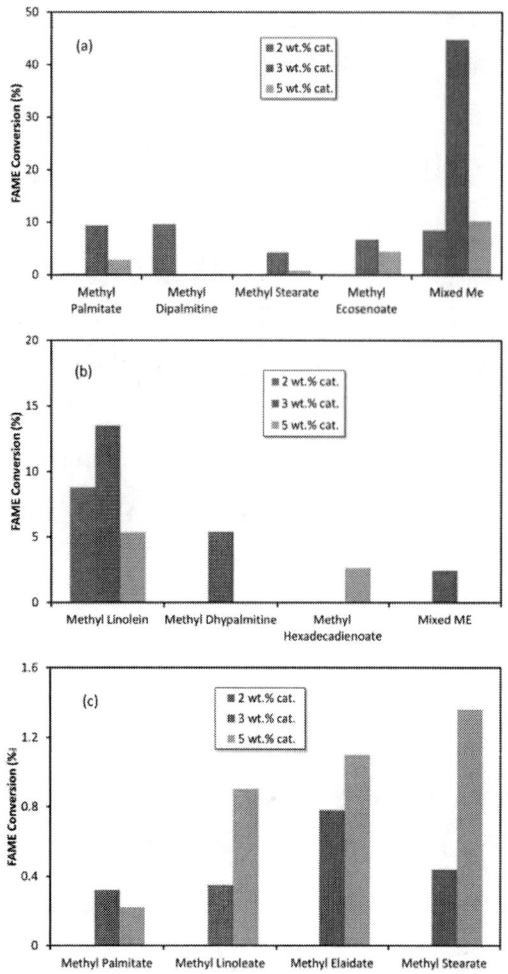

Fig. 5 FAME conversion from soybean oil in the electrolysis process using kaolinite thermal 800 °C (K800) (a), kaolinite acid 800 °C (KA800) (b) and kaolinite 600 °C (K600) as a catalyst.

4 Conclusions

The enhancement of FAME yield from the (trans)esterification of used-cooking oil using electrolysis process with acid kaolinite (KA800) as a catalyst had been evaluated. Acid kaolinite (KA800) as a catalyst could be a better material to enhance the electrolysis process. High catalyst concentration (3 wt. %) of KA800 with 60 minutes of electrolysis time at room temperature was required in order to get a high FAME yield as much as 18.85 % compared with thermal kaolinite K600 and K800 as a catalyst. In contrast, the results were not significantly on the FAME yield without catalyst in the electrolysis process.

Authors would like thank to KEMENRISTEKDIKTI, the Republic of Indonesia for financial support through Penelitian Produk Terapan (PPT) 2017. Furthermore, additional support from AIPT 2016 and Directorate of Talent Development and Student Welfare of Universitas Islam Indonesia to M.A.., R.I.K. and W.S. are also gratefully acknowledged.

References

1. R.L. Naylor, M.M. Higgins, Renewable and Sustainable Energy Review, **77**, 695-705 (2017)
2. G.F. Zagonel, P. Peralta-Zamora, L.P. Ramos, Talanta, **63**(4), 1021–5 (2004)
3. R.S. Putra, P. Hartono, T. Julianto, Energy Procedia, **65**, 309–316 (2015)
4. G. Guan, K. Kusabe, Chem. Eng, **153**, 159–163 (2009)
5. J.R.Ortiz, J.M. Valtierra, M.M. Rosales, J.Eng. Tech., **56**, 977-978 (2011)
6. C. Belver, M.A.B. Munoz, M.A. Vicente, Chem. Mater, **14**, 2033–2043 (2002)
7. S. Kumar, A.K. Panda, R. K. Singh.Bull, Chem. React. Eng., Catal., **8** (1), 61– 69 (2013)
8. R. Delhez, T.H. Keijser, E.J. Mittemeijer, Fresenius, J. Anal. Chem, **312**, 1–10 (1982)

The effect of inlet curvature and flange on wind turbine diffuser performance

Muhammad Nasir[1,*], *Ignasius* Bayu Dwianditya [1], *Kutut* Suryopratomo[1], and *Sihana*[1]

[1] Nuclear Engineering and Engineering Physics Department Universitas Gadjah Mada, Yogyakarta, Indonesia

Abstract. The modification is required to the conventional wind turbine in order to increase the power produced by the wind turbine located in the low wind velocity area. Attaching a diffuser to the conventional wind turbine is one of the ways to increase the wind velocity of the approaching wind thus increasing the extracted power. This research discussed the influence of geometry variation of diffuser augmented wind turbine equipped with inlet curvature and flanged. The analyzed geometry parameters were the length (L), the expansion angle (θ), the angle of inlet curvature (β), the radius of inlet curvature (R), and the flange height (h). In this study, experimental investigations were carried out to measure wind velocity ratio U/U_∞ (wind velocity on the central axis/free stream wind velocity) of various diffuser geometry designs with an installed rotor/wind turbine in it. The present diffuser augmented wind turbine provided higher wind velocity at the inlet section reaching the maximum velocity ratio U_{max}/U_∞ of 2.01 compared to a bare wind turbine.

1 Introduction

The depletion of fossil fuels and global warming issues have led to a reawakening of the interest in renewable energy development. One of the renewable energy sources as the alternative to substitute the utilization of fossil fuels is the wind energy. Wind energy technologies have been developing rapidly in the recent years. However, despite the fast growth of wind energy technologies, the extraction of the energy from the low wind velocity is still a major problem. Therefore, several research groups have tried to find artificial ways to collect and accelerate the approaching wind effectively.

The study on this topic is mainly encouraged by the physical law, which states that the wind power generated by the wind turbine is proportional to the wind velocity cubed approaching a wind turbine. It means that a slight acceleration of the wind gives a large impact on the higher energy generation [1]. This gives a better approach to resolve the low-wind velocity problem.

Attaching a diffuser to the conventional wind turbine is one of the modifications to increase the wind velocity. The diffuser is able to collect and accelerate the approaching wind thus the obtained power is increasing. The use of the diffuser as the wind power augmentation is also known as Diffuser Augmented Wind Turbine (DAWT). Various types and shapes of diffusers have been studying since 1979 [2-5]. One of the very interesting diffuser shapes is the flanged diffuser developed by Ohya et al. [6]. Despite proposing the similar diffuser-shaped structure like others previously developed, the characteristic distinctions of their diffuser feature were the additional curvature-shaped at the inlet section and a flange encircled at its exit section. Its performance is strongly linked to its geometric parameters. Those geometric parameters are the length (L), the expansion angle (θ), the inlet curvature angle (β), the radius of inlet curvature (R) and the flange height (h) (Fig. 1). When a diffuser's shape has the appropriate dimension, a maximum wind velocity ratio U_{max}/U_∞ (maximum wind velocity on the central axis/free stream wind velocity) of about 1.7 and slightly more could be reached [6].

Previous works have shown that the optimum expansion angle (θ) at which the wind speed increase, reached a maximum, was considered to be around 4° [6] and 6° [7] for an empty diffuser. Furthermore, for a larger expansion angle (θ) (10° or more) applied to DAWT with an installed rotor in it, higher wind turbine performance was obtained [8]. It was also shown that there is an optimal ratio flange height/diffuser inlet section diameter (h/D) of 0.25 for the diffuser without the wind turbine in it and 0.5 for the wind turbine installed in it [6]. Those research had clearly explained the influence of flange at the exit section of the diffuser. However, the influence of the curvature shroud at the inlet had not been discussed. Therefore, this research proposed to study the influence of the inlet curvature and the flange when both are attached at the diffuser. Further research experiments that study the influence of inlet curvature and flange are required since suitable data on this topic are hardly available in the published literature and mainly are derived from numerical simulations.

* Corresponding author: muhammad.nasir1995@mail.ugm.ac.id

 https://doi.org/10.1051/e3sconf/20184301025

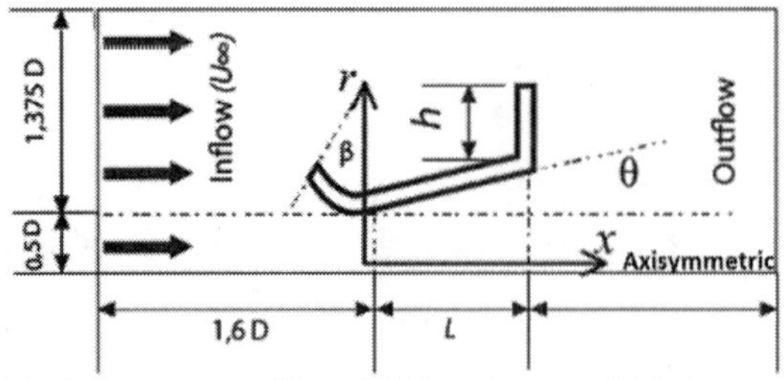

Fig. 1. Geometry condition of Diffuser Augmented Wind Turbine.

2 Method

Having considered the above background, this research discussed the influence of geometry variation of DAWT equipped with inlet curvature and flange. The analyzed geometry parameters were the length (L), the expansion angle (θ), the angle of inlet curvature (β), the radius of inlet curvature (R), and the flange height (h). In this study, experimental investigations were carried out to measure wind velocity ratio U/U_∞ (wind velocity on the central axis/free stream wind velocity) of various diffuser geometry designs with an installed rotor/wind turbine in it. By processing the obtained data, the influence of geometry variations was analyzed.

All the experiments were performed at Renewable Energy Laboratory of Nuclear Engineering and Engineering Physics Department, Universitas Gadjah Mada. The source of the wind is generated by a three blades fan with the diameter of 75 cm. The free stream velocities produced by the fan were 1.1 m/s, 1.7 m/s, and 2 m/s. The wind from the fan was turbulent, therefore the flow conditioner is required to minimize the turbulent flow. The diameter of flow conditioner was 80 cm. The flow conditioner was made from the set of straws with the length of 18 cm and the diameter of 0.8 cm for each straw.

In general, the structures of the diffuser are composed of the main body of the diffuser, a curvature-shaped shroud attached at the inlet, and a circular flange section encircled at the rear of the diffuser. The diameter of the diffuser throat was set constant 20 cm. The length of the diffuser (L) was 30 cm and the flange height (h) was 10 cm. The expansion angles (θ) of the diffusers were 4°, 8°, and 12°. While the radius (R) and the angle (β) of the inlet curvature were 5 cm, 8 cm and 40°, 45°, and 50° respectively. The rotor/wind turbine that was installed inside the diffuser has 13 bladed-rotor with the diameter of 17.5 cm and the diameter of the center hub was 4 cm.

The wind velocity on the central axis of the diffuser was measured based on the rotational velocity of the installed rotor inside the diffuser, which was derived by using a tachometer. The rotational velocity was then compared to the velocity measured by an anemometer. The tachometer used in this experiment was a digital tachometer DT-2234C. The resolution of this tachometer was 0.1 RPM (2.5-999.9 RPM) and 1 RPM (>1000 RPM). The accuracy of this tachometer was ± (0.05% + 1 digit) with the sampling time of 0.8 seconds. The Anemometer used in this experiment was a digital anemometer AM 4200 with the measurement range of 0.8 – 30.0 m/s. The resolution of this anemometer was 0.1 m/s and the accuracy was ± (3% + 2 digit).

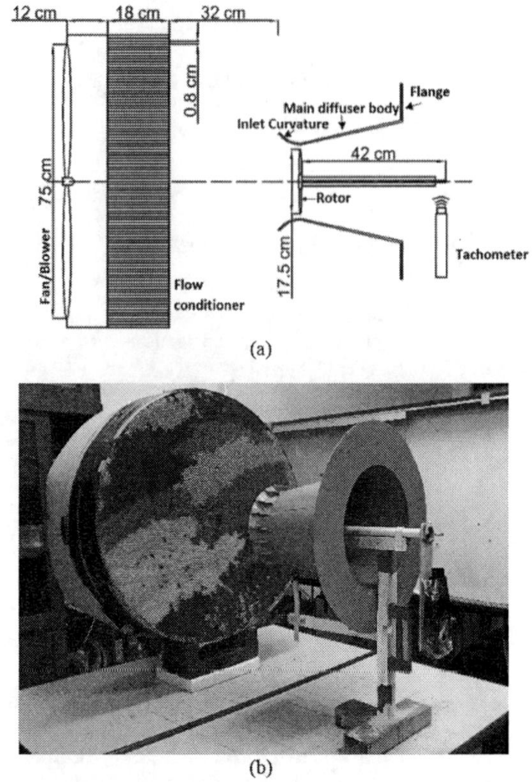

Fig. 2. (a) Detail information of experimental design; (b) The real appearance of the experimental design.

3 Results and Discussion

3.1 The overall behaviour

Six configuration models of the diffuser were tested in this experiment. The experiment was conducted at the same wind velocity conditions. The measurement was obtained from the centerline of the diffuser. By plotting the wind velocity ratio (U/U_∞) derived from experiment against the dimensionless length (X/L), we found that the maximum wind velocity ratio was obtained where the value of dimensionless length (X/L) ranging from 0.03 to 0.10. This range is located in the center line of the diffuser near the inlet. It can be seen from Fig. 3 that the maximum velocity was obtained near the inlet of the diffuser. The maximum wind velocity ratio (U/U_∞) reached values of 2.01 for the diffuser equipped with inlet curvature and flange that having the diameter inlet (D) of 20 cm; L/D ratio of 1.5; expansion angle (θ) of 8°; inlet curvature angle (β) of 50°; inlet curvature radius (R) of 8 cm; and h/D ratio of 0.5. This result showed that the wind velocity ratio was greater than the wind velocity ratio provided by the diffuser that only consisted of flange without inlet curvature shroud designed by Ohya et al [6].

In a diffuser, the flow expands along inside wall unless a massive separation occurs in the near wall region, resulting in the increase of static pressure and the decrease of the velocity towards the diffuser exit. This condition suggests that the flow have to accelerate near the inlet in order to cover such a deceleration that happens inside the body of the diffuser toward the exit. This is the reason why the approaching flow accelerates and the maximum velocity was obtained near the inlet.

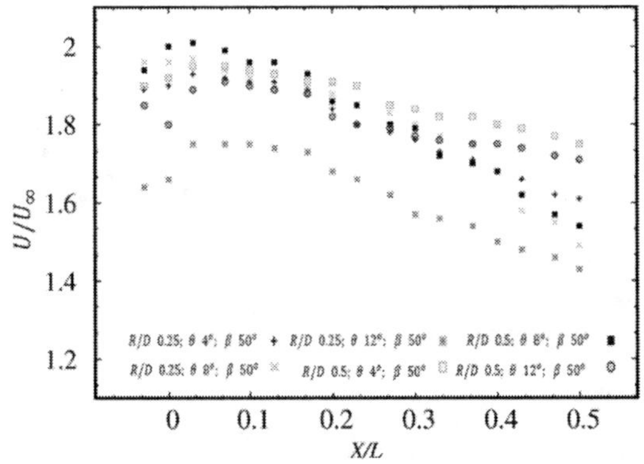

Fig. 3. The wind velocity distribution at the center line of the six diffusion configuration models.

3.2 The effect of inlet curvature and flange

The previous studies showed that as the body of the diffuser becomes longer, the wind velocity obtained will increase [6]. However, in terms of field application, it is not preferable to use a long body diffuser. Therefore, shorter diffuser body that mainly has an L/D of less than 2 is more desirable when it is able to give the similar performance the same as the long diffuser body performance.

In our experiment, we utilized a short diffuser body with the L/D of 1.5. The attachment of a curvature-shaped inlet shroud and the flange were the ideas aiming to improve the performance of the diffuser. During the experiment, we study the performance of the inlet shroud by varying its size which consisted of two variables, the radius (R) and the angle (β) (see Fig. 1). The size of the flange was made the same with the one designed by Ohya et al [6]. It was found by Ohya et al [6] that the h/D ratio of 0.5 provided the maximum wind velocity when the turbine was installed in it.

As a result, when the flange is employed, a higher wind velocity ratio (U/U_∞) was derived compare to the diffuser model only. A greater increased obtained when both curvature-shaped inlet and flange were attached to the diffuser. The derived value noticeably exceeded the case of diffuser model with flange only.

Our work showed that the additional inlet curvature with the specification R/D of 0.25 and β of 50° generated the wind velocity ratio $\pm$7.82 % compared to the diffuser that only has the flange. This result showed that the inlet curvature acted to increase the mass flow through the rotor and, therefore, to increase the wind velocity ratio. The curvature-shaped inlet practically provides a better shape for a smooth flow of the wind to enter the diffuser and well prevents the flow separation to happen near the entrance.

The increase of the wind velocity when a flange attached at the outer periphery exit of a diffuser is mainly due to the emergence of a low-pressure area in the near wake of the diffuser. Therefore, owing to the effect, the vortex is formed at the exit of the diffuser and the flow coming into the diffuser can be effectively concentrated and accelerated.

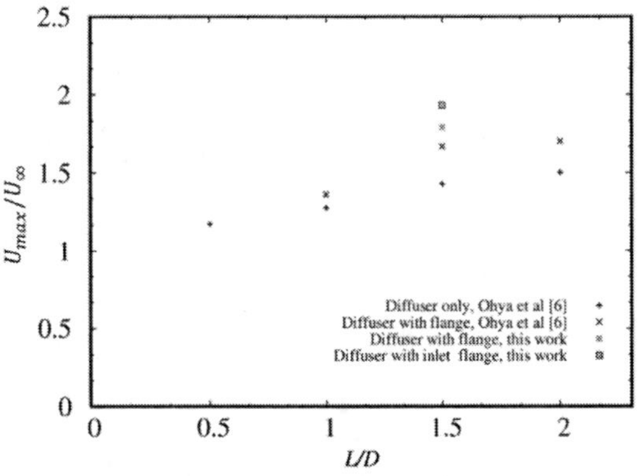

Fig. 4. The increase of the maximum wind velocity due to the diffuser model structure improvement.

3.3 The effect of the expansion angle (θ)

Expansion angle (θ) or also known as the open angle is also a very influential variable to the performance of the diffuser. Therefore, it is very important to understand the effect of the expansion angle (θ) toward the optimization of the diffuser utilization. We found that the wind velocity ratio (U/U_∞) increases with a bigger expansion angle (θ). However, when the expansion angle (θ) bigger than 12°, the wind velocity ratio (U/U_∞) tends to decrease again.

Comparing our result with those Matsushima et al. [7], there is a good synchronization concerning the optimal expansion angle (θ) that was reached in their study which approximately was about 6°. This can be explained by the different method and configuration used in this paper and the work done by Matsushima et al. [7] which was simulated based on 3D thermos hydrodynamic analysis software, I-DEAS, without an installed rotor inside the diffuser and curvature at the inlet section. By doing another comparison with work done by R. Chaker [8], the fairly well relationship is also found. From their simulation and experimental work, it appeared that the maximum expansion angle (θ) is about $\pm$10°. All of these results showed a similarity that optimum wind velocity ratio (U/U_∞) will be derived when the expansion angle (θ) of the diffuser was ranging about 6° - 12°. When the expansion angle (θ) is out of this range, the velocity of the wind tends to decrease. These results suggest that the expansion angle (θ) of the diffuser should not be out of this range.

The remarkable phenomena of the inclination in generating a lower wind velocity ratio (U/U_∞) when the

https://doi.org/10.1051/e3sconf/20184301025

expansion angle (θ) bigger than $\pm$ 12° mainly corresponds to the separation of the boundary layer. The detachment of the boundary layer gives a real impact on the aerodynamic performance of the diffuser.

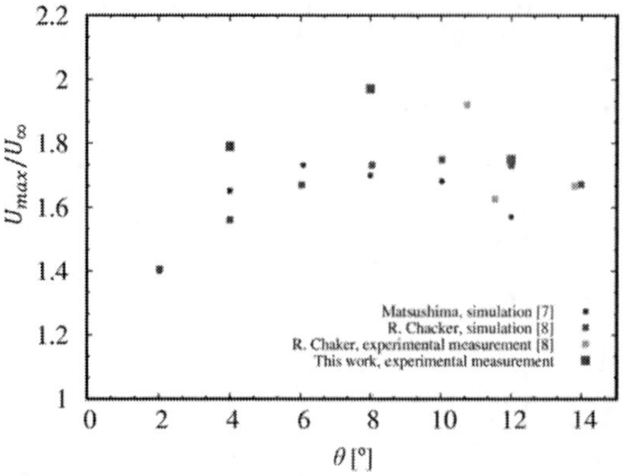

Fig. 5. The relationship between the expansion angle (θ) and maximum velocity ratio.

3.4 Recommended improvement of inlet curvature angle (β) and radius (R)

In order to have a better shape of the inlet shroud, in this experiment, we also tried to improve the inlet curvature radius (R) and the angle (β). As a result of our work, the relationship between the radius of the inlet shroud (R) with the wind velocity ratio (U/U_∞) can be studied from Fig. 3. It shows that the increase of inlet shroud radius (R) offers an increase of the wind velocity ratio (U/U_∞). However, this phenomenon was partially observed as the radius of the inlet shroud (R) provided in this experiment was limited only to two different values. It will be very crucial to work on another experiment providing more variations of the inlet shroud radius (R) to study the relationship between these two variables.

To study the relationship between the inlet curvature angle (β), three configurations of different inlet curvature angle (β) were tested in this experiment. The condition of these three models was the same. The value of R/D was 0.5; the expansion angle (θ) was 8°; and flange h/D was 0.5. Our result showed that as the angle of the inlet curvature (β) increased, the wind velocity ratio (U/U_∞) also increased (see Fig. 6). The inlet curvature angle (β) 50° generated higher wind velocity at the inlet section compared to inlet curvature angle (β) 45° and 40°. This phenomenon occurred due to an increase of the mass flow of the wind as the inlet curvature angle (β) becomes bigger. However, the same as the relationship between the radius (R) of the inlet shroud with the wind velocity ratio (U/U_∞), the influence of the inlet curvature angle (β) to the wind velocity ratio (U/U_∞) was also partially observed as we have not tested the performances of the diffuser when the inlet curvature angle (β) is bigger than 50°. Therefore, it will be very important to work on another experiment with a bigger inlet curvature angle (β) to enhance the performance of the diffuser.

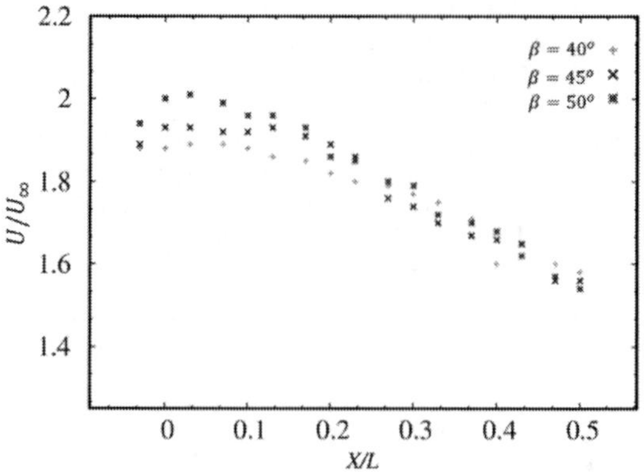

Fig. 6. The effect of the inlet curvature angle (β) to the wind velocity ratio.

3.4 Review of diffusion augmented wind turbine field application

The study of the diffuser augmented wind turbine for the field application had been conducted by several researchers [6,7,9,10]. One of the interesting projects was a project conducted in a desert region of northwest China. This project consisted of six units of 5kW diffuser augmented wind turbines equipped with the flanges that were installed in order to construct an irrigation plant. The greenery inspection on this project had been conducted and it was concluded to be effective. The other three 5kW diffuser augmented wind turbines had also been applied in the seashore Fukuoka city, Japan [10].

A 500 W class prototype of diffuser augmented wind turbine with flange had also been tested in the field. The diameter of the rotor was 0.7 m, while the diameter of the diffuser (D) was 0.72 m. The length and the height of the flange were 1.25D and 0.5D respectively. The field data of 10 min averaged values of the wind speed (m/s) and the power produced by the wind turbine were plotted (see Fig. 7). The result showed that the data derived from the experiment conducted in the field was similar to the data obtained from the wind tunnel experiment where the power curve of C_w = 1.4. The dashed line represented the power produced by the wind turbine that was not augmented with the diffuser. This result also demonstrated that the power generated by the tested diffuser augmented wind turbine with flange was about 4-5 times compared to the conventional wind turbine. During this field experiment, it was also found that the flange was able to rotate the body of the diffuser towards the direction of the approaching wind. Thus, the additional instrument to control the movement of the diffuser towards the wind direction was not required. In addition, the existence of the diffuser was not only generating an increase of the wind velocity ratio (U/U_∞) but also played a very important role in order to reduce the noise produced by the wind turbine. Moreover, it also provided an enhancement of the safety in case of the failure of the blades occurred [6].

 https://doi.org/10.1051/e3sconf/20184301025

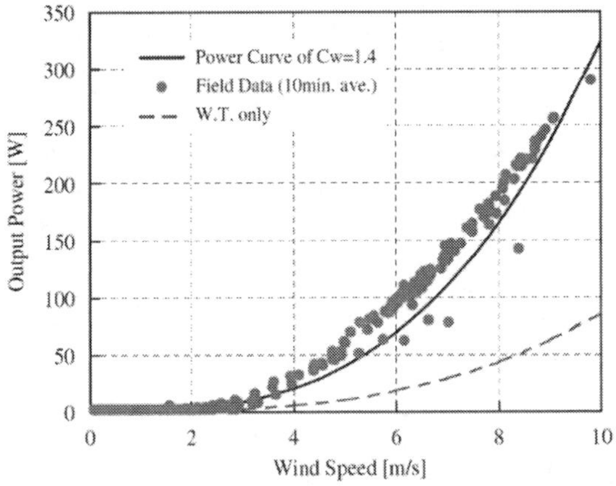

Fig. 7. The output power generated by different wind speed for 500 W class of diffuser augmented wind turbine [6].

4 Conclusion

In order to improve the performance of the diffuser augmented wind turbine, we examined the influence of the additional structure namely a curvature-shaped inlet shroud that was attached at the entrance of the diffuser and a flange that was encircled at the outer periphery of the diffuser. The present study showed that the modified diffuser noticeably provided a higher wind velocity ratio (U/U_∞) compare to conventional wind turbine as well as to the wind turbine that only equipped with the flange. The curvature-shaped of the inlet shroud allows a smooth flow of the wind to enter the diffuser. It also plays a role to prevents the flow separation to happen near the entrance. The utilization of the flange creates a low-pressure area that consequently draws and accelerates the wind into the diffuser. Our study also showed that optimum wind velocity ratio (U/U_∞) is derived when the expansion angle (θ) of the diffuser was ranging about 6° - 12°. The separation of the boundary layer is likely to happen when the value of the expansion angle (θ) is bigger than this range. This separation of the boundary layer causes a decrease of the wind velocity ratio (U/U_∞). The study of the relationship between the radius (R) of the inlet shroud with the wind velocity ratio (U/U_∞) confirmed that the increase of inlet shroud radius (R) offers an increase of the wind velocity ratio (U/U_∞). The phenomenon of an increase of the wind velocity ratio (U/U_∞) also observed when the inlet curvature angle (β) increases. However, both of the relations need a further investigation as the work done in this experiment was limited by the variation of the models that were used to observe.

References

1. Y. Ohya and K. Abe, *An Investigation of Flow Fields around Flanged Diffusers Using CFD*, J. Wind. Eng. Indus. Aero., vol. **92**, no. 3-4, pp. 315 – 330 (2004)
2. M. O. L. Hansen, N. N. Sorensen and R. G. J. Flay, *Effect of Placing a Diffuser around a Wind Turbine*, J. Wind Energy, vol. **3**, no. 4, pp. 207-213 (2000)
3. G. J. W. Van Bussel, *The Science of Making More Torque from Wind: Diffuser Experiments and Theory Revisited*, J. Physics.: Conf. Ser., vol. **75**.
4. K. Abe, M. Nishida, A. Sakurai, Y. Ohya, H. Kihara, E. Wada and K. Sato, *Experimental and Numerical Investigations Flow Fields Behind a Small Wind Turbine with a Flanged Diffuser*, J. Wind Eng. Ind. Aerodyn., vol. **93**, no. 12, pp. 951-970 (2005)
5. D. G. Phillips, P. J. Richards, and R. G. J. Flay, *CFD Modelling and the Development of the Diffuser Augmented Wind Turbine*, Wind Stuct., vol. **5**, pp. 267-276 (2002)
6. Y. Ohya, T. Karasudani, A. Sakurai, A. Ken-ichi and M. Inoue, *Development of a Shrouded Wind Turbine with Flanged Diffuser*, J. Wind Eng. Ind. Aerodyn., pp. 524-539 (2008)
7. T. Matsushima, S. Takagi, and S. Muroyama, *Characteristics of the highly efficient propeller type small wind turbine with a diffuser*, Renew. Energy., vol. **31**, pp. 315-330 (2006)
8. R. Chaker, M. Kardous, F. Aloui and S. Ben Nasrallah, *The Relationship Between Open Angle and Aerodynamics Performances of DAWT*, in The 4th International. Renew. Energy. Congress, Tunisia (2012)
9. P. Ten Hoopen, *Master of Science Thesis: An Experimental and Computational*, Delft University of Technology (2009)
10. Y. Ohya, Takashi Karasudani, *A Shrouded Wind Turbine Generating High Output Power with Wind-lens Technology*, Energies., vol. **3**, pp. 634-649 (2010)

Sustainable Innovation System Using Process Of Bagasse Become Sodium Lignosulfonate Surfactant For Enhanced Oil Recovery

Rini Setiati [1*], *Septoratno* Siregar[2], *Taufan* Marhaendrajana[3], and *Deana* Wahyuningrum[4]

[1] Petroleum Engineering, Trisakti University , PhD Student Petroleum Engineering. ITB, Bandung, Indonesia
[2] Professor of Petroleum Engineering, ITB, Bandung, Indonesia
[3] Assoc.Prof. of Petroleum Engineering, ITB, Bandung, Indonesia
[4] Assoc.Prof. of Chemistry, ITB, Bandung, Indonesia

Abstract. The purpose of this research is to get new product innovation process from bagasse, that is Sodium LignoSulfonate surfactant. Lignosulfonates surfactants in petroleum engineering are used as injection fluids into oil reservoirs to increase oil recovery, which is known as Enhanced Oil Recovery process. Lignosulfonates is made of lignin as raw material, which can be extracted from bagasse as one of its sources. Bagasse contains 24 - 25% lignin, so it is sufficient to be processed into lignosulfonates. Today, bagasse is one of the biomass resources widely used as a boiler fuel in sugar factory, source of animal feed, material for paper, cement and brick reinforcement .This study presents an innovation of bagasse utilization. This innovation involves two scientific application fields, firstly, chemistry in the processing of bagasse into sodium lignosulfonates surfactant and secondly, petroleum engineering in the effort of using sodium lignosulfonates surfactant to increase oil production from the reservoir. The last stage in this process is injection of the sodium lignosulfonates surfactant into a synthetic core in laboratory scale use water and surfactant injection.. The amount of oil that is produced from the injected core shows the increase in oil yield from the sodium lignosulfonates surfactant injection.

1 Introduction

In specific, bagasse is scientifically defined as a waste of sugar cane liquid extraction after milling process and is in a fibrous form. The amount of bagasse production each year are abundant, easily obtained, and economical. Based on the data from Indonesian Sugar Farm Research Center (P3GI), bagasse amounts to approximately 32% of milled sugar cane weight or about 10,2 million ton/year all around Indonesia.Furthermore, Bagasse contains 48-52% water, sugar at an average of 3,3%, and fiber at an average of 47,7%. Also, bagasse fiber is undissolvable in water and it consists of cellulose, pentosan and lignin. In essence, Lignin is the raw material, which is processed through sulfonation before subsequently turning to surfactant. Thus, the product is classified as Surfactant Sodium Lignosulfonate. In addition, Lignosulfonate is one of the variants of anionic surfactant that is often utilized in a chemical injection process of EOR (Enhanced Oil Recovery) in the oil industry. High lignin content is what makes bagasse eligible candidate for surfactant. Today, bagasse is one of the biomass resources widely used as a boiler fuel in sugar factory, source of animal feed, material of paper, cement and brick reinforcement material. Sustainable innovation process from bagasse can be done by isolation process and sulphonation process into SLS surfactant. The result of chemical process can be applied as an injection fluid to increase oil recovery.

2 Methods

In this study, the innovation of bagasse processing is made in a gradual and sustainable system. This innovation includes the processing of bagasse into sodium lignosulfonates surfactant and using sodium lignosulfonates surfactant as an injection fluid in enhanced oil recovery . Figure 1 show the processing of bagasse into lignosulfonates can be done in two stages: lignin separation process from bagasse (isolation process) and lignin sulfonation process into lignosulfonates. In the lignin isolation stage, bagasse and 2% sodium hydroxide was refluxed directly for five hours, followed by diluting the filtrate sodium hydroxide with distilled water at 1:1 volume ratio. Then, H_2SO_4 is added in drops until pH equals 2 and the resulting solution is precipitated for at least eight hours and the precipitate is filtered and dried, and the solid obtained is lignin. The bagasse lignin is then compared with the commercial lignin (from Aldrich and Kraft) and characterized by FTIR spectrophotometry,which identifies the functional groups contained in the lignin from the isolation process.

The next process is sulfonating lignin into lignosulfonates using 0.1 M sodium bisulfite and refluxed at 150 °C for five hours while stirring with a magnetic stirrer. Then the precipitate is filtered, and the solid is dried to obtain sodium lignosulfonates surfactant. The next process is testing the surfactant with FTIR and NMR to determine the structure of the sulfonate group.The test is performed as a screening of sodium lignosulfonates surfactant, which is followed by characterization of the sodium lignosulfonates surfactant consisting of compatibility test, phase behavior test, middle emulsion test, adsorption test, IFT test and wettability test.

* Corresponding author: rinisetiatidgm@gmail.com

 https://doi.org/10.1051/e3sconf/20184301026

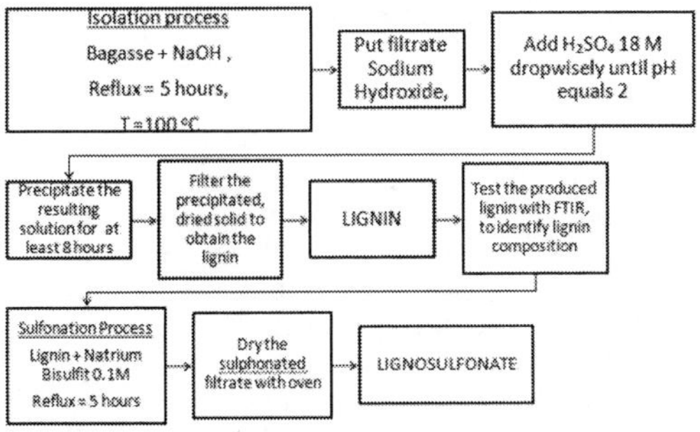

Fig. 1. Schematic processing of bagasse into SLS surfactant

Lignosulfonates surfactants in petroleum engineering are used as injection fluids into oil reservoirs to increase oil recovery, which is known as Enhanced Oil Recovery process. Currently, the lignosulfonates surfactants which are widely used have petroleum base, known as petroleum sulfonate. Since it uses petroleum as raw material, its price also follows oil price which is relatively expensive. Lignosulfonates is made of lignin as raw material, which can be extracted from bagasse as one of its sources. Bagasse contains 24 - 25% lignin, so it is sufficient to be processed into lignosulfonates. High lignin content is what makes bagasse potential candidate for lignosulfonates surfactant. Today, bagasse is one of the biomass resources widely used as a boiler fuel in sugar factory, source of animal feed, material for paper, cement and brick reinforcement .This study presents an innovation of bagasse utilization. This innovation involves two scientific application fields, firstly, chemistry in the processing of bagasse into sodium lignosulfonates surfactant and secondly, petroleum engineering in the effort of using sodium lignosulfonates surfactant to increase oil production from the reservoir. This innovation process is optimized gradually and continuously in order to obtain maximum results.

Based on the characteristic test, followed by core injection process. The last stage in this process is injection of the sodium lignosulfonates surfactant into a synthetic core in laboratory scale. The core injection process consists of two stages: water injection, which is followed by sodium lignosulfonates surfactant injection (surfactant flooding). The amount of oil that is produced from the injected core by water and surfactant shows the increase in oil yield from the sodium lignosulfonates surfactant injection. Figure 2 showed scheme of surfactant injection for EOR process

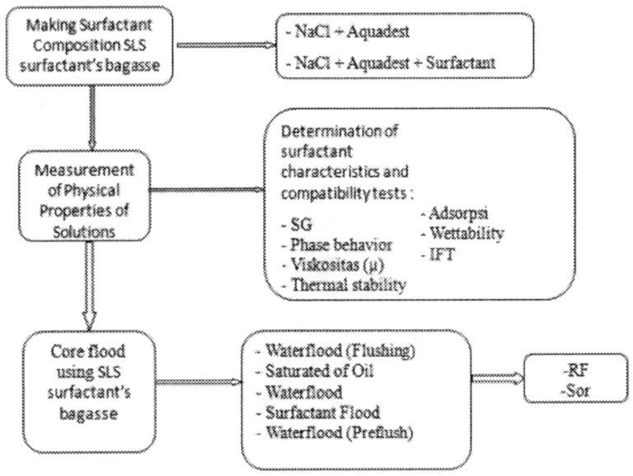

Fig. 2. Scheme of surfactant injection for EOR process

Figure 3 below showed process activity of bagasse into SLS surfactant and its application as injection fluid in EOR process.

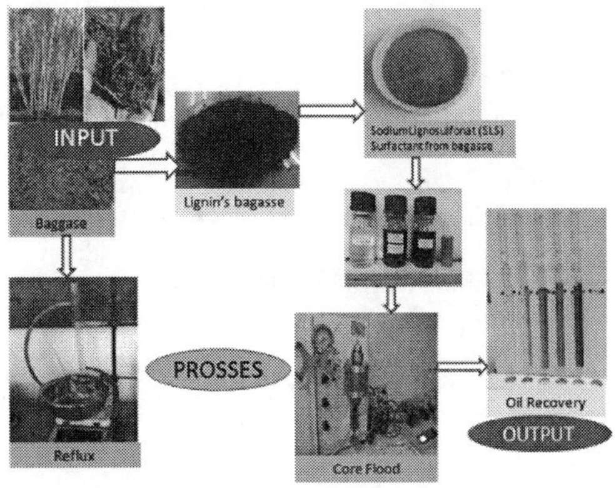

Fig. 3. Process activity of bagasse into SLS surfactant and its application in EOR process.

3 Result

The FTIR test on lignin isolation shows that bagasse lignin contains O-H phenolic functional group, aliphatic and aromatic group -CH-, C = O group, arena-functional group -C = C-, amine C-N groups and alkyl C-H groups. . The results of comparing bagasse lignin with the commercial standard lignin of Aldrich and Kraft products show that the bagasse has the same qualification as the standard lignin , so it supports the option to continue to the next stage, ie sulfonation process.

The results of the FTIR test on the sodium lignosulfonates from bagasse show that it contains components comprising of C = C alkene, S = O sulphat, C = O carboxylic acids and S-OR esters elements, with wavelength spectral values approaching standard wavelength spectrum. From table 1 showed that FTIR results and comparisons with existing standard products, both in lignin and lignosulfonates, suggest that the isolation lignin and sulfonation processes is complete and comparable with standard products.

Table 1. Componen of SLS surfactant's bagasse.

	Cmponen of SLS Surfactant	Wave number (cm^{-1})		
		Standar SLS Surfactant (Patricia)	Standar SLS Surfactant (Aldrich)	SLS Surfactant (Bagasse)
1.	Alkena -C=C-	1630 - 1680	1608,34	1635,34
2.	Sulfonate S=O	1350	1365	1384,64
3.	carboxylates C=O	1000 - 1300	1187,94	1114,64
4.	Ester S-OR	500 - 540	499,831	462,832

The calculation of Hydropilic-Lipophilic Balance (HLB) values of sodium lignosulfonates surfactant from bagasse shows HLB value of 11.62 which means the product is qualified as Oil-Water emulsion, therefore can function as emulsion-forming fluid.

Table 2. Characteristics of SLS surfactant's bagasse.

The empirical formula	$(C_{11}H_{16}O_8S)_n$
Component of SLS Surfactant	Alkena -C=C-
	Sulfonate S=O
	carboxylates C=O
	Ester S-OR
Molecular weight	308.06
HLB	11.62

The characterization test of the sodium lignosulfonates surfactant from bagasse (SLS surfactant's bagasse) at various concentrations and brine salinities shows that it is stable in the aqueous stability test, forming the middle-phase emulsion (microemulsion) with light oil component and having a stable 1 - 10% microemulsion. With light oil, surfactant has IFT value range from 6.81 - 1.091mN/m. The adsorption of the surfactant on the core surface ranges from 77.78 - 205.33 mg/g, and the wettability test showes contact angle ranging from 28°-48°.

Table 3. The result characterization test of the SLS surfactant's bagasse and the result of injection core

No.	Salinity (ppm)	Surfactant concentration (%)	Micro emulssion (%)	IFT (mN/m)	Thermal stability	RF WF (%)	RF SF (%)
1.	5.000	1.5	1.25	6.810	Unstabel	31.00	7.00
2.	10.000	1.5	10.00	2.728	Stabel	16.00	9.25
3.	10.000	3.0	7.50	1.675	Stabel	24.00	9.50
4.	20.000	1.5	5.00	4.131	Stabel	20.00	8.55
5.	20.000	4.5	0.00	1.272	Stabel	21.00	1.05
6.	40.000	1.5	6.00	4.107	Unstabel	20.00	1.80
7.	40.000	4.0	0.00	2.724	Unstabel	20.08	1.16
8.	80.000	1.5	1.25	3.614	Stabel	18.00	10.71
9.	80.000	4.0	1.00	1.091	Unstabel	20.00	5.60

The final test result, which is injection of with sodium lignosulfonates surfactant into a synthetic core containing light oil shows an oil gain of 1.05 - 10.71% which is also done at various concentrations of dissolved sodium lignosulfonates surfactant at various brine salinities.

4 Conclusion

1. Bagasse into lignosulfonates process is done in two stages: lignin separation process from bagasse (isolation process) and lignin sulfonation process into lignosulfonates
2. The lignosulfonates is complete and comparable with standard products and therefore can be used as injection fluid .
3. Sodium lignosulfonates surfactant of bagasse shows HLB value of 11.61 which means the product is classified as O/W emulsion, therefore can function as emulsion-forming fluid.
4. Characterization of sodium lignosulfonates surfactant shows that it is stable in the aqueous stability test, forming the middle-phase emulsion (microemulsion) with light oil component and having a stable 1 - 10% volume microemulsion, has an IFT value range from 6.81 - 1.091mN/m.
5. Sodium lignosulfonates surfactant injection into synthetic core containing light oil gives an increase of oil recovery of 1.05 - 10.71%.
6. The innovation to process bagasse into sodium lignosulfonates surfactant which has been completed involves two scientific application fields, chemistry and petroleum engineering to increase oil production from a reservoir. This innovation process should be optimized gradually and continuously in order to obtain maximum results.

This research was facilitated by the Research Consortium OGRINDO at ITB, Trisakti University, and the Indonesian Directorate General of Higher Education Decentralization Fund for PUPT Fund 2014 Usakti – Dikti, Decentralization Fund 2014 ITB, PUPT Fund 2015 ITB – Dikti and PUPT Fund 2016 ITB – Dikti. We also thank to Astechnova 2017 who publishes this article, which in turn will benefit the society.

References

1. Chakar, *Review of Current and Future Softwood Kraft Lignin Process Chemistry*, International Lignin Institute Conference, 131 – 141 (2004)
2. J. F. Craig, *The Reservoir Engineering Aspects Of Waterflooding*, Tulsa: Society of Petroleum Engineers (1993)
3. O. ElMofty, *Surfactant enhanced oil recovery by wettability alteration in sandstone reservoirs*, Missouri University of Science and Technology, Missouri: Student Research & Creative Works at Scholars' Mine (2012)
4. D. Fengel, G. Wegener, *Wood: Chemistry, Ultrastructure, Reactions*, Berlin: Walter de Gruyter (1989)
5. T. Fujimoto, *New Introduction to Surface Active Agent*, Kyoto: Sanyo Chemical Industri (1985)
6. D.W. Green, G. Willhite, *Enhanced Oil Recovery*, Texas: SPE textbook (1998)
7. E. Hambali, *Proses Pembuatan Surfaktan Metil Ester Sulfonat (MES) Untuk Aplikasi Industri Migas*, IPB, Surfactant and Bioenergi Research Centre (2008)
8. L. L.Schramm, *Surfactant: Fundamentals andApplication in the Petroleum Industry*, Cambrigde: University of Cambrige (2000)
9. Lacey, *Moulding of Sugar Cane Bagasse*, Annals of Applied Biology, **76(1)**, 63-76 (1974)
10. L. Lake, *Enhanced Oil Recovery*, Austin: University of Texas (2003)
11. D. Myers, *Surfactant Science and Technology*, New Yersey: Wiley Interscience (2006)

 https://doi.org/10.1051/e3sconf/20184301026

12. R. J. Patricia, *Relationship between the structure of Fe-Lignosulfonate complexes determined by FTIR spectroscopy and their reduction by the leaf Fe reductase*, The Proceedings of the International Plant Nutrition Colloquium XVI, hal.1 (2009)

The wireless energy transfer (WET) using ultra high frequency (UHF) for human body implant recharging

Fahmy Rinanda Saputri[1], *Sunarno*[2*], *Memory* Motivanisman Waruwu[3], *Rony* Wijaya[4]

[1]Master Student in Department of Nuclear Engineering and Engineering Physics, Faculty of Engineering, Universitas Gadjah Mada
[2,3]Department of Nuclear Engineering and Engineering Physics, Faculty of Engineering, Universitas Gadjah Mada
[4]Production House Unit, Faculty of Medicine, Universitas Gadjah Mada

Abstract. The wireless energy transfer (wet) is widely used in many fields. In particularly for medical, this technique can be implemented to an electronic human body implant recharging. The possible size and shape of the receiver antenna in the experiment before still become a problem. Using Yagi-Uda model antenna for the transmitter unit in ultra-high frequency (UHF) band, this paper refers to design a receiver antenna for implemented as a human implant recharging. The receiver antenna has successfully designed in minimal with its size and shape of the designed antenna is about 1.44 cm^3.

1 Introduction

The wireless energy transfer (wet) has been carried out in various field. Sunarno has designed the basic system for the wireless energy transfer based on ultra high frequency (UHF) band [1]. The transmitter and receiver system was using parabolic metal reflector system. This system able to transfer energy and fill the rechargeable battery with the efficiency was about 1%.

Sunarno and the team have also succeeded in their experiment about transferring energy using wireless technology based on UHF band with Yagi-Uda Directional antenna as a transmitter antenna. From the experiment can be concluded that distance and angle of transmitter antenna influenced for the amount of energy received. In this case, the amount of energy received is proportional with the current and voltage measured [2].

Reference [3] gave an experiment about designing ultra wideband antenna. It's application for a short distance communication system. The performance of the antenna is good with its fractional bandwidth of more than 108.2% and the range antenna resonance frequency from 3.1 to 10.5 GHz. In [4], the author presented elaborated prototype of energy harvesting in the ultra-high frequency band. It is used for excess energy and conducting radio communication process.

For implantable Bioelectronics, wireless energy transfer is possible to be applied. Reference [5] reviews the theory, link designed, and different wireless strategy ranging from the near-field inductive and capacitive coupling to the ultrasonics, mid-field and far field for the neural implants. The author considered that wirelessly powered implants are efficient and integrable if it is implemented as implantation device.

Recent various contributions in the area of energy harvesting are summarized in [6]. Based on the author's reviews, there is interdisciplinary nature from the devices that wireless transfer and harvest energy. Additionally, in reference [7], the authors discuss issue and challenge in future implementing the integration of wireless charging technologies.

Exiting the various applications of the wireless energy transfer, this paper that continuing the research before [2] was conducting research specifically for the medical area. In the medical, the wet can be utilized as an electronic human body implant recharging. The methods have already done in the experiments before, but there are problems, such as the size of the receiver unit doesn't match if it is implemented as an implant and can't be applied as an energy harvesting yet. An energy harvesting technique is very advantageous because recharging energy can occur automatically if the people who use implant are in the radiation electromagnetic field and should not be done intensively in the hospital.

The paper is organized into five sections. Section II discusses various terms explanation for understanding this research topic. The setup and methodological of the system concerns specifically in Section III. Furthermore, in Section IV, we discuss with a brief result clarifications based on the final graphs and data. We end with a conclusion and future research in the real receiver application for human body implant in Section V.

2 Theoretical Background

2.1. Ultra high-frequency antenna

Ultra high frequency is electromagnetic waves in the frequency of 300 MHz to 3.000 MHz. The range of wave length is 10 cm to 100 cm [10]. The popular UHF antenna is Yagi-Uda. Yagi-Uda antenna is used to transfers or receives the radio waves and has a directional function, meaning it only can take or receives the signal in single direction i.e. front. Yagi-Uda antenna has range of gain is 3 to 20 dB. The elements of this antenna are driven, reflector, director, and boom [10]. Driven is the feed point to perform the transfer of power of the antenna [2]. The length of the driven antenna is 0.5λ. A reflector is utilized for reflecting the signal. The position is in the behind. The length of a reflector is about 0.55λ. Director as a referrer has the shorter length than driven. Adding director will increase the gain,

* Corresponding author: sunarno@ugm.ac.id

 https://doi.org/10.1051/e3sconf/20184301027

however, the beam pattern will more narrow. The boom is part of the antenna made of the wood and metal stem for placing driven, reflector, and director. The antenna has a space between elements is 0.1λ of the frequency.

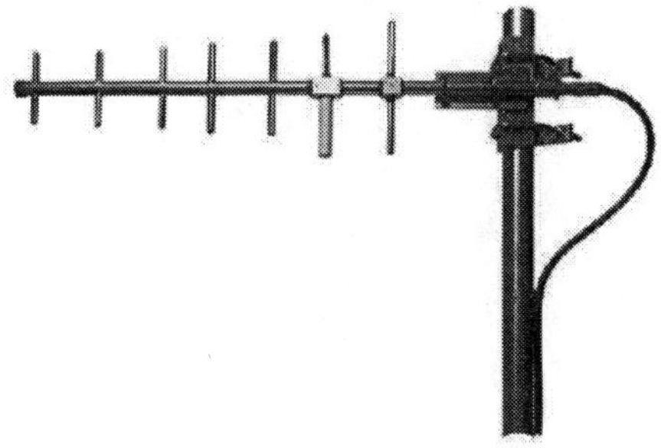

Fig. 1. The Yagi-Uda antenna [10]

2.2 Wireless charging

Wireless charging is the technology that enables a power to transmit electromagnetic energy to an electrical load without cable. Wireless charging has many advantages [7]: wireless charging improves user friendliness as without cable for implementation. better product durability renders the smaller devices design and fabrication and is more flexible and energy-efficient in an on-demand fashion. There are two major wireless charging technologies, i.e. radio frequency based wireless charging and coupling-based wireless charging. Wireless charging is attracting a many application, such as smartphone battery charging, vehicle charging, wireless communication, and biomedical devices [2]. For biomedical devices as an implantation, the coupling-based wireless charging has been conducted [11].
white, and this should be taken into account when preparing them.

3 Experimental setup

This study proposed that the use of the Yagi-Uda model antenna using ultra high frequency (UHF) is able to transfer energy with the wireless energy transfer technique. The wireless technique can be applied in many fields, especially in the medical field, that is to recharging electronic implant.

The methodology and technique of research have been conducted in previous research that was held by Sunarno *et.al*. For the transmitter unit, we also used Yagi-Uda model antenna. The beaming pattern of Yagi-Uda antenna is the reason. It is very focused, moreover in the narrow spot area. The transmitter antenna is able to changes electron wave to electromagnetic wave to the receiver antenna.

The wet system has designed by using ultra-high frequency band, that is 430 MHz as an energy carrier. Firstcom FR-488 instrument is used for measuring measured energy [2]. The power transmitter unit is 55 watt and 12 volts. The value of standing wave ratio in this research is 1.12.

In this system, the same electronic scheme, the Yagi-Uda Antenna, and the beam antenna were used. In previous research, that components were conducted by Sunarno *et.al*, are shown in Fig. 2, 3, 4, and 5.

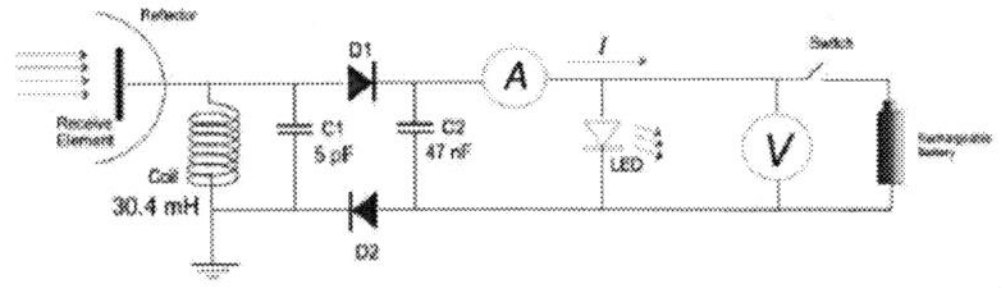

Fig. 2. The scheme of receiver circuit unit [1]

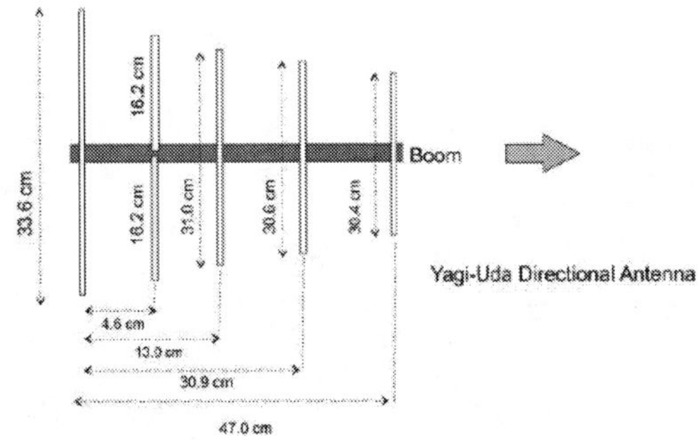

Fig. 3. The Yagi-Uda model antenna [2]

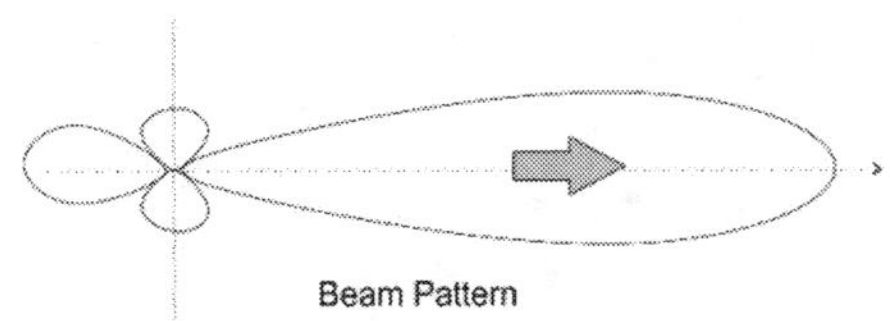

Fig. 4. The beam pattern of Yagi-Uda model antenna [2]

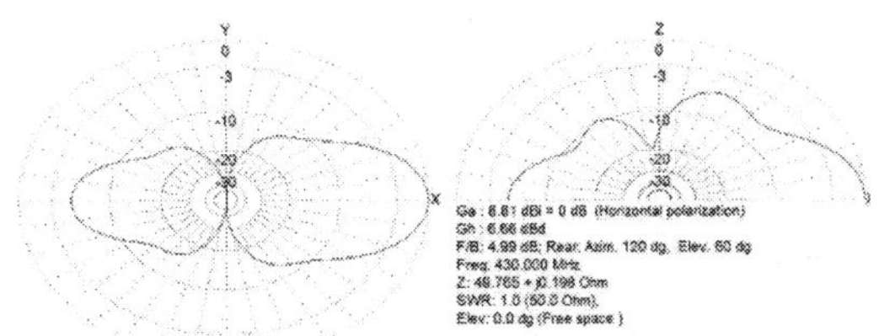

Fig. 5. The 2D prediction of beam pattern [2]

https://doi.org/10.1051/e3sconf/20184301027

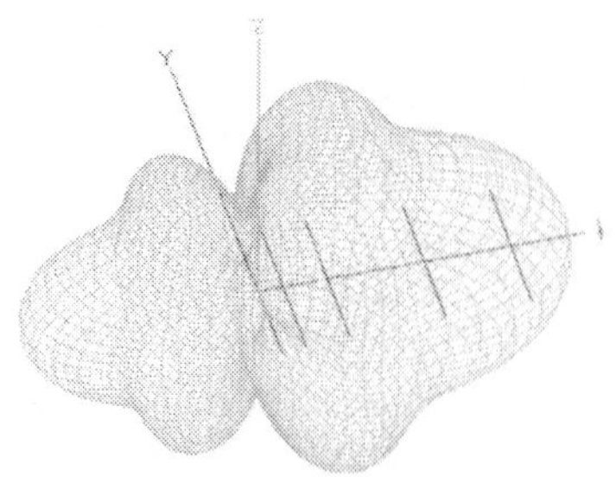

Fig. 6. The 3D prediction of beam pattern [2]

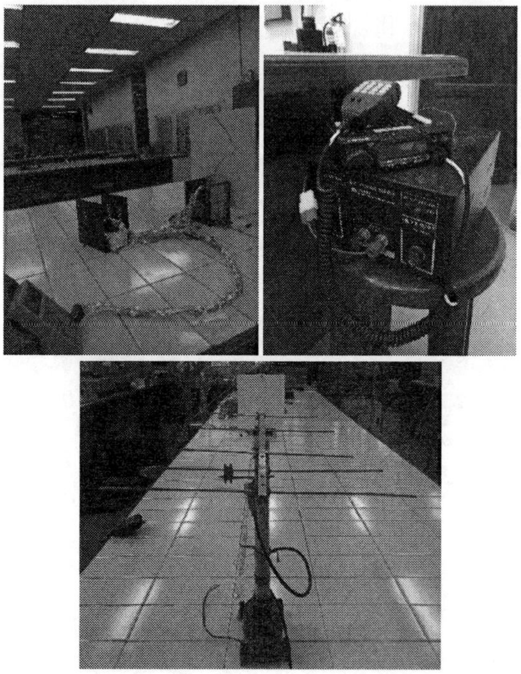

Fig. 7. The research devices

Fig. 6 shown the devices of wireless energy transfer system for this research. We used Yagi-Uda antenna of the transmitter unit, Firstcom FR-488 for measuring transmitter unit, voltmeter and ampere meter for measuring received energy and use the modified antenna as a receiver antenna. Because the implementation of the receiver antenna is for human implant electronic for recharging system, we design the minimal antenna. The size of the antenna is about 1.44 cm^3. The size is able enough to insert in the body as human impant electronic. The receiver antenna is shown in Fig. 7.

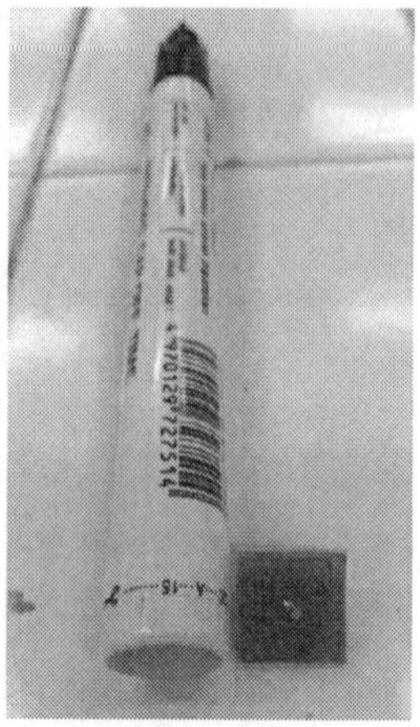

Fig. 8. The receiver antenna

4 Result and discussion

The study focused on the analysis of the unit receiver antenna design and doing experiment with varying the distance between transmitter and receiver antenna. The energy received based on varying distance is also analyzed. The measured result of the system based on varying distance is shown in Fig. 8.

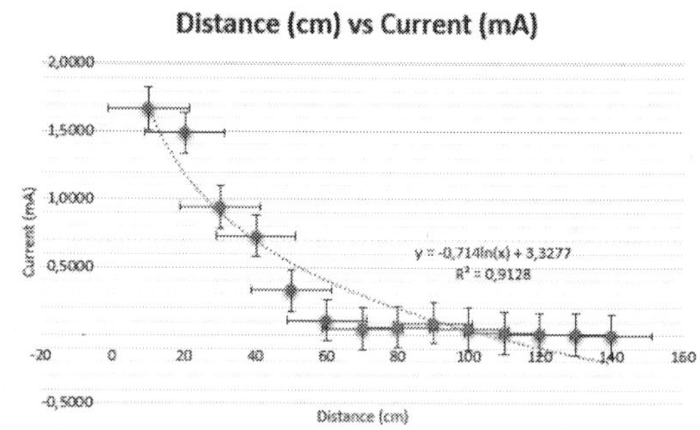

Fig. 9. Capture current versus distance with 55-watt energy source

The results show that the most of the current decreases when the distance enlarged between transmitter and receiver unit. The received power can be calculated by a formula,

$$P = V \times I \tag{1}$$

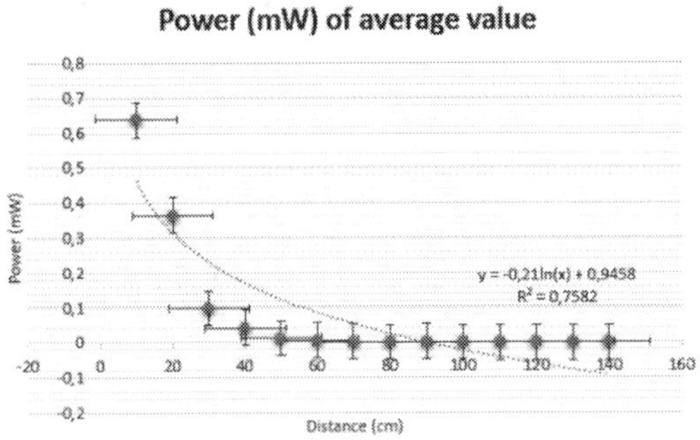

Fig. 10. Capture power versus distance with 55-watt energy source

The result of power in mW can be calculated by the measured voltage times the measured current in the receiver antenna. Fig. 9 is the graph capture of the received power. The technique was applied to transferring energy. The size of the receiver antenna that has been designed is 1.44 cm^3. The technique can recharge implant device without doing surgery. The antenna receiver design is successful to receiving energy in the ultra high-frequency band with 0.00116% of the efficiency, 10 cm of the distance, using Yagi-Uda model antenna as a transmitter. The energy received by the constructed system was also higher with the short distance between transmitter and receiver antenna compared to the far distance.

In the next research, received antenna will be implemented to the deaf. Fig. 10 shows wireless energy transfer for human implantation system diagram that will be used in the next research, specifically for bone vibrator implantation system as in Fig. 11.

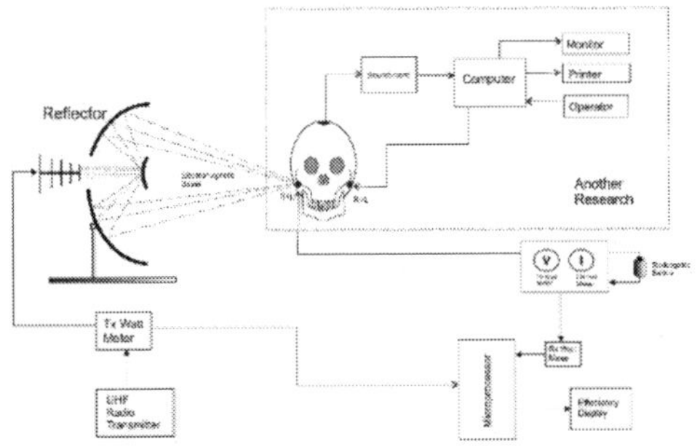

Fig. 11. The wireless energy transfer for human implantation research system diagram experiment system diagram

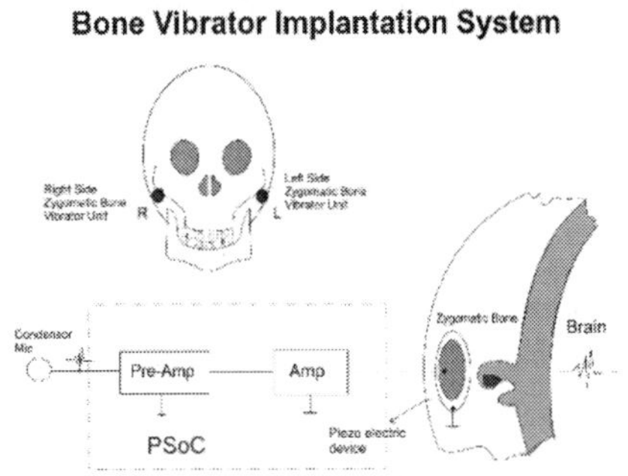

Fig. 12. Bone vibrator implantation system

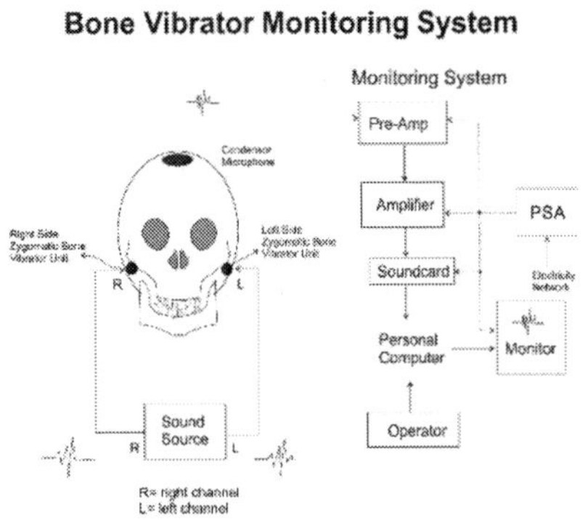

Fig. 13. Capture power versus distance with 55-watt energy source

Fig. 12 is the monitoring system of the final research. The final of the future research is implementation the wireless energy transfer using ultra-high frequency for the implantation electronic to the deaf. This electronic is able to help the deaf to feel the vibration of the sound and can monitor the vibrator.

4 Conclusion

The energy transfer was successfully performed using a wireless energy transfer technique using the ultra-high frequency (UHF) band with the Yagi-Uda model antenna as a transmitter and a small receiver antenna design, which is about 1.44 cm^3. The research has succeeded transfer energy with the minimal design od the receiver antenna. This minimal antenna received size allows it to be implanted into the human body as an electronic implant recharging.

The authors would like to thank Sensor and Telecontrol System Laboratory, Department of Nuclear Engineering and Engineering Physics, Faculty of Engineering, Universitas Gadjah Mada for providing the lab facilities to complete this research.

References

1. Sunarno, *Astechnova*, 62-70 (2014)
2. Sunarno, F.R. Saputri, M.M. Waruwu, R. Wijaya, *International Conference on Science and Technology* (2017)
3. V.K. Rai, M. Kumar, M.D. Upadhayay, *International Conference on Computational Intelligence and Communication Networks*, 16-19 (2015)
4. L. Bernacki, R. Gozdur, N. Salamon, *Conference Series*, 709 (2016)
5. K. Agarwal, R. Jegadeesan, Y-X. Guo, N. V. Thakor, IEEE Reviews in Biomedical Engineering, 10 (2017)
6. S. Ulukus, A. Yener, E. Erkip, O. Simeone, M. Zorzi, P. Grover, K. Huang, IEEE J. Selected Areas in Communications, 33 (2015)
7. X. Lu, P. Wang, D. Niyato, D. I. Kim, Z. Han, IEEE Comm. Surveys & Tutorials, 18, 1413-1452 (2016)
8. L. Xie, Y. Shi, Y. T. Hou, W. Lou, IEEE Wireless Communications, 140-145 (2013)
9. P. Nepa, H. Rogier, IEEE Antennas & Propagation Magazine, 30-52 (2015)
10. Sunarno, Diktat Mata Kuliah TKF 3515-Komunikasi Data, Jurusan Teknik Fisika Fakultas Teknik Universitas Gadjah Mada (2013)
11. S. C. Tang, T. L. T. Lun, Z. Guo, K. W. Kwok, N. J. McDannold, IEEE Trans. Power Electronics **32**, 3844-3857 (2017)

AUTHOR INDEX